CONCEPTS OF
PROBABILITY THEORY

PAUL E. PFEIFFER

Department of Mathematical Sciences
Rice University

Second Revised Edition

DOVER PUBLICATIONS, INC.
NEW YORK

to Mamma and Daddy

*Who contributed to this book
by teaching me that hard work can be
satisfying if the task is worthwhile.*

Published in Canada by General Publishing
Company, Ltd., 30 Lesmill Road, Don Mills,
Toronto, Ontario.
Published in the United Kingdom by Con-
stable and Company, Ltd., 10 Orange Street,
London WC2H 7EG.

This Dover edition, first published in 1978,
is an unabridged republication of the work
originally published by McGraw-Hill Book
Company, New York, in 1965. This Dover edition has been extensively revised and updated
by the author.

International Standard Book Number: 0-486-63677-1
Library of Congress Catalog Card Number: 78-53757

Manufactured in the United States of America
Dover Publications, Inc.
180 Varick Street
New York, N. Y. 10014

Preface

My purpose in writing this book is to provide for students of science, engineering, and mathematics a course at the junior or senior level which lays a firm theoretical foundation for the application of probability theory to physical and other real-world problems. Mastery of the material presented should provide an excellent background for further study in such rapidly developing areas as statistical decision theory, reliability theory, dynamic programming, statistical game theory, coding and information theory, communication and control in the presence of noise, etc., as well as in classical sampling statistics.

My teaching experience has shown that students of engineering and science can master the essential features of a precise mathematical model in a way that clarifies their thinking and extends their ability to make significant applications of the theory. While the ultimate aim of the development in this book is the application of probability theory to problems of practical import and interest, the central task undertaken is an exposition of the basic concepts of probability theory, including a substantial introduction to the idea of a random process. Rather than provide a treatise on applications and techniques, I have attempted to provide a clear development of the fundamental concepts and theoretical perspectives which guide the formulation of problems and the discovery of methods of solution. Considerable attention is given to the task of translating real-world problems into the precise concepts of the model, so that the problem is stated unambiguously and may be attacked with all the resources provided by the mathematical theory as well as physical insight.

The rich theory of probability may be constructed on the essentially simple conceptual framework of that mathematical model generally known as the Kolmogorov model. The central features of this model may be grasped with the aid of certain graphical, mechanical, and notational representations which facilitate the formulation and visualization of concepts and relationships. Highly sophisticated techniques are seen as the means of performing conceptually simple tasks. The whole theory is formulated in a way that makes contact with both the literature of applications and the literature on pure mathematics. At many points I have borrowed specifically and explicitly from one or the other; the resulting treatment points the way to extend such use of the vast reservoir of knowledge in this important and rapidly growing field.

In introducing the basic model, I have appealed to the notion of probability as an idealization of the concept of the relative frequency of occurrence of an event in a large number of repeated trials. In most places, the primary interpretation of probability has been in terms of this familiar concept. I have also made considerable use of the idea that probability indicates the uncertainty regarding the outcome of a trial before the result is known. Various thinkers are currently advocating other approaches to formulating the mathematical model, and hence to interpreting its features. These alternative ways of developing and interpreting the model do not alter its character or the strategies and techniques for dealing with it. They do serve to increase confidence in the usefulness and "naturalness" of the model and to point to the desirability of achieving a mastery of the theory based upon it.

The background assumed in the book is provided in the freshman and sophomore mathematics courses in many universities. A knowledge of limits, differentiation, and integration is essential. Some acquaintance with the rudiments of set theory is assumed in the text, but an Appendix provides a brief treatment of the necessary material to aid the student who does not have the required background. Many students from the high schools offering instruction in the so-called *new mathematics* will be familiar with most of the material on sets before entering the university. Although some applications are made to physical problems, very little technical background is needed to understand these applications. The book should be suitable for a course offered in an engineering or science department, or for a course offered by a department of mathematics for students of engineering or science. The practicing engineer or scientist whose formal education did not provide a satisfactory course in probability theory should be able to use the book for self-study.

It has been my personal experience, as well as my observation of others, that success in dealing with abstract systems rests in large part on the ability to find concrete mental images and constructs which serve as aids in visualizing, remembering, and relating the abstract concepts and ideas. This being so, success in teaching abstract systems depends in similar measure on making explicit use of the most satisfactory images, diagrams and other conceptual aids in the act of communicating ideas.

The literature on probability—both works on pure mathematics and on practical applications—contains a number of such aids to clear thinking, but these aids have not always been exploited fully and efficiently. I can lay little claim to originality in the sense of novelty of ideas or results. Yet I believe the synthesis presented in this book, with its systematic exploitation of several ideas and techniques which ordinarily play only a marginal role in the literature known to me, provides an approach to probability theory which has some definite pedagogical advantages.

Among the features of this presentation which may deserve mention are:

1. A full exploitation of the concept of probability as mass; in particular, the idea that a random variable produces a point-by-point mass transfer from the basic probability space is introduced and utilized in a manner that has proved helpful.

2. Exploitation of minterm maps, minterm expansions, binary designators, and other notions and techniques from the theory of switching, or logic, networks as an aid to systematizing the handling of compound events.

3. Use of the indicator function for events (sets) to provide analytical expressions for discrete-valued random variables.

4. Development of the basic ideas of integration on an abstract space to give unity to the various expressions for mathematical expectation. The mass picture is exploited here in a very significant way to make these ideas comprehensible.

5. Development of a calculus of mathematical expectations which simplifies many arguments that are otherwise burdened with unnecessary details.

None of these ideas or approaches is new. Some have been utilized to good advantage in the literature. What may be claimed is that the systematic exploitation in the manner of this book has provided a treatment of the topic that seems to communicate to my students in a way that no other treatment with which I am familiar has been able to do. This work is written in the hope that this treatment may be equally helpful to others who are seeking a more adequate grasp of this fascinating and powerful subject.

Acknowledgments

My indebtedness to the current literature is so great that it surely must be apparent. I am certainly aware of that debt.

Among the many helpful comments and suggestions by students and colleagues, I am particularly grateful for my discussions with Dr. A. J. Welch and Dr. Shu Lin, graduate students at the time of writing. The critical reviews—from different points of view—by Dr. H. D. Brunk, my former professor, and Dr. John G. Truxal, as well as those by several unnamed reviewers, have aided and stimulated me to produce a better book than I could have written without such counsel.

The patient and dedicated work of Mrs. Arlene McCourt and Mrs. Velma T. Goodwin has been an immeasurable aid in producing a manuscript. I can only hope that the pride and care they showed in their part of the work is matched in some measure by the quality of the contents.

PAUL E. PFEIFFER

Contents

chapter **1**

Introduction

As is true of so much of mathematics, probability theory has a long history whose beginnings are largely unknown or obscure. In this chapter we examine very briefly the classical concept of probability which arose in early investigations and which still remains the basis of many applications. It seems reasonably certain that the principal impetus for the development of probability theory came from an interest in games of chance. Interest in gambling is ancient and widespread; games of chance involve an element of "randomness"; it is, in fact, puzzling that the idea of randomness and the attempt to describe it mathematically did not develop earlier. David [1962] discusses this cultural enigma in an interesting study of the early gropings toward a theory of probability and of the early work in the field.

The rudiments of a mathematical theory probably took shape in the sixteenth century. Some evidence of this is provided by a short note written in the early seventeenth century by the famous mathematician and astronomer Galileo Galilei. In a fragment known as *Thoughts about Dice Games*, Galileo dealt with certain problems posed to him by a gambler whose identity is not now known. One of the points of interest in this note (cf. David [1962, pp. 65ff.]) is that Galileo seems to assume that his reader would know how to calculate certain elementary probabilities.

The celebrated correspondence between Blaise Pascal and Pierre de Fermat in 1654, the treatise by Christianus Huygens, 1657, entitled *De Ratiociniis in Aleae Ludo*, and the work *Ars Conjectandi* by James Bernoulli (published posthumously in 1713, but probably written some time about 1690) are landmarks in the formulation and development of the classical theory. The fundamental definition of probability which was accepted in this period, vaguely assumed when not explicitly stated, remained the classical definition until the modern formulations developed in this century provided important extensions and generalizations.

We shall simply examine the classical concept, note some of its limitations, and try to identify the fundamental properties which underlie modern axiomatic formulations. It turns out that the key properties are extremely simple. All the results of the classical theory are obtained as easily in the more general system which we study in this book. Because of the success of the more general system, we shall not examine separately the extensive mathematical system developed upon the classical base. For such a development, one may consult the treatise by Uspensky [1937].

1-1 Basic ideas and the classical definition

The interest in games of chance which stimulated early work in probability not only provided the motivation for that work but also influenced the character of the emerging theory. Almost instinctively, it seems, the best minds attempted to analyze the probability situations into sets of possible outcomes of a gaming operation. These possibilities were then assumed to be "equally likely." The success of the analysis in predicting "chances" led eventually to the precise definition of probability, which remained the classical definition until early in the present century.

Definition 1-1a   Classical Probability

A trial is made in which the outcome is one of N equally likely possible outcomes. If, among these N possible outcomes, there are N_A possible outcomes which result in the occurrence of the event A, the *probability of the event A* is defined by

$$P(A) = \frac{N_A}{N}$$

This definition seems to be motivated by two factors:

1. The intuitive idea of *equally likely* possible outcomes
2. The empirical fact of the *statistical regularity of the relative frequencies* of the occurrence of events to be studied

The statistical regularity of the relative frequencies of the occurrence of various events in gambling games has long been observed. In fact, some of the problems posed to noted mathematicians were the result of small variations of observed frequencies from those anticipated by the gamblers (David [1962, pp. 66, 89]). The character of the games was such that the notion of "equally likely" led to successful predictions. So natural did this concept seem that it has been defended vigorously upon philosophical grounds.

Once the definition is made—for whatever reasons—no further appeal to intuition or philosophy is needed. A situation is given; two questions must be answered:

1. How many possible outcomes are there (i.e., what is the value of N)?

2. How many of the possible outcomes result in the occurrence of event A (i.e., what is the value of N_A)?

Once these questions are answered, the probability is determined by the ratio specified in the definition. The problem is to determine answers to these two questions. This, in turn, is a problem of counting the possibilities.

Consider a simple example. Two dice are thrown. Suppose the event A is the event that "a six is thrown." This means that the pair of numbers which appear (in the form of spots) must add to six. What is the probability of throwing a six? First, we must identify the equally likely possible outcomes. Then we must perform the appropriate counting operations. If the dice are "fair," it seems that it is equally likely that any one of the six sides of either of the dice will appear. It is thus natural to consider the appearance of each of the 36 possible pairs of sides of the two dice as equally likely. These various possibilities may be represented simply by pairs of numbers. The first number, being one of the integers 1 through 6, represents the corresponding side of one of the dice. The second number represents the corresponding side of the second die. Thus the number pair $(3, 2)$ indicates the appearance of side 3 of the first die and of side 2 of the second die. The sides are usually numbered according to the number of spots thereon.

We have said there are 36 such pairs. For each possibility on the first die there are six possibilities on the second die. Thus, for the rolling of the dice, we take N to be 36. To determine N_A, we may in this case simply enumerate those pairs for which the sum is 6. These are the pairs $(1, 5)$, $(2, 4)$, $(3, 3)$, $(4, 2)$, and $(5, 1)$. There are five such outcomes, so that $N_A = 5$. The desired probability is thus, *by definition*, $5/36$.

It should not be assumed from the simple example just discussed that probability theory is trivial. Counting, in complex situations, can be a very sophisticated matter, as references to the literature will show (cf. Uspensky [1937] or Feller [1957]).   Much of the classical probability theory is devoted to the development of counting techniques.   The principal tool is the theory of permutations and combinations.   A brief summary of some of the more elementary results is given in Appendix A. An excellent introductory treatment is given in Goldberg [1960, chap. 3]; a more extensive treatment is given in Feller [1957, chaps. 2 through 4].

Upon this simple base a magnificent mathematical structure has been erected.   Introduction of the laws of compound probability and of the concepts of conditional probability, random variables, and mathematical expectation have provided a mathematical system rich in content and powerful in its application.   As an example of the range of such theory, one should examine a work such as the classical treatise by J. V. Uspensky [1937], entitled *Introduction to Mathematical Probability*. So successful was this development that Uspensky could venture the opinion that modern attempts to provide an axiomatic foundation would result in interesting mental exercises but would have little value for application [*op. cit.*, p. 8].

The classical theory suffers some inherent limitations that inhibit its applications to many problems.   Moreover, the success of modern mathematical models in extending the classical theory has provided a more flexible base for applications.   Thus it seems desirable, both for applications and for purely mathematical investigations, to move beyond the classical model.

1-2   Motivation for a more general theory

There are two rather obvious limitations of classical probability theory. For one thing, it is limited to situations in which there is only a finite set of possible outcomes.   Very simple situations arise, even in classical gambling problems, in which a finite set of possibilities is not adequate. Suppose a game is played until one player is successful in performing a given act (i.e., until he "wins").   Any particular sequence of plays is likely to terminate in a finite number of trials.   But there is no a priori assurance that this will happen.   A man could conceivably flip a coin indefinitely without ever turning up a head.   At any rate, no one can determine a number large enough to include all possible sequences ending in a successful toss.   Other simple gaming operations can be conceived in which the game goes on endlessly.   In order to account for these possibilities, there must be a model in which the possibilities are not limited to any finite number.

It is also desirable, both for theoretical and practical reasons, to extend the theory to situations in which there is a continuum of possibilities.   In such situations, some physical variable may be observed: the height of an individual, the value of an electric current in a wire, the amount of water in a tank, etc. Each of the continuum of possible values of these variables is to be considered a possible outcome.

A second limitation inherent in the classical theory is the assumption of equally likely outcomes. It is noted above that the classical theory seems to be rooted in the two concepts of (1) equally likely outcomes and (2) statistical regularity of relative frequencies. It often occurs that these two concepts do not lead to the same definition. A simple example is the loaded die. For a die which is asymmetrical in terms of mass or shape, it is not intuitively expected that each side will turn up with equal frequency; as a matter of fact, both experience and intuition agree that the relative frequencies will not be the same for the different sides. But *it is expected that the relative frequencies will show statistical regularity.* Experience bears this out in many situations, of which the loaded die is a simple example.

These considerations suggest that the extension of the definition of probability should preserve the essential characteristics of relative frequencies. Two properties prove to be satisfactory for the extension:

1. If f_A is the relative frequency of occurrence of an event A, then $0 \leq f_A \leq 1$.

2. If A and B are mutually exclusive events and C is the event which occurs iffi (if and only if) either A or B occurs, then $f_C = f_A + f_B$.

In the next chapter we begin the development of a theory which defines probability as a function of events; the characteristic properties of the probability function are (1) that it takes values between zero and one and (2) that it has a fundamental *additivity property* for the probability of mutually exclusive events.

The idea of the relative frequency of the occurrence of events plays such an important role in motivating the concept of probability and in interpreting the meaning of the mathematical results that some competent mathematicians have developed mathematical models in which probability is defined as a limit of a relative frequency. This approach has the advantage of tying the fundamental concepts closely to the experiential basis for the introduction of the theoretical model. It has the disadvantage, however, of introducing certain complications into the formulation of the basic definitions and axioms.

It seems far more fruitful to postulate the existence of probabilities which have the simple fundamental properties discussed above. When these probabilities are *interpreted* as relative frequencies, the behavior of the mathematical model can be compared with the behavior of the physical (or other) system that it is intended to represent. The frequency interpretation is aided by the development of certain theorems known under the generic title of *the law of large numbers.* The high degree of correlation between suitable models based on this approach and the observed behavior of many practical systems have provided grounds for confidence in the suitability of such models.   This approach is based philosophically on the view that one cannot "prove" anything about the physical world in terms of a mathematical model.   One constructs a model, studies its "behavior," uses the results to predict phenomena in the "real world," and evaluates the usefulness of his model in terms of the degree to which the behavior of the mathematical model corresponds to the behavior of the real-world system.   "The proof is in the pudding."   The growing literature on applications in a wide variety of fields indicates the extent to which such models have been successful (cf. the article by S. S. Shu in Bogdanoff and Kozin [1963] for a brief survey of the history of applications of probability theory in physics and engineering).

Because of these considerations, we do not attempt to examine the theory constructed upon the foundation of the classical definition of probability; instead, we turn immediately to the more general model. Not only does this general theory include the classical theory as a special case; it is often simpler to develop the more general concepts—in spite of certain abstractions—and then examine specific problems from the vantage point provided by this general approach.   More elegant solutions and more satisfactory interpretations of problems and solutions are often obtainable with a smaller total effort.

Selected references

DAVID [1962]: "Games, Gods, and Gambling."   This interesting work deals with "the origins and history of probability and statistical ideas from the earliest times to the Newtonian era."   A readable treatment, with many interesting personal and historical sidelights.   The author has a keen interest in the history of ideas as well as in the development of the technical aspects of probability theory in its early stages.

FELLER [1957]: "An Introduction to Probability Theory and Its Applications," vol. 1, 2d ed.   An introduction and an extensive treatment of probability theory in the case of a finite or countably infinite number of possible outcomes.   Chapters 2, 3, and 4 provide a rather extensive treatment of the problem of counting the ways an event can occur.

GOLDBERG [1960]: "Probability: An Introduction."   A lucid treatment of the modern point of view, which is mathematically easy because the author deals only with the case of a finite number of possible outcomes.   Chapter 3 provides an excellent introduction to the theory of permutations and combinations needed for many probability problems, both in the classical and in the more general case.

USPENSKY [1937]: "Introduction to Mathematical Probability."   A classical treatment of classical probability.   This work is still a major reference for many aspects of the mathematical theory and its applications, although its author takes a dim view of the modern axiomatic model which the present work attempts to expound.   Available in a paperback edition, it probably should be on the bookshelf of any person having a serious interest in probability theory.

chapter 2

A mathematical model for probability

The discussion in Chap. 1 has shown that the classical theory of probability, based upon a finite set of equally likely possible outcomes of a trial, has severe limitations which make it inadequate for many applications. This is not to dismiss the classical case as trivial, for an extensive mathematical theory and a wide range of applications are based upon this model. It has been possible, by the use of various strategies, to extend the classical case in such a way that the restriction to equally likely outcomes is greatly relaxed. So widespread is the use of the classical model and so ingrained is it in the thinking of those who use it that many people have difficulty in understanding that there can be any other model. In fact, there is a tendency to suppose that one is dealing with physical reality itself, rather than with a model which represents certain aspects of that reality. In spite of this appeal of the classical model, with both its conceptual simplicity and its theoretical power, there are many situations in which it does not provide a suitable theoretical framework for dealing with problems arising in practice. What is needed is a generalization of the notion of probability in a manner that preserves the essential properties of the classical model, but which allows the freedom to encompass a much broader class of phenomena.

In the attempt to develop a more satisfactory theory, we shall seek in a deliberate way to describe a *mathematical model* whose essential features may be correlated with the appropriate features of real-world problems.   The history of probability theory (as is true of most theories) is marked both by brilliant intuition and discovery and by confusion and controversy.   Until certain patterns had emerged to form the basis of a clear-cut theoretical model, investigators could not formulate problems with precision, and reason about them with mathematical assurance. Long experience was required before the essential patterns were dis-covered and abstracted.   We stand in the fortunate position of having the fruits of this experience distilled in the formulation of a remarkably successful mathematical model.

A mathematical model shares common features with any other type of model.   Consider, for example, the type of model, or "mock-up," used extensively in the design of automobiles or aircraft.   These models display various essential features: shape, proportion, aerodynamic characteristics, interrelation of certain component parts, etc.   Other features, such as weight, details of steering mechanism, and specific materials, may not be incorporated into the particular model used.   Such a model is not equivalent to the entity it represents.   Its usefulness depends on how well it displays the features it is designed to portray; that is, *its value depends upon how successfully the appropriate features of the model may be related to the "real-life" situation, system, or entity modeled.* To develop a model, one must be aware of its limitations as well as its useful properties.

What we seek, in developing a *mathematical model* of probability, is a *mathematical system* whose concepts and relationships correspond to the appropriate concepts and relationships in the "real world."   Once we set up the model (i.e., the mathematical system), we shall study its mathematical behavior in the hope that the patterns revealed in the mathematical system will help in identifying and understanding the cor-responding features in real life.

We must be clear about the fact that the mathematical model cannot be used to *prove* anything about the real world, although a study of the model may help us to *discover* important facts about the real world.   A model is not true or false; rather, a model fits (i.e., corresponds properly to) or does not fit the real-life situation.   A model is useful, or it is not. A model is useful if the three following conditions are met:

1. Problems and situations in the real world can be *translated* into problems and situations in the mathematical model.

2. The model can be studied as a mathematical system to obtain solutions to the *model problems* which are formulated by the translation of real-world problems.

3. The solutions of a model problem can be correlated with or *interpreted* in terms of the corresponding real-world problem.

The mathematical model must be a consistent mathematical system. As such, it has a "life of its own." It may be studied by the mathematician without reference to the translation of real-world problems or the interpretation of its features in terms of real-world counterparts. To be useful from the standpoint of applications, however, not only must it be mathematically sound, but also its results must be physically meaningful when proper interpretation is made. Put negatively, a model is considered unsatisfactory if either (1) the solutions of model problems lead to unrealistic solutions of real-world problems or (2) the model is incomplete or inconsistent mathematically.

Although long experience was needed to produce a satisfactory theory, we need not retrace and relive the mistakes and fumblings which delayed the discovery of an appropriate model. Once the model has been discovered, studied, and refined, it becomes possible for ordinary minds to grasp, in reasonably short time, a pattern which took decades of effort and the insight of genius to develop in the first place.

The most successful model known at present is characterized by considerable mathematical abstractness. A complete study of all the important mathematical questions raised in the process of establishing this system would require a mathematical sophistication and a budget of time and energy not properly to be expected of those whose primary interest is in application (i.e., in solutions to real-world problems). Two facts motivate the study begun in this chapter:

1. Although the details of the mathematics may be sophisticated and difficult, *the central ideas are simple* and *the essential results are often plausible*, even when difficult to prove.

2. A mastery of the ideas and a reasonable skill in translating real-world problems into model problems make it possible to grasp and solve problems which otherwise are difficult, if not impossible, to solve. *Mastery of this model extends considerably one's ability to deal with real-world problems.*

In addition to developing the fundamental mathematical model, we shall develop certain auxiliary representations which facilitate the grasp of the mathematical model and aid in discovering strategies of solution for problems posed in its terms. We may refer to the combination of these auxiliary representations as the *auxiliary model.*

Although the primary goal of this study is the ability to solve real-world problems, success in achieving this goal requires a reasonable mastery of the mathematical model and of the strategies and techniques of solution of problems posed in terms of this model. Thus considerable attention must be given to the model itself. As we have already noted, the model may be studied as a thing in itself, with a "life of its own." This means that we shall be engaged in developing a mathematical theory. The study of this mathematics can be an interesting and challenging game in itself, with important dividends in training in analytical thought. At times we must be content to play the game, until a stage is reached at which we may attempt a new correlation of the model with the real world. But as we reach these points in the development of the theory, repeated success in the act of interpretation will serve to increase our confidence in the model and to make it easier to comprehend its character and see its implications for the real world.

The model to be developed is essentially the axiomatic system described by the mathematician A. N. Kolmogorov (1903–   ), who brought together in a classical monograph [1933] many streams of development. This monograph is now available in English translation under the title *Foundations of the Theory of Probability* [1956]. The Kolmogorov model presents mathematical probability as a special case of abstract measure theory. Our exposition utilizes some concrete but essentially sound representations to aid in grasping the abstract concepts and relations of this model. We present the concepts and their relations with considerable precision, although we do not always attempt to give the most general formulation. At many places we borrow mathematical theorems without proof. We sometimes note critical questions without making a detailed examination; we merely indicate how they have been resolved. Emphasis is on concepts, content of theorems, interpretations, and strategies of problem solution suggested by a grasp of the essential content of the theory. Applications emphasize the translation of physical assumptions into statements involving the precise concepts of the mathematical model.

It is assumed in this chapter that the reader is reasonably familiar with the elements of set theory and the elementary operations with sets. Adequate treatments of this material are readily available in the literature. A sketch of some of these ideas is given in Appendix B, for ready reference. Some specialized results, which have been developed largely in connection with the application of set theory and boolean algebra to switching circuits, are summarized in Sec. 2–6. A number of references for supplementary reading are listed at the end of this chapter.

Sets, Events, and Switching [1964]. A number of references for supplementary reading are listed at the end of this chapter.

2-1   In search of a model

The discussion in the previous introductory paragraphs has indicated that, to establish a mathematical model, we must first identify the sig-

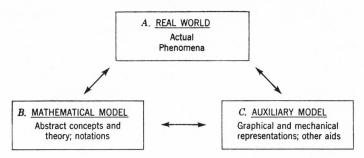

Fig. 2-1-1   Diagrammatic representation of the relationships between the "real world" and the models.

nificant concepts, patterns, relations, and entities in the "real world" which we wish to represent.   Once these features are identified, we must seek appropriate mathematical counterparts.   These mathematical counterparts involve concepts and relations which must be defined or postulated and given appropriate names and symbolic representations.

In order to be clear about the situation that exists when we utilize mathematical models, let us make a diagrammatic representation as in Fig. 2-1-1.   In this diagram, we analyze the object of our investigation into three component parts:

A. The *real world* of actual phenomena, known to us through the various means of experiencing these phenomena.

B. The imaginary world of the *mathematical model*, with its abstract concepts and theory.   An important feature of this model is the use of symbolic notational schemes which enable us to state relationships and facts with great precision and economy.

C. An *auxiliary model*, consisting of various graphical, mechanical, and other aids to visualization, remembering, and even discovering important features about the mathematical model.   It seems likely that even the purest of mathematicians, dealing with the most abstract mathematical systems, employ, consciously or unconsciously, concrete mental images as the carriers of their thought patterns.   We shall develop explicitly some concrete representations to aid in thinking about the abstract mathematical model; these in turn will help us to think clearly and systematically about the patterns abstracted from (i.e., lifted out of) our experience of the real world of phenomena.

Much of our attention and effort will be devoted to establishing the mathematical model *B* and to a study of its characteristics.   In doing this, we shall be concerned to relate the various aspects of the mathematical model to corresponding aspects of the auxiliary model *C*, as an aid to learning and remembering the important characteristics of the mathematical model.   Our real goal as engineers and scientists, however, is to use our knowledge of the mathematical model as an aid in dealing with problems in the real world.   This means that we must be able to move from one part of our system to another with freedom and insight. For clarity and emphasis, we may find it helpful to indicate the important transitions in the following manner:

$A \rightarrow B$: *Translation* of real-world concepts, relations, and problems into terms of the concepts of the mathematical model.

$B \rightarrow A$: *Interpretation* of the mathematical concepts and results in terms of real-world phenomena.   This may be referred to as the *primary interpretation*.

$B \rightarrow C$: *Interpretation* of the mathematical concepts and results in terms of various concrete representations (mass picture, mapping concepts, etc.).   This may be referred to as a *secondary interpretation*.

$C \rightarrow B$: The movement from the auxiliary model to the mathematical model exploits the concrete imagery of the former to aid in discovering new results, remembering and extending previously discovered results, and evolving strategies for the solution of model problems.

$A \leftrightarrow C$: The correlation of features in A and C often aids both the translation of real-world problems into model problems and the interpretation of the mathematical results.   In other words, the best path from A to B or from B to A may be through C.

The first element to be modeled is the relative frequency of the occurrence of an event.   It is an *empirical fact* that in many investigations the relative frequencies of occurrence of various events exhibit a remarkable *statistical regularity* when a large number of trials are made. This feature of many games of chance served (as we noted in Chap. 1) to motivate much of the early development of probability theory.   In fact, many of the questions posed by gamblers to the mathematicians of their time were evoked by the fact that observed frequencies deviated slightly from that which they expected.

This phenomenon of constant relative frequencies is in no way limited to games of chance.   Modern statistical communication theory, for example, makes considerable use of the remarkable statistical regularities which characterize languages.   The relative frequencies of occurrence of symbols of the alphabet (including all symbols such as space, numbers, punctuation, etc.), of symbol pairs, triples, etc., and of words, word combinations, etc., have been studied extensively and are known to be quite stable.   Of course, exceptions are known.   For example, Pierce [1961] quotes a paragraph from a novel which is written without the use of a single letter E in its entire 267 pages.   In ordinary English, the letter

E is used more frequently than any other letter of the alphabet. Such marked deviations from the usual patterns require special effort or indicate unusual situations. In the normal course of affairs one may expect rather close adherence to the common patterns of statistical regularity.

The whole life insurance industry is based upon so-called mortality tables, which predict the relative frequency of deaths at various ages (or give information equivalent to specifying these frequencies). In a similar manner, reliability engineering makes extensive use of the life expectancy of articles manufactured by a given process. These life expectancies are based on the equivalent of mortality tables for the products manufactured. Closely related is the theory of errors, in which the relative frequencies of errors of various magnitudes may be assumed to follow quite closely certain well-known patterns.

The empirical fact of a stable (i.e., constant) relative frequency serves as the basis in experience for choosing a mathematical model of probability. This empirical fact cannot in any way imply logically the existence of a probability number. We set up a model by *postulating* the existence of an ideal number, which we call the *probability* of an event. If this is a sound model and we have chosen the number properly, we may expect the relative frequency of occurrence of the event in a large number of trials to lie quite close to this number. But we cannot "prove" the existence in the real world of such an ideal limiting frequency. We simply set up a model. We then examine the "behavior" of the model, *interpret* the probabilities as relative frequencies, check or *test* these probabilities against observed frequencies (where possible), and *try to determine experimentally whether the model fits the real-world situation to be modeled. We can build up confidence in our model; we cannot prove or disprove our model.* If it works for enough different problems, we move with considerable confidence to new problems. Continued success tends to increase our confidence, so that we come to place a large degree of reliance upon the adequacy of the model to predict behavior in the real world.

If the feature of the real world to be represented by probability is the relative frequency of occurrence of a given event in a large number of trials, then the probability must be a number associated with this event. These numbers must obey the same laws as relative frequencies. We can readily list several elementary properties which must therefore be possessed by the probability model.

1. If an event is sure to occur, its relative frequency, and hence its probability, must be unity. Similarly, if an event cannot possibly occur, its probability must be 0.

2. Probabilities are real numbers, lying between 0 and 1.

3. If two events are mutually exclusive (i.e., cannot both happen on any one trial), the probability of the compound event that either one or

the other of the original events will occur is the sum of the individual probabilities.

4. The probability that an event will not occur is 1 minus the probability of the occurrence of the event.

5. If the occurrence of one event implies the occurrence of a second event, the relative frequency of occurrence of the second event must be at least as great as that of the first event.   Thus the probability of the second event must be at least as great as that of the first event.

Many more such properties could be enumerated.   One of the concerns of the model maker is to discover the most basic list, in the sense that the properties included in this list imply as logical consequences all the other desirable or necessary properties.   When we come to the formal presentation of our mathematical model in Sec. 2-3, we shall see that the basic list desired is contained in the list of properties above.

In order to realize an economy of expression, we shall need to introduce an appropriate notational scheme to be used in formulating probability statements.   Before attempting to do this, however, we should take note of the fact that we have used the term *event* in the previous discussion without attempting to characterize or represent this concept. Since probability theory deals in a fundamental way with events and various combinations of events, it is desirable that we give attention to a precise formulation of the concept of event in a manner that will be mathematically useful and precise and that will be meaningful in terms of real-world phenomena.   We turn our attention to that problem in the next section.

2-2   A model for events and their occurrence

In order to produce a mathematical model of an event, it is necessary to have a clear idea of the situations to which probability theory may be applied.   Historically, the ideas developed principally in connection with games of chance.   In such games the concept of some sort of trial or test or experiment is fundamental.   One draws a card from a deck, throws a pair of dice, or selects a colored ball from a jar.   The trial may be composite.   A single composite trial may consist of drawing, successively, five cards from a deck to form a hand, or of flipping a coin ten times and noting the sequence of heads and tails that turn up.   But *in any case, there must be a well-defined trial and a set of possible outcomes.*   In the coin-flipping trial, for example, it is physically conceivable that the coin may land neither "heads" nor "tails."   It may stand on edge.   It is necessary to decide whether the latter eventuality is to be considered a satisfactory performance of the experiment of flipping the coin.

It is perhaps helpful to realize that each of the situations described above is equivalent to drawing balls from a jar or urn. Each throw of a pair of dice, for instance, results in an identifiable outcome. Each such outcome can be represented by a single ball, appropriately marked. Throwing the dice is then equivalent to choosing at random a ball from the urn. Similarly, the composite trial of drawing five cards from a deck may be equivalent to drawing a single ball. Suppose we were to lay out each of the possible hands of five cards, with the cards arranged in the order of their drawing; a picture could be taken of each possible hand; the set of all pictures (or a set of balls, each with one of the pictures of a hand printed thereon) could be put in a large urn. Drawing a single picture (or ball) from the urn would then be equivalent to the composite trial of drawing five cards from the deck to form a hand.

The concept of the outcome of a trial, which is so natural to a game of chance, is readily extended to situations of more practical and scientific importance. One may be sampling opinions or recording the ages at death of members of a population. One may be making a physical measurement in which the precise outcome is uncertain. Or one may be receiving a message on a communication system. In each case, the result observed may be considered to be the result of a trial which yields one of a set of possible outcomes. This set of possible outcomes is clearly defined, in principle, at least. Even here we may use the mental image of drawing balls from a jar, provided we are willing to suppose our jar may contain an infinite number of balls. Making a trial, no matter what physical procedure is required, is then equivalent to choosing a ball from the jar.

We have a simple mathematical counterpart of a jar full of balls. This is the fundamental mathematical notion of an abstract set of elements. We consider a basic set S of elements ξ. Each element ξ represents one of the possible outcomes of the trial. The *basic set*, or *basic space* (also commonly called the *sample space*), S represents the collection of *all* allowed outcomes. The single element ξ is referred to as an *elementary outcome*. Regardless of the nature of the physical trial referred to, *the performance of a trial is represented mathematically as the choice of an element ξ.* In many considerations it is, in fact, convenient to refer to the element ξ as the *choice variable.*

Events

Probability theory refers to the *occurrence of events*. Let us now see how this idea may be expressed in terms of sets and elements. Suppose, for example, a trial consists of drawing a hand of five cards from an ordinary deck of playing cards. We may think in terms of our equivalent urn experiment in which each ball has the picture of one of the possible hands, with one ball for each such hand. Selecting a ball is equivalent to drawing the hand pictured thereon. What is meant by the statement "A hand with one or more aces is drawn"? This can mean but one thing. The ball drawn is one of those having a picture of a hand with one or more aces. But the set of balls satisfying this condition constitutes a well-defined *subset* of the set of all the balls in the jar.

Our mathematical counterpart of the jar full of balls is the abstract basic space S. There is an element ξ corresponding to each ball in the urn. We suppose that it is possible, in principle at least, to identify each element in terms of the properties of the ball (and hence the hand) it represents. *The occurrence of the event is the selection of one of the elements in the subset of those elements which have the properties determining the event.* In our example, this is the property of having one or more aces. Viewed in this way, it seems natural to identify the *event* with the *subset* of elements (balls or hands) having the desired properties. The *occurrence* of the event is the act of choosing one of the elements in this subset. In probability investigations, we suppose the choice is "random," in the sense that it is uncertain which of the balls or elements will be chosen before the actual trial is made. It should be apparent that the occurrence of events is not dependent upon this concept of randomness. A deliberate choice—say, one in which the balls are examined in the process of selection—would still lead to the occurrence or nonoccurrence of an event.

In general, an event may be defined by a proposition. The event occurs whenever the proposition is true about the outcome of a trial. On the other hand, a proposition about elements defines a set in the following manner. Suppose the symbol $\pi_A(\cdot)$ represents a proposition about elements ξ in the basic set S. In the example above, the symbol $\pi_A(\xi)$ is understood to mean "the hand represented by ξ has one or more aces." This statement may or may not be true, depending upon which element ξ is being considered. Let us consider the set A of those elements ξ for which the proposition $\pi_A(\xi)$ is true. We use the symbolic representation $A = \{\xi : \pi_A(\xi)\}$ to mean "A is the set of those ξ for which the proposition $\pi_A(\xi)$ is true." Thus, in our example, A is the set of those hands which have one or more aces. *We identify this set with the event A. The event A occurs iff (if and only if) the ξ chosen is a member of the set A.*

Let us illustrate further. Consider the result of flipping a coin three times. Each resulting sequence may be represented by writing a sequence of H's and T's, corresponding to throws of heads or tails, respectively. Thus the symbol THT indicates a sequence in which a tail appears on the first and third throws and a head appears on the second throw. There are eight elementary outcomes, each of which is represented by an element.   We may list them as follows:

$$\xi_0 = TTT \qquad \xi_1 = TTH \qquad \xi_2 = THT \qquad \xi_3 = THH$$
$$\xi_4 = HTT \qquad \xi_5 = HTH \qquad \xi_6 = HHT \qquad \xi_7 = HHH$$

Consider the event that a tail turns up on the second throw.   This is the event $T_2 = \{\xi_0, \xi_1, \xi_4, \xi_5\}$.   This event occurs if any one of the elements in this list is chosen, which means that any one of the sequences represented by ξ_0, ξ_1, ξ_4, or ξ_5 is thrown.   Similarly, the set $H_1 = \{\xi_4, \xi_5, \xi_6, \xi_7\}$ corresponds to the event that a head is obtained on the first throw. Many other events could be defined and the elements belonging to them listed in a similar manner.

We have made considerable progress in developing a mathematical model for probability theory.   The essential features may be outlined as follows:

Real world	Mathematical model	Auxiliary model
1. Relative frequency of occurrence of events	Probability of events	
2. Set of possible outcomes of a trial	Basic set S of elements ξ	Urn filled with balls
3. Events	Subsets of S	Subsets of the balls
4. Occurrence of an event	Selection of an element from the appropriate subset	Selection of a ball from the appropriate subset

Some aspects of the formulation of the model have not been made precise as yet.   Some of the flexibility and possibilities of the representation of events as sets have not yet been demonstrated.   But considerable groundwork has been laid.

Special events and combinations of events

In probability theory we are concerned not only with single events and their probabilities, but also with various combinations of events and relations between them.   These combinations and relations are precisely those developed in elementary set theory.   Thus, having introduced the concept of an event as a set, we have at our disposal an important mathematical resource in the theory of sets.   In the following treatment, it is assumed that the reader is reasonably familiar with elementary topics in set theory.   A brief exposition of the facts needed is given in Appendix B, for ready reference.

Before considering combinations of events, it may be well to note the following fact.   When a trial or choice is made, one and only one element ξ is chosen—i.e., only one elementary outcome is observed—but a large number of events may have occurred.   Suppose, in throwing a pair of

dice, the pair of numbers 2, 4 turn up.   Only one of the possible outcomes
of the trial has been selected.   But the following events (among many
others) have occurred:

1. A "six" is thrown.
2. A number less than seven is thrown.
3. An even number is thrown.
4. Both of the numbers appearing are even numbers.
5. The larger number in the pair thrown is twice the smaller.

These are distinct events, and not just different names for the same event,
as may be verified by enumerating the elements in each of the events
described.   Each of these events has occurred; yet only one outcome has
been observed.

It is important to distinguish between the *elementary outcome* ξ and
the *elementary event* $\{\xi\}$ whose only member is the elementary outcome ξ.
Whenever the result of a trial is the elementary outcome ξ, the elementary
event $\{\xi\}$ occurs; so also, in general, do many other events.   We have
here an example of the necessity in set theory of distinguishing logically
between an element ξ and the single-element set $\{\xi\}$.   The reader should
be warned of an unfortunate anomaly in terminology found in much of
the literature.   In his fundamental work, Kolmogorov [1933, 1956] used
the term elementary event for the elementary outcome ξ; he used no
specific term for the event $\{\xi\}$.   Although he does not confuse logically
the elementary outcome ξ with the event $\{\xi\}$, his terminology is inconsistent at this point.   We shall attempt to be consistent in our usage.   A
little care will prevent confusion in reading the literature which follows
Kolmogorov's usage.

Let us consider now several special events, combinations of events,
and relations between events.   We shall illustrate these by reference to
the coin-flipping experiment described earlier in this section.   We let
H_k be the event of a head on the kth throw in the sequence (k is 1, 2, or
3 in this experiment) and T_k be the event of a tail on the kth throw.   For
convenience, we list the corresponding sets as follows:

$$H_1 = \{\xi_4, \xi_5, \xi_6, \xi_7\} \qquad T_1 = \{\xi_0, \xi_1, \xi_2, \xi_3\}$$
$$H_2 = \{\xi_2, \xi_3, \xi_6, \xi_7\} \qquad T_2 = \{\xi_0, \xi_1, \xi_4, \xi_5\}$$
$$H_3 = \{\xi_1, \xi_3, \xi_5, \xi_7\} \qquad T_3 = \{\xi_0, \xi_2, \xi_4, \xi_6\}$$

Suppose we are interested in the compound event of "a head on the
first throw *or* a tail on the second throw."   To what set of elementary
outcomes does this correspond?   The event "a head on the first throw" is
the set of elementary outcomes H_1; the event "a tail on the second throw"
is the set of elementary outcomes T_2.   The compound event occurs iffi an
element from either of these sets is chosen.   Thus the desired event is the

set of elementary outcomes $H_1 \cup T_2$, that is, the set of all those elementary outcomes which are in at least one of the events H_1, T_2. The argument carried out for this illustration could be repeated without essential change for any two events A and B. Thus, for any pair of events A, B, the event "A *or* B" is the event $A \cup B$. We commonly refer to this as the *or event;* also, we may use the language of sets and refer to this as the *union of the events* A, B.

In a similar way, the event of "a head on the first throw *and* a tail on the second throw" can be represented by the intersection $H_1 T_2 = H_1 \cap T_2$. From the point of view of set theory, this is the set of those elements which are in both H_1 and T_2. This is precisely what is required by the joint occurrence of H_1 and T_2; the element chosen must belong to both of the events (sets) H_1 and T_2, which is the same as saying that it belongs to $H_1 T_2$. We may refer to this as the *joint event*, or as the *intersection of the events*, H_1 and T_2. Again, the argument does not depend upon the specific illustration, and we may speak of the joint event AB for any two events A and B.

If we consider the intersection $H_1 T_1$ of the events H_1 and T_1, we find that it contains no element; i.e., there is no sequence which has both a head on the first throw and a tail on the first throw. As a set, the intersection $H_1 T_1$ is the *empty set* $\emptyset$; as an event, the intersection is *impossible*, since no possible outcome can meet the conditions which determine the event. It is immediately evident that the impossible event always corresponds to the empty set. In abstract set theory, the symbol $\emptyset$ is commonly used to represent the empty set; we shall also use it to indicate the impossible event. When two events A,B have the relation that their joint occurrence is impossible, they are commonly called *mutually exclusive events*. Thus two mutually exclusive events are characterized by having an empty intersection; in the language of sets, we may refer to these as *disjoint events*.

It may be noted that the set T_1 has as members all the elements that are not in H_1. T_1 is thus the set known as the *complement of H_1*, designated $H_1{}^c$. The event T_1 occurs whenever H_1 fails to occur—if a head does not appear on the first throw, a tail must appear there. We speak of the event $H_1{}^c$ as the *complementary event* for H_1. Inspection of the list above shows that, for each $k = 1$, 2, or 3, we must have $H_k{}^c = T_k$ and $T_k{}^c = H_k$, as common sense requires. If we take the event $H_k \cup T_k$ for any k, we have the set of all possible outcomes. This means that the event $H_k \cup T_k$ is sure to happen. Every sequence must have either a head or a tail (but not both) in the kth place. The event S, corresponding to the *basic set* of all possible outcomes, is thus naturally referred to as the *sure event*.

We have noted that two events A and B may stand in the relation of being mutually exclusive, in the sense that if one occurs the other cannot occur on the same trial. At the other extreme, we may have the situation in which the occurrence of the event A implies the occurrence of the event B. This, in terms of set theory, means that set A is contained in set B, a relationship indicated by the notation $A \subset B$. We shall utilize the same notation for events; if the occurrence of event A requires the occurrence of event B (this means that every element ξ in A is also in B), we shall indicate the fact by writing $A \subset B$. In keeping with the custom in set theory, we shall suppose the impossible event $\emptyset$ implies any event A. It certainly is true that every element which is in event $\emptyset$ is also in event A. The convention involved in this seemingly pointless statement, denoted in symbols by $\emptyset \subset A$ for any $A \subset S$, is often useful in the symbolic manipulation of events.

Because of the essential identity of the notions of sets and events, as we have formulated our model, we shall find it convenient to use the language of sets and that of events interchangeably, or even in a manner which mixes the terminology. We have, in fact, already done so in the discussions of the preceding paragraphs. The resulting usage is so natural and obvious that no confusion is likely to result.

A systematic tabulation of the terminology and concepts discussed above may be useful for emphasis and for reference. We emphasize the

Real world	Mathematical model
1. Sure event S.	Basic space (sample space) S.
2. Impossible event $\emptyset$.	Empty set (null set) $\emptyset$.
3. Complementary event A^c: A^c is the event that occurs whenever the event A does not occur.	Complement of the set A: A^c is the set of those elements which are not in A.
4. Event A *and* B, usually referred to as the joint event AB or $A \cap B$: AB is the event that occurs iffi both A occurs and B occurs on a given trial.	Intersection $AB = A \cap B$: AB is the set of elements that belong to both A and B.
5. Mutually exclusive events: The joint occurrence of mutually exclusive events is impossible.	Disjoint sets: $AB = \emptyset$; that is, there are no elements common to sets A and B.
6. Event A *or* B, designated $A \cup B$: $A \cup B$ is the event that occurs whenever at least one of the events A, B occurs.	Union $A \cup B$: $A \cup B$ is the set of those elements which belong to at least one of the sets A, B.
7. Event A implies event B, designated $A \subset B$: $A \subset B$ iffi the occurrence of event A requires the occurrence of the event B.	Set A is contained in set B, designated $A \subset B$: $A \subset B$ iffi every element in set A is also in set B.

fact that the language of events comes largely from our ordinary experience by listing various terms in this language in the "real-world" column. In the same column, however, we include symbols which are taken from the set theoretic formulation of our concept of events. In spite of some anomalies here, the tabulation may be useful in making the tie between the real world, where events actually happen, and the mathematical model, in which they are represented in an extremely useful way.

We have extended our mathematical model by the use of the theory of sets. An important addition to our auxiliary model is provided by the *Venn diagrams* (Appendix B), used as an aid in visualizing set combinations and relations. These now afford a means of visualizing combinations of events and relations among them. The Venn diagram, in fact, provides a somewhat simplified version of the urn or jar model. Points in the appropriate regions on the Venn diagram correspond to the balls in the urn. Various subsets may be designated by designating appropriate regions. Determining the outcome of a trial corresponds to picking a point on the Venn diagram. The occurrence of an event A corresponds to the choice of one of that set of points determined by the region delineating the event A. We shall illustrate the use of Venn diagrams in the following discussion.

Classes of events

In the foregoing discussion, events are defined as sets which are subsets of the basic space S; relationships between pairs of events are specified in terms of the appropriate concepts of set theory. These relations and concepts may be extended to much larger (finite or infinite) classes of events in a straightforward manner, utilizing the appropriate features of abstract set theory. Thus we may speak of finite classes of events, sequences of events, mutually exclusive (or disjoint) classes of events, monotone classes of events, etc. The *and* and *or* (intersection and union) combinations may be extended to general classes of events. Where necessary or desirable, we make the modifications of terminology required to adapt to the language of events.

Notation for classes

It frequently occurs that the most difficult part of a mathematical development is to find a way of stating precisely and concisely various conditions or relationships pertinent to the problem. To this end, skill in the use of mathematical notation may play a key role. In dealing with classes of events (sets), we shall often find it expedient to exploit the notational devices described below.

A class of events is a set of events. We shall ordinarily use capital letters A, B, C, etc., to designate events and shall use capital script letters

α, $\mathfrak{B}$, $\mathfrak{C}$, etc., to indicate classes of events. To designate the membership of a class, we use an adaptation of the notation employed for subsets of the basic space S. For example,

$$\alpha = \{A, B, C, D\}$$

is the class having the four member events listed inside the braces.

One of the common means of designating the member events in a class of events is to use the same letter for all events in the class and to distinguish between the various events in the class by appropriate subscripts (or perhaps in some cases, superscripts). When this scheme is used, the membership of the class may be designated by an appropriate specification of which indices are to be allowed. This means that we designate the membership of the class when we specify the *index set*, i.e., the set of all those indices such that the corresponding event is a member of the class. For example, suppose α is the class consisting of four member events A_1, A_2, A_3, and A_4. Then we may write

$$\begin{aligned}
\alpha = \{A_1, A_2, A_3, A_4\} &= \{A_i : i = 1, 2, 3, 4\} \\
&= \{A_i : 1 \leq i \leq 4\} \\
&= \{A_i : i \in J\} \qquad \text{where } J = \{1, 2, 3, 4\}
\end{aligned}$$

We may thus (1) list the member events, (2) list the indices in the index set, (3) give a condition determining the indices in the index set, or (4) simply indicate symbolically the index set, which is then described elsewhere. It is obvious that the last two schemes may be used to advantage when the class has a very large (possibly infinite) number of members. Also, the symbolic designation of the index set J is useful in expressions in which the index set may be any of a large class of index sets. For example, we may simply require that J be any finite index set. We say class α is *countable* iffi the number of member sets is either finite or countably infinite, in which case the index set J is a finite or countably infinite set.

Suppose we wish to consider the *union* of the events in the class α just described. This is the event that occurs iffi at least one of the member events occurs. We speak of this event as the *union of the class* and designate it in symbols in one of the following ways:

$$A_1 \cup A_2 \cup A_3 \cup A_4 = \bigcup_{i=1}^{4} A_i = \bigcup_{i \in J} A_i$$

The latter notation is particularly useful when we may want to consider a general statement true for a large class of different index sets.

If the class of events under consideration is a *mutually exclusive* class, we may designate the union of the class by replacing the symbol $\cup$ by the symbol $\uplus$. Use of the latter symbol not only indicates the union of the class, but implies the stipulation "the events are mutually exclusive." Thus, when we write

$$\biguplus_{i \in J} A_i$$

we are thereby adding the requirement that the A_i be mutually exclusive. It is not incorrect in such a case, however, to use the ordinary union symbol $\cup$ if we are not concerned to display the fact that the events are mutually exclusive.

In a similar way, we may deal with the *intersection* of the events of the class $\mathcal{C}$. This is the event that occurs iff all the member events occur. We speak of this event as the *intersection of the class* (or sometimes as the *joint event* of the class) and designate it in symbols in one of the following ways:

$$A_1 \cap A_2 \cap A_3 \cap A_4 = \bigcap_{i=1}^{4} A_i = \bigcap_{i \in J} A_i$$

Extensions of this notational scheme to various finite and infinite cases are immediate and should be obvious.

Use of the index-set notation is also useful in the case of dealing with sums of numbers. Suppose a_1, a_2, a_3, and a_4 are numbers. We may designate the sum of these numbers in any of the following ways:

$$a_1 + a_2 + a_3 + a_4 = \sum_{i=1}^{4} a_i = \sum_{i \in J} a_i \quad \text{where } J = \{1, 2, 3, 4\}$$

Here we have used the conventional sigma Σ to indicate summation. Extensions to sums of larger classes of numbers parallels the extension of the convention for unions of events to larger classes. We shall find this notational scheme extremely useful in later developments.

Partitions

As our theory develops, we shall find that expressing an event as the union of a class of mutually exclusive events plays a particularly important role. In anticipation of this fact, we introduce and illustrate some concepts and terminology which will be quite useful. As a first step, we make the following

Definition 2-2a

A class of events is said to form a *complete system of events* if at least one of them is sure to occur.

It is immediately apparent that the class $\mathcal{C} = \{A_1, A_2, \ldots, A_n\}$ consisting of the indicated n events is a complete system iff

$$A_1 \cup A_2 \cup \cdots \cup A_n = \bigcup_{i=1}^{n} A_i = S$$

If any element is chosen, it must be in at least one of the A_i (it could be in more than one). A class thus forms a complete system iff the union of the class is the sure event.

One of the simplest complete systems is formed by taking any event A and its complement A^c. One of these two events is sure to occur. Suppose, for example, that during a storm a photographer sets up his camera and opens the shutter for 10 seconds. Let the event A be the event that one or more lightning strokes are recorded. Then the event A^c is the event that no lightning stroke is recorded. One of the two events A or A^c must occur, for $A \cup A^c = S$, the sure event.

As a more complicated complete system, suppose we take a sample of 10 electrical resistors from a large stock. We suppose this choice of 10 resistors corresponds to one trial. The resistors are tested individually to see that their resistance values lie within specified limits of the nominal value marked upon them. Let A_k, for any integer k between 0 and 10, be the event that, for the sample taken, k or more of the resistors meet specification. Then the 11 events $A_0, A_1, \ldots, A_{10}$ form a complete system of events. In any sample taken, some number (possibly zero) of resistors must meet specification. Suppose in a given sample this number is three. Then the element ξ representing this particular sample is an element of each of the events A_3, A_2, A_1, and A_0 and hence is an element of the union of all the events. We may note that we have a class of events satisfying $A_0 \supset A_1 \supset \cdots \supset A_{10}$; that is, we have a monotone-decreasing class. As a matter of fact, the event A_0 is the sure event, for every sample must have zero or more resistors which meet specification.

A more interesting class of events in the resistor-sampling problem is the class $\mathcal{B} = \{B_k : 0 \leq k \leq 10\}$, where for each k the event B_k is the event that the sample has exactly k resistors which meet specification. In this case, the different events in the class are mutually exclusive, since no sample can have exactly k and also exactly $j \neq k$ resistors which meet tolerance. Thus $\mathcal{B}$ is a mutually exclusive class. Now it is apparent that the class $\mathcal{B}$ forms a complete system of events, for every element corresponding to a sample must belong to one (and only one) of the B_k. The particular k in each case is determined by counting the number of resistors in the sample which meet specification.

Complete systems of events in which the members are mutually exclusive are so important that we find it convenient to give them a name as follows:

Definition 2-2b

A class of events is said to form a *partition* iffi one and only one of the member events must occur on each trial.

It is apparent that the class of events $\mathcal{B} = \{B_k : 0 \leq k \leq 10\}$ defined in the resistor-sampling problem above is a partition; so also is the complete system of events $\{A, A^c\}$ illustrated in the photography example. It is

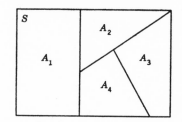

Fig. 2-2-1   Venn-diagram representation of a
partition.

also apparent that a partition is characterized by the two properties

1. The class is mutually exclusive.
2. The union of the class is the sure event S.

The name partition is suggested by the fact that a partition divides the
Venn diagram representing the basic space S into nonoverlapping sets.
While the members of the partition may be quite complicated sets
(representing events), the essential features of a partition are exhibited in
a schematic representation such as that in Fig. 2-2-1.   One deficiency of
such a representation is that it is not clear to which set the points on the
boundary are assigned.   If it is kept in mind that disjoint sets are being
represented, this need cause no difficulty in interpretation.

We may extend the notion of a partition in the following useful way:

Definition 2-2c

A mutually exclusive class $\mathcal{C}$ of events whose union is the event A
is called a *partition of the event A* (or sometimes a *decomposition of A*).

It is obvious that a partition of the sure event S is a partition in the sense
of Def. 2-2*b*.   A Venn diagram of a partition of an event A is illustrated
in Fig. 2-2-2.

As an illustration of the partition of an event, we consider some
partitions of the union of two events.   A certain type of rocket is known
to fail for one of two reasons: (1) failure of the rocket engine because
the fuel does not burn evenly or (2) failure of the guidance system.
Let the elementary outcome ξ correspond to the firing of a rocket.   We
let A be the event the rocket fails because of engine malfunction and

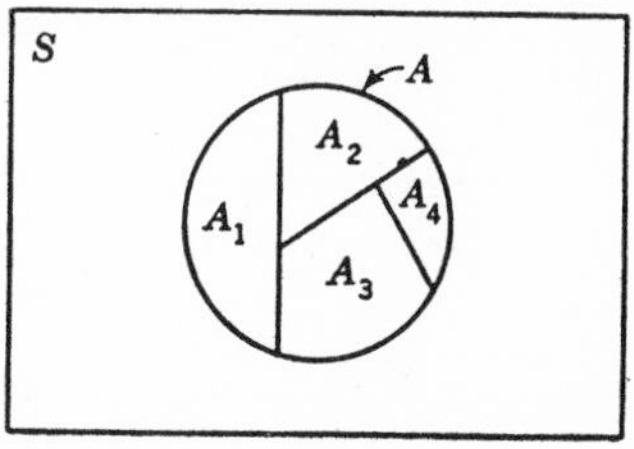

Fig. 2-2-2   Venn-diagram representation of a
partition of an event A.

B be the event the rocket fails because of guidance failure. The event F of a failure of the rocket is thus given by $F = A \cup B$. Now we cannot assert that events A and B are mutually exclusive. We may assert, however, that

$$F = A \uplus A^c B \qquad \text{(engine fails or engine operates and guidance fails)}$$

$$= B \uplus B^c A \qquad \text{(guidance fails or guidance operates and engine fails)}$$

$$= A^c B \uplus AB \uplus AB^c \qquad \text{(engine operates and guidance fails, or both fail, or engine fails and guidance operates)}$$

We thus have three partitions of $A \cup B$, namely, $\{A, A^c B\}$, $\{B, AB^c\}$, and $\{A^c B, AB, AB^c\}$. The first and third of these are illustrated in Fig. 2-2-3.

As a second example, which is quite important in probability theory, consider a mutually exclusive class of events

$$\mathfrak{B} = \{B_1, B_2, B_3, B_4\} = \{B_i : i \in J\}$$

where in this case $J = \{1, 2, 3, 4\}$. Suppose

$$A \subset B_1 \uplus B_2 \uplus B_3 \uplus B_4 = \underset{i \in J}{\uplus} B_i$$

Then the class $\{AB_1, AB_2, AB_3, AB_4\} = \{AB_i : i \in J\}$ is a mutually exclusive class whose union is the event A. Thus $\{AB_i : i \in J\}$ is a partition of the event A. If the class $\mathfrak{B}$ is a partition (of the sure event S), then every A is contained in the union of the class. This manner of producing a partition of an event A by taking the intersections of A with the members of a partition is illustrated in the Venn diagram of Fig. 2-2-4. Note that some of the members of the partition may be

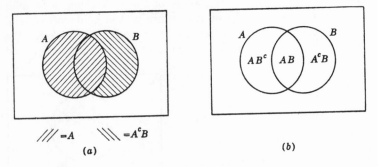

$$/\!/\!/ = A \qquad \backslash\!\backslash\!\backslash = A^c B$$

(a)   (b)

Fig. 2-2-3   Partitions of the union of two events A and B.

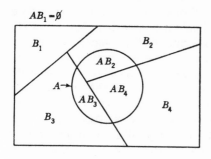

Fig. 2-2-4   A partition of an event A produced by the intersection with a partition of the sure event S.

impossible events (empty sets). These may be ignored when the fact is known. They do not affect the union if included, since they contribute no elementary outcomes. The use of the general-index-set notation in the argument above shows that the result is not limited to the special choice of the mutually exclusive sets.

Because of its importance, we state the general case as

Theorem 2-2A

Let $\mathfrak{B} = \{B_i \colon i \in J\}$ be any mutually exclusive (disjoint) class of events. If the event A is such that $A \subset \biguplus_{i \in J} B_i$, then the class $\{AB_i \colon i \in J\}$ is a partition of A, so that $A = \biguplus_{i \in J} AB_i$.

We shall return to the important topic of partitions in Sec. 2-7.

Sigma fields of events

We have defined events as subsets of the basic space S and have developed a considerable mathematical system based on these ideas. Our discussion has proceeded as if every subset of S could be considered an event, and hence as if the class $\mathcal{E}$ of events could be considered the class of all subsets of S. In the case of a finite number of elementary outcomes, this is the usual procedure. In many investigations in which the basic space is infinite, there are technical mathematical reasons for defining a less extensive class of subsets as the class of events. Fortunately, in applications it is usually not important to examine the question of which of the subsets of S are to be considered as events. For the purposes of this book, we shall simply suppose that in each case a suitable class of subsets of S makes up the class of events. It is usually important to describe carefully the various individual events of interest, but once these are described, they are assumed, without further examination, to belong to an appropriate class.

Development of the mathematical foundations has shown that the class of events must be a *sigma field* of sets.   A brief discussion of such classes is given in Appendix B.   We simply note here that such classes are closed under the formation of unions, intersections, and complements. This means that if the class of events includes a countable class $\alpha = \{A_i : i \in J\}$, then the union of α, the intersection of α, and the complements of any of the A_i must be events also.   This provides the mathematical flexibility required for building up composite events from various events defined in formulating a problem, as subsequent developments show.

Having noted some of the general properties which probability numbers must have and having examined the concept of an event in considerable detail, we are now ready to make a formal definition of probability.

2-3   A formal definition of probability

In Sec. 2-1 we discussed informally several properties of probability which must hold if we are to consider the probability of an event as the relative frequency of occurrences of the event to be expected in a large number of trials.   This informal discussion was restricted by the fact that the concept of an event had not been formulated in a precise way. The subsequent development in Sec. 2-2 was devoted to formulating carefully the concept of an event and relating the resulting mathematical model to the real-world phenomena which it represents.   An event is represented as a set of elements from a basic space; each element of that space represents one of the possible outcomes of a trial or experiment under study; the occurrence of an event A corresponds to the choice of an element from that subset of the basic space which represents the event.   As a result of this development, we have adopted the following mathematical apparatus:

1. A *basic space* S, consisting of *elements* ξ (referred to as *elementary outcomes*)

2. *Events* which are *subsets* of S (more precisely, members of a suitable class of subsets of S)

In this model, the *occurrence* of the event A is the *choice* of an element ξ in the set A (thus ξ is sometimes referred to as the *choice variable*).

We now wish to introduce a third entity into this model: a mathematical representation of probability.   It is desired to do this in a manner that extends the classical probability model and includes it as a special case.   In particular, it is desired to preserve the properties which make possible the relative-frequency interpretation.   We note first that probability is associated with events rather than with individual outcomes.   In the classical case, the probability of an event A is the ratio of the number of elements in A (i.e., the number of ways that A can occur) to the total number of elements in the basic space S (i.e., the total number of possible outcomes).   Thus, in our model, probability must be a *function of sets;* that is, it must associate a number with a set of elements rather than with an individual element.

An examination of the list of properties which probability must have, as developed in Sec. 2-1, shows that probability has the formal properties characterizing a class of functions of sets known in mathematics as *measures.*   If we are able to use this well-known class of set functions as our probability model, we shall be able to appropriate a rich and extensive mathematical theory.   It is precisely this notion of probability as a measure function which characterizes the highly successful Kolmogorov model for probability.   An examination of the theory of measure shows that it is not necessary to postulate all the properties of probability discussed and listed in Sec. 2-1.   Several of these properties can be taken as basic or axiomatic, and the others deduced from them. This trimming of the list of basic properties to a minimum or near minimum often results in an economy of effort in establishing the fact that a set function is a probability measure.   It is also an aid in grasping the essential character of the function under study.   At this point, we shall simply introduce the definition of probability in a formal manner; the justification for this choice rests upon the many mathematical consequences which follow and upon the success in relating these mathematical consequences in a meaningful way to phenomena of interest in the real world.

Probability systems

We are now ready to make the following

Definition 2-3a

A *probability system* (or probability space) consists of the triple:

  1. A *basic space S of elementary outcomes* (elements) ξ
  2. A *class $\mathcal{E}$ of events* (a sigma field of subsets of S)
  3. A *probability measure $P(\cdot)$* defined for each event A in the class $\mathcal{E}$ and having the following properties:

(P1) $P(S) = 1$ (probability of the sure event is unity)
(P2) $P(A) \geq 0$ (probability of an event is nonnegative)
(P3)   If $\mathcal{C} = \{A_i : i \in J\}$ is a countable partition of A (i.e., a mutually exclusive class whose union is A), then

$$P(A) = \sum_{i \in J} P(A_i) \qquad \text{(additivity property)}$$

The *additivity property* (P3) provides an essential characterization of a probability measure $P(\cdot)$. It may be stated in words by saying that *the probability of the union of mutually exclusive events is the sum of the probabilities of the individual events.* We have seen how this property is required for a finite number of events in order to preserve a fundamental property of relative frequencies. The extension in the mathematical model to an infinity of events is needed to allow the analytical flexibility required in many common problems. For technical mathematical reasons, this property is limited to a *countable* infinity of events. A countably infinite class is one whose members may be put into a one-to-one correspondence with the positive integers. It is in this sense that the infinity is "countable." As an example of a simple situation in which an infinity of mutually exclusive events is required for analysis, consider

Example 2-3-1

Suppose a game is played in which two players alternately toss a coin. The player who first tosses a "head" is the winner. In principle, the game as defined may continue indefinitely, for it is not possible to specify in advance how many tosses will be required before one player wins. Let A_k be the event that the first player (i.e., the player who makes the first toss in the sequence) wins on the kth toss. We note that a win is impossible for this player on the even-numbered throws. Thus $A_k = \emptyset$ for k even. We note that any two events A_k and A_j for different j and k must be mutually exclusive, since we cannot have a head appear for the first time in a given sequence on both the kth and the jth toss. Thus the class $\mathfrak{C} = \{A_1, A_3, A_5, \ldots\} = \{A_i : i \in J_0\}$, where J_0 is the set of odd positive integers, is a mutually exclusive class. Also, the event A of the first man's winning is the union of the class $\mathfrak{C}$. This means that the first man wins iff he wins on one of the odd-numbered throws in the sequence which corresponds to the compound trial defining the game. We thus have

$$A = \biguplus_{i \in J_0} A_i$$

If we can calculate the probabilities $P(A_i)$, we can determine the probability of the event A by the relation

$$P(A) = \sum_{i \in J_0} P(A_i)$$

We shall see later that the $P(A_i)$ may be determined quite easily under certain natural assumptions. ∎

We shall encounter a great many problems in which it is desirable to be able to deal with an infinity of events.

The example above illustrates a *basic strategy* in dealing with probability problems. It is desired to calculate the probability $P(A)$ that the first man wins the game. The event A can be expressed as the union of the mutually exclusive events A_k. Under the usual conditions and assumptions (to be discussed later), it is easy to evaluate the probabilities $P(A_k)$. Evaluation of $P(A)$ is then made by use of the fundamental additivity rule.   The key to the strategy consists in finding a partition of the event A such that the probabilities of the member events of the partition may be determined readily.

Before considering some of the properties of the probability measure $P(\cdot)$ which follow as consequences of the three axiomatic properties used in the definition, let us verify the fact that the new definition includes the classical probability as a special case.   In order to do this, we must show that the three basic properties hold for the classical probability measure.   In the classical case, the basic space has a finite number of elements and the class of events is the class of all subsets.   Let the symbol $n(A)$ indicate the number of elements in event A; that is, $n(A)$ is the number of ways in which event A can occur.   For disjoint (mutually exclusive) events, $n(A \uplus B) = n(A) + n(B)$.   The classical probability measure is defined by the expression $P(A) = n(A)/n(S)$.   From this it follows easily that

(P1) $P(S) = \dfrac{n(S)}{n(S)} = 1$

(P2) $P(A) = \dfrac{n(A)}{n(S)} \geq 0$

(P3) $P(A \uplus B) = \dfrac{n(A \uplus B)}{n(S)} = \dfrac{n(A) + n(B)}{n(S)} = P(A) + P(B)$

The property (P3) can easily be extended to any finite number of mutually exclusive events.   In the case of a finite basic space, there can be only a finite number of nonempty, mutually exclusive events.   The demonstration shows that the axioms are consistent, in the sense that there is at least one probability system satisfying the axioms.   As we shall see, there are many others.

Other elementary properties

In order that the probability of an event be an indicator of the relative frequency of occurrence of that event in a large number of trials, it is immediately evident that several properties are required of the probability measure.   Several of these are listed near the end of Sec. 2-1. In defining probability, we chose from this list three properties which are taken as basic or axiomatic.   We now wish to see that the remaining properties in this list may be derived from the three basic properties. In addition, we shall obtain several other elementary properties which are fundamental to the development of the theory and to its application to real-world problems.

Property 4 in the list in Sec. 2-1 states that the probability that an event will not occur is 1 minus the probability of the occurrence of the event.   This follows very easily from the fact that for any event A, the pair $\{A, A^c\}$ forms a partition; that is, $A \uplus A^c = S$.   From properties (P1) and (P3) we have immediately $P(A) + P(A^c) = P(S) = 1$, so that we may assert

(P4) $P(A^c) = 1 - P(A)$

As a special case, we may let $A = S$, the sure event, so that $A^c = \emptyset$. As a result we have

(P5) $P(\emptyset) = 0$

The probability of the impossible event is zero.   This with property (P1) completes property 1 in Sec. 2-1.

We have now included in our model all the properties listed in Sec. 2-1 except property 5.   This property states that if the occurrence of one event A implies the occurrence of a second event B on the same trial (i.e., if $A \subset B$), then $P(B) \geq P(A)$.   We first note that the condition $A \subset B$ implies $A \cup B = B$, so that $B = A \uplus A^c B$.   Since every element in A is also in B, the elements in B must be those which are either in A or in that part of B not in A.   By property (P3) we have $P(B) = P(A) + P(A^c B)$.   Since the second term in the sum is non-negative by property (P2), it follows immediately that $P(A) \leq P(B)$. For reference, we state this property formally as follows:

(P6)   If $A \subset B$, then $P(A) \leq P(B)$.

As an illustration of the implications of this fact, consider the following

Example 2-3-2   *Serial Systems in Reliability Theory*

In reliability theory one is concerned with the success or failure of a system or device. Usually the system is analyzed into several subsystems whose success or failure affects the success or failure of the larger system.   The larger system is called a *serial system* if the failure of any one subsystem causes the complete system to fail. Such may be the case for a complex missile system with more than one stage.   Failure of any stage results in failure of the entire system.   If we let

A = event entire system succeeds
A_i = event ith subsystem is successful

then

$P(A) = R$ is called *reliability of the system*
$P(A_i) = R_i$ is called *reliability of ith subsystem*

A serial system composed of n subsystems is characterized by

$$A = A_1 A_2 \cdots A_n = \bigcap_{i=1}^{n} A_i$$

or equivalently by

$$A^c = A_1{}^c \cup A_2{}^c \cup \cdots \cup A_n{}^c = \bigcup_{i=1}^{n} A_i{}^c$$

That is, the system succeeds iffi all the subsystems succeed, or the system fails iffi any one or more of the subsystems fail. This relationship requires $A \subset A_i$ for each $i = 1, 2, \ldots, n$. By property (P6) we must therefore have $R \leq R_i$ for each i. That is, the serial system reliability is no greater than the reliability of any subsystem. ∎

We now wish to develop several other properties of the probability measure which are used repeatedly in both theory and applications. In Sec. 2-2 we discussed in terms of an example the following partitions of the event $A \cup B$: $\{A, A^cB\}$, $\{B, AB^c\}$, and $\{A^cB, AB^c, AB\}$. The first and third of these are illustrated in Fig. 2-2-3. Use of property (P3) gives several alternative expressions for the probability of $A \cup B$.

$$\textbf{(P7)} \quad \begin{aligned} P(A \cup B) &= P(A) + P(A^cB) \\ &= P(B) + P(AB^c) \\ &= P(A^cB) + P(AB^c) + P(AB) \\ &= P(A) + P(B) - P(AB) \end{aligned}$$

The first three expressions are obtained by direct application of property (P3) to the successive partitions. The fourth expression is obtained by an algebraic combination of the first three in the form $P(A \cup B) + P(A \cup B) - P(A \cup B)$. These expressions may be visualized with the aid of a Venn diagram, as in Fig. 2-3-1. Probabilities of nonoverlapping sets add to give the probability of the union. If one simply adds $P(A) + P(B)$, he includes $P(AB)$ twice; hence it is necessary to subtract $P(AB)$ from the sum, as in the last expression in (P7). If the events A and B are mutually exclusive (i.e., if the sets are disjoint), $AB = \emptyset$ and, by property (P5), $P(AB) = 0$. Thus, for mutually exclusive events, property (P7) reduces to a special case of the additivity rule (P3).

The next example provides a simple illustration of a property which is frequently useful in analysis.

Example 2-3-3

A manufacturing process utilizes a basic raw material supplied by companies 1, 2, and 3. A batch of this material is drawn from stock and used in the manufacturing process. The material is selected at random but is packaged in such a way that the batch drawn is from only one of the suppliers. Let A be the event that the finished product meets specifications. Let B_k be the event that the raw material chosen is supplied by company numbered k ($k = 1, 2,$ or 3). When event A occurs, one and

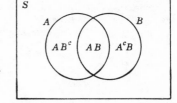

Fig. 2-3-1   A partition of the union $A \cup B$ of two events.

only one of the events B_1, B_2, or B_3 occurs. We thus have the situation formalized in Theorem 2-2A. The B_k form a mutually exclusive class. Occurrence of the event A implies the occurrence of one of the B_k. That is, $A \subset B_1 \uplus B_2 \uplus B_3$. From this it follows that $A = AB_1 \uplus AB_2 \uplus AB_3$. By property (P3) we have $P(A) = P(AB_1) + P(AB_2) + P(AB_3)$. If the information about the selection process and the characteristics of the material supplied by the various companies is sufficient, it may be possible to evaluate each of the joint probabilities $P(AB_k)$. A discussion of the manner in which this is usually done must await the introduction of the idea of conditional probability, in Sec. 2-5. ∎

The pattern illustrated in this example is quite general. We may utilize the formalism of Theorem 2-2A to give a general statement of this property as follows.

(P8)   Let $\mathfrak{B} = \{B_i : i \in J\}$ be any countable class of mutually exclusive events. If the occurrence of the event A implies the occurrence of one of the B_i (i.e., if $A \subset \underset{i \in J}{\uplus} B_i$), then $P(A) = \displaystyle\sum_{i \in J} P(AB_i)$.

It should be noted that if the class $\mathfrak{B}$ is a partition (of the whole space), then the occurrence of any event A implies the occurrence of one of the B_i, so that the theorem is applicable. The validity of the general theorem follows from Theorem 2-2A, which states that the class $\{AB_i : i \in J\}$ must be a partition of A, and from the additivity property (P3).

Example 2-3-4

Let us return to the coin-tossing game considered in Example 2-3-1. We let A_k be the event that the first man wins by tossing the first head in the sequence on the kth toss. We may let k be any positive integer; the fact that $A_k = \emptyset$ for k even will not affect our argument. Let us now consider the event B_k that the first man wins on the kth toss or sooner. We have a number of interesting relationships. For one thing, $B_k \subset B_{k+1}$ for any k since, if the man wins on the kth toss or sooner, he certainly wins on the $(k + 1)$st toss or sooner. Also, $B_k = \underset{i=1}{\overset{k}{\uplus}} A_i$ since the man wins on the kth toss or sooner iffi he tosses a head for the first time on some toss between the first and the kth. We note also that $A_k = B_k B_{k-1}^c$; that is, the man tosses a head for the first time on the kth toss in the sequence iffi he tosses it on the kth toss or sooner but not on the $(k - 1)$st toss or sooner. We must also have the event A that the first man wins expressible as

$$A = \underset{i=1}{\overset{\infty}{\uplus}} A_i = \bigcup_{k=1}^{\infty} B_k$$

We note that the second union is not the union of a disjoint class. In fact, since the B_k form an increasing sequence, we may assert (see the discussion of limits of sequences in Appendix B) that $A = \underset{k}{\lim} B_k$. By the additivity property,

$$P(A) = \sum_{i=1}^{\infty} P(A_i) = \lim_{k \to \infty} \sum_{i=1}^{k} P(A_i)$$

The last expression is the definition of the infinite sum given in the second expression. Again, by the additivity property,

$$P(B_k) = \sum_{i=1}^{k} P(A_i)$$

From this we infer the interesting fact that

$$P(A) = P(\lim_k B_k) = \lim_{k \to \infty} P(B_k) \blacksquare$$

An examination of the argument in the example above shows that the last result does not depend upon the particular situation described. It depends upon the fact that the B_k form an increasing sequence and that the union of the B_k can be expressed as the disjoint union of the A_k in the manner shown. Further examination shows that if we had started with any increasing sequence $\{B_k: 1 \le k < \infty\}$ and had defined A_k by the relation $A_k = B_k B_{k-1}^c$, the A_k and B_k would have the properties and relationships utilized in the problem. Thus the final result holds generally for increasing sequences.

One can carry out a similar argument for a decreasing sequence. If $\{C_k: 1 \le k < \infty\}$ is a sequence such that $C_i \supset C_{i+1}$ for every i, we have

$$C = \lim_k C_k = \bigcap_{k=1}^{\infty} C_k.$$ By well-known rules for complements,

$$C^c = \bigcup_{k=1}^{\infty} C_k{}^c$$

and the events $C_k{}^c$ form an increasing sequence. We may use the result to assert

$$P(C^c) = 1 - P(C) = \lim_{k \to \infty} P(C_k{}^c) = 1 - \lim_{k \to \infty} P(C_k)$$

From this it follows that $P(\lim_k C_k) = \lim_{k \to \infty} P(C_k)$. We may summarize these results in the following statement.

(**P9**) If $\{A_n: 1 \le n < \infty\}$ is a decreasing or an increasing sequence of events whose limit is the event A, then

$$\lim_{n \to \infty} P(A_n) = P(A).$$

We add one more property which is frequently useful in making probability estimates. For the present we simply state it as a mathematical theorem whose usefulness remains to be demonstrated.

(**P10**) Let $\mathfrak{a} = \{A_i: i \in J\}$ be any finite or countably infinite class of events, and let A be the union of the class. Then

$$P(A) \le \sum_{i \in J} P(A_i) \qquad \text{(subadditivity property)}$$

PROOF We consider the case of an infinite sequence of events; the finite case may be considered as a special case of the infinite case by setting all $A_i = \emptyset$ for $i > n$. We may use the following device to write A as a disjoint union. Let $B_i = A_i \cap \bigcap_{j=1}^{i-1} A_j^c$; then, for any $i \neq j$, the events B_i and B_j must be mutually exclusive. A careful reading of the following expressions shows that

$$\bigcup_{i=1}^{n} A_i = \biguplus_{i=1}^{n} B_i \quad \text{for any } n \quad \text{and} \quad \bigcup_{i=1}^{\infty} A_i = \biguplus_{i=1}^{\infty} B_i$$

By property (P6), $P(B_i) \leq P(A_i)$ for each i, since $B_i \subset A_i$. By the additivity property (P3) and general rules on inequalities,

$$P(A) = \sum_{i=1}^{\infty} P(B_i) \leq \sum_{i=1}^{\infty} P(A_i) \quad \blacksquare$$

For the most part, the properties listed above have been discussed in terms of the frequency interpretation or have been illustrated in terms of real-world situations translated into the mathematical model. Their full significance for probability theory, however, will become evident only as the theory is developed further. Some of the properties have fairly obvious significance in terms of the relative-frequency interpretation; others find their principal importance as analytical tools for developing further special properties and for carrying out solutions of model problems in a systematic manner. The list of properties is certainly not exhaustive, but both the properties themselves and the arguments employed to verify them are useful in further development of the theory.

Determination of probabilities

The discerning reader will have noted that the mathematical model for probability introduced in this section does not tell how to determine the probabilities of events, except for those very special events the sure event S and the impossible event $\emptyset$. This is characteristic of the theory. It tells how the probabilities of various events may be related when the events themselves stand in certain logical relationships. But the theory does not tell what the probability of a given event may be. In the classical case, to be sure, a specific rule is given for determining the probability of any event, once that event is characterized in terms of its membership, i.e., once it is known which elementary outcomes result in the occurrence of the event. But empirical evidence has shown that this rule is not always applicable in a good model. Generally, it is necessary to resort to empirical evidence to determine how probabilities are distributed among events of interest. Such evidence can never be conclusive. One must be content with assumptions or estimates based on experience. Even when the classical model is employed, we can merely assume and test against experience the fact that this is the best model available (or is at least a satisfactory approximation).

The nature of probability theory is such that *if* the probabilities can be determined for certain particular events (say, by assuming some rule of distribution of probabilities among these events, as in the classical case), then the probabilities for various other events of interest may be calculated. It often happens that it is difficult to observe directly the empirical evidence for the probability assumed for some particular event (or class of events). There may be reasons to guess what values these probabilities should have. From these assumed values it may then be possible to infer the consequent probabilities of other events which are more easily tested empirically. If the inferred probabilities agree sufficiently well with the empirically observed frequencies of occurrence, then one may use the originally assumed probabilities with a considerable measure of confidence. This is precisely the character of much of modern statistical theory. A distribution of probabilities over a given class of events is assumed; deductions are made concerning the probabilities of related events; these probabilities are tested against statistical evidence; if the "fit" is good enough, one proceeds to carry out the analysis on the basis of the assumed probabilities. Technique in these matters may become quite sophisticated and difficult.

We shall touch only briefly at a few points on the problem of determining empirically the correct probabilities to assume. For the most part, we shall assume that in any given problem it is possible to arrive at reasonable values for the required probabilities. We devote most of our attention—in so far as we attempt to solve problems—to the matter of determining other probabilities implied by these assumptions. We may rely on intuition, experience, or just exploratory guesses to obtain the basic probabilities. We shall be concerned, however, to formulate our problems so that it is plain what probabilities and what conditions among events are assumed. It is desirable to make these assumptions as simple and as natural as possible, and they should be stated in such a way that their validity is most easily tested against experience and judgment based on experience.

2-4 An auxiliary model—probability as mass

In this section we present an auxiliary model which serves as an aid in grasping the essential features of the basic mathematical model for probability. This auxiliary model provides concrete imagery to aid in thinking about the patterns and relations that exist in the abstract model. A probability measure has been characterized as a completely additive, nonnegative set function. Such measures seem to be somewhat strange and remote from the ordinary experience of life. As a matter of fact, the concept of measure arose in dealing with quite familiar physical and geometrical quantities. In our ordinary physical experience, we are quite familiar with several important quantities which may be represented by functions of this type.

At the most elementary level is the concept of the *volume* of a region in space. The set is the set of geometrical points usually taken to represent such regions. Volume is a number associated with such a set; it is nonnegative, and the volume of the combination (union) of any finite number of nonoverlapping regions is the sum of the volumes of the separate regions. For analytical purposes, it is usually assumed that this additivity extends to countably infinite combinations.

In a similar manner we may consider the *mass* located in a region of space. Again, the set is the set of geometrical points representing the region. A nonnegative number called the mass is associated with each such set of points. The mass in any two nonoverlapping (i.e., disjoint) regions is the sum of the masses in the two regions. This is the additivity property, which is usually assumed in physical theory to extend to masses in any countable infinity of nonoverlapping regions.

Mass may be viewed as continuously distributed. In such a situation, the mass associated with any region consisting of a single point is zero. But sets of points with positive volume may have nonzero masses. In addition to continuously distributed masses, there may be mass concentrations in the neighborhood of a point or along a line or over a surface; these may be idealized into point masses, line distributions of mass, or surface distributions of mass.

A similar concept is that of the electric charge in a region in space. The physical picture is similar to that of mass, except that the charge may be negative as well as positive.

The mathematical concepts corresponding to these physical concepts and, in fact, abstracted from them are:

1. *Measure* as a nonnegative, completely additive set function. In physical expressions, the basic space is the euclidean space, which is an idealization of physical, geometrical space. The more general mathematical model has been obtained by extending the domain of the measure function to abstract spaces, in which ordinary geometrical concepts of volume, continuity, and the like are not applicable. The essential patterns are, for the most part, those which are encountered in representations of ordinary physical situations.

2. *Signed measure*, which is the extension of the concept of measure obtained by removing the restriction to nonnegative values. It serves as the mathematical model of the classical concept of charge, when the

space is euclidean.   But mathematically, the space may be quite abstract, as in the case of ordinary measure functions.

A probability measure $P(\cdot)$ is a measure function on an abstract space.   It is characterized by the fact that the value assigned to the whole space S is unity.   If the space S is visualized as a set of points in ordinary physical space, as in the case of Venn diagrams, *the probability of any event* (set) *may be visualized as a mass associated with the set of points*.   The total mass has unit value.

This physical picture is quite valid and serves as an important means of visualizing abstract relationships.   To make the physical picture quite concrete, one may think of the Venn diagram as consisting of a thin sheet of material of variable density.   When a set of points is determined by marking out a region, the mass associated with this region coincides with the usual physical picture.   Thus, if probability is viewed as mass, we have a concrete physical picture of the abstract concept of a probability measure.

We shall frequently appeal to the mental image of probability as mass by referring to the *probability mass* associated with a given set of points corresponding to a specified event.   We may exploit this conceptual image as an aid in visualizing or even discovering abstract relations.   When the relations so visualized are precisely formulated, their validity may be established without dependence upon the geometrical picture.   However, the construction of an analytical proof is often aided by proper mental visualization.

It should be apparent that a wide variety of mass distributions are possible on any given basic space.   This is precisely in keeping with the fact that many probability measures may be defined on the same basic space.   In order to illustrate the usefulness of the mass picture, we consider an important special class of probability systems.

Discréte probability spaces

In many investigations, only a finite number of outcomes of a trial have nonzero probabilities; these probabilities total to unity.   In such a case, the probability mass is concentrated at a finite number of points on the basic space.   For purposes of probability theory, any other points in the basic space may be ignored, so that the basic space is considered to be a finite space.   In other investigations, the number of outcomes may be extended to a countable infinity of possibilities, with probabilities concentrated on a discrete set of points in the basic space.   Such discrete probability spaces are quite important in applications.

We consider a basic space S which has a finite or countably infinite number of elements.   Then $S = \{\xi_i : i \in J\}$, where J is a finite or countably infinite index set.   As the class of events $\mathcal{E}$, we take the *class of all subsets* of S.   We note that if S is finite with N elements, $\mathcal{E}$ has 2^N sets as members.   It is easy to see that this is a completely additive class.

In order to complete the probability space (or system), it is only necessary to assign a probability function.   This is done by the simple expedient of associating with each point ξ_i in the space a probability mass p_i.   Strictly speaking, the probability mass is to be associated with the elementary event $\{\xi_i\}$ rather than with the elementary outcome ξ_i, but this need cause no confusion.   To each event (i.e., to each subset of the basic space) is associated the probability mass carried by the totality of the points in the set.   The sum of the p_i must be unity, and each number p_i must be nonnegative (in fact, we often assume each p_i to be strictly positive).   It is sometimes helpful to consider the basic space as a collection of balls.   Each ball has an appropriate mass.   An event corresponds to an appropriate subcollection, or subset, of the balls.   The probability of an event A is the total mass assigned to the balls in that subset.   Since the mass of all the balls in the whole collection corresponding to the basic space is unity, the mass associated with the event A is equal to the fraction of the total mass assigned to the balls in the subset.

Example 2-4-1

The best-known example of such a space is the *classical probability* model, for which $J = \{1, 2, \ldots, N\}$ and $p_i = 1/N$.   The number of elements $n(A)$ in an event A is the number of elements $n(J_A)$ in its index set J_A.   The probability assigned to any event A is, in these terms, $P(A) = n(J_A)/N = n(A)/N$.

Example 2-4-2

Consider the results of throwing a pair of dice.   Suppose (1) the dice are not distinguished and (2) that only the sum of the numbers of spots appearing is of interest. The possible numbers appearing are the integers 2 through 12.   It is convenient to think of each of these as corresponding to one of the elements $\xi_2, \xi_3, \ldots, \xi_{12}$.   The index set J consists of the integers 2 through 12.   The elementary event $\{\xi_7\}$, for example, is the event that a seven is thrown.   To each of these elementary events can be assigned the probabilities calculated from the classical model, under the usual assumptions. Thus $p_2 = \frac{1}{36}$, $p_7 = \frac{1}{6}$, etc.   When these numbers are determined (or assumed), the probability measure $P(\cdot)$ is defined for all subsets of the basic space. ∎

The foregoing example is of some interest because it shows that there may be no unique mathematical model for a given situation.   The classical model could have been used for this situation.   In fact, we derived the present model from the classical case.   For some purposes, however, the model just described is the more convenient to use.

We may summarize once more the model we have developed in the following table.

Real world	Mathematical model	Auxiliary model
1. Outcomes of trials	Elements of basic space S	Points on Venn diagram or balls in urn
2. Events	Subsets of elements	Portions of Venn diagram or sets of balls in urn
3. Relative frequency of occurrence of events	Probabilities	Mass of set of points or set of balls

2-5 Conditional probability

Much of the success of the mathematical theory of probability is made possible by an extension of the basic model to include the idea of conditional probability. As we have developed the theory, a basic set S of possible outcomes is assumed. It frequently occurs, however, that one has information about a trial which assures the occurrence of a particular event E. If this condition exists, the set of possible outcomes is in effect modified and one would change his point of view in assessing the "chances" of the occurrence of an event A. If it is given that the event E must occur (or has occurred), the only possible elementary outcomes are those which are in the set E. Thus, if the event A occurs, the elementary outcome ξ chosen must be in both A and E; that is, the event AE must occur. In such a situation, how should probabilities be assigned?

Again we may turn to the concept of relative frequency for the pattern to be incorporated into the mathematical model. If the outcome of a trial is to be conditioned by the requirement that the event E must occur, one would often be interested in the relative frequency of occurrences of the event A among the outcomes ξ for which E also occurs. Suppose in a large number n of trials the event E occurs n_E times; and suppose that among those trials resulting in the event E the event A also occurs n_{AE} times. The number n_{AE} is the number of occurrences of the joint event AE. The relative frequency of interest is n_{AE}/n_E. For sufficiently large n, we should expect this ratio to lie close to some ideal number, which we call the *conditional probability* of the event A, given the event E. Now we note that the relative frequency of occurrence of event E is $f_E = n_E/n$ and that of event AE is $f_{AE} = n_{AE}/n$. Simple algebra shows that

$$\frac{n_{AE}}{n_E} = \frac{n_{AE}/n}{n_E/n} = \frac{f_{AE}}{f_E}$$

Since the relative frequency f_{AE} is expected to lie close to the probability

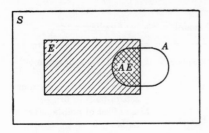

<figure>*Fig.* 2-5-1   Diagram indicating the interpretation of conditional probability in terms of the mass picture.</figure>

$P(AE)$ and the relative frequency f_E is expected to lie close to the probability $P(E)$, the ratio n_{AE}/n_E should lie close to the ratio $P(AE)/P(E)$.

On the basis of this argument, it would seem natural to introduce into the mathematical model the concept of conditional probability in the following manner:

Definition 2-5a

If E is an event with positive probability, the *conditional probability of the event* A, *given* E, written $P(A|E)$, is defined by the relation

$$P(A|E) = \frac{P(AE)}{P(E)}$$

The analytical reason for requiring $P(E) > 0$ is evident from the definition. Generally, when we write a conditional probability, we shall assume tacitly that the conditioning event has positive probability. For any fixed E with positive probability, it is apparent that $P(\cdot|E)$ is a function defined for every event A.

We may examine this function in terms of the mass picture for probability. Suppose E is any fixed event with positive probability, as represented on the Venn diagram of Fig. 2-5-1. This event has assigned to it probability mass $P(E)$. The event AE has assigned to it probability mass $P(AE)$. The conditional probability of A, given E, is thus the fraction of the probability mass assigned to event E which is also assigned to event A. The probability $P(A)$, on the other hand, is the fraction of the total probability mass (assigned to the sure event S) which is assigned to event A. In this regard, we note that

$$P(A|S) = P(A)$$

In fact, we require only that $P(E) = 1$ to ensure $P(A|E) = P(A)$. To show that this is not a trivial observation, we must await further development of the theory.

On the basis of the mass picture, a number of properties of the conditional probability function $P(\cdot|E)$ are readily visualized. The conditional probability of an event is a nonnegative number which never exceeds unity. It must have the value 1 when $A = S$. Also, the conditional probability function must have the additivity property. This fact may be verified by the following analytical argument, although it should be apparent from the mass picture:

$$P(\underset{i}{\uplus} A_i | E) = \frac{P(E \cap [\underset{i}{\uplus} A_i])}{P(E)} = \frac{P(\underset{i}{\uplus} EA_i)}{P(E)} = \frac{\underset{i}{\sum} P(EA_i)}{P(E)}$$

$$= \sum_i P(A_i | E)$$

The properties just noted are precisely the fundamental properties (P1), (P2), and (P3) which characterize a probability measure; therefore we conclude that *a conditional probability function is a probability measure*. As such, the conditional probability measure must have all the properties common to other probability measures. In addition, it has special properties which follow from the fact that it is derived from the original probability measure $P(\cdot)$. In order to distinguish between the original probability measure $P(\cdot)$ and the conditional probability measure $P(\cdot|E)$, we shall refer to $P(\cdot)$ as a *total probability measure* and to its values as total probabilities.

Since a conditional probability is in fact a probability measure, it is sometimes difficult to know from a statement of a problem whether the total probability or a conditional probability is to be determined. In fact, in formulating a real-world problem it is not always clear which probability should be obtained to answer the question actually posed by the real-world situation. The following example presents such a dilemma.

Example 2-5-1

Three of five prisoners are to be shot. Their names are selected by drawing slips of paper from a hat. Prisoner 1 has figured the odds on his being selected. He then persuades the guard to point out one of the persons (other than himself) whose name has been chosen. Prisoner 3 is designated. Prisoner 1 then proceeds to recalculate his chances of being among those selected for execution. What probability does he calculate? Does the information that 3 is to be shot give him grounds for more hope that he will escape a similar fate?

SOLUTION   The classical situation of equally likely choices is assumed for the selection process. We let A_1 be the event prisoner 1 will be shot and A_3 be the event prisoner 3 will be shot. It is apparent that the first probability calculated (before knowledge of the occurrence of event A_3) is $P(A_1)$. Event A_1 can occur in C_2^4 ways, since this is the number of ways two objects can be chosen from among four (the fifth one having been fixed). The total number of selections of the three prisoners is C_3^5. Hence, using the classical model, we have $P(A_1) = C_2^4/C_3^5 = \frac{3}{5}$. Once the fact of the occurrence of A_3 is known, the question arises: What probability should serve as the basis of estimating the prisoner's chances of being shot? Some would argue that nothing has been changed about the selection by the knowledge that A_3 has occurred. Such an argument would suppose that $P(A_1)$ is the proper probability to use. A second possible argument is that since A_3 has occurred, it is the event A_1A_3 that is of interest; thus the desired probability is $P(A_1A_3)$. This may be calculated to be $C_1{}^3/C_3{}^5 = \frac{3}{10}$. The third point of view—and the one most commonly taken in applications—argues that knowledge of the occurrence of A_3 in effect sets up a new basic space. Thus the desired probability is $P(A_1|A_3)$. This may be calculated as follows:

$$P(A_1|A_3) = \frac{n(A_1A_3)}{n(A_3)} = \frac{C_1{}^3}{C_2{}^4} = \frac{1}{2}$$

On the basis of this probability, prisoner 1 would be justified in having a little more hope for escape than before he learned of the fate of prisoner 3. This approach has been criticized on the ground that it is meaningless or improper to deal with the event A_3. It is argued that some assumption must be made (or knowledge obtained) as to *how* the guard decides to tell prisoner 1 that prisoner 3 is to be shot. If, as is customary, it is assumed that the prisoner knows that the guard has selected at random from among the prisoners to be shot, after eliminating number 1 if present, then we have a different conditioning event B_3. Under the usual assumptions about the meaning of "at random," this gives

$$P(B_3) = 1/4, \ P(A_1B_3) = 3/20, \text{ and hence } P(A_1|B_3) = 3/5 = P(A_1)$$

Under this condition, knowledge of the occurrence of B_3 does not change the likelihood of A_1, interpreted as probability conditioned by the information given. We are thus confronted with a second critical issue in formulating the problem. Exactly what information is given; hence, what conditioning event is determined? When clarity on this point is attained and a decision as to which probability is meaningful is reached, the purely mathematical problem posed admits of an unambiguous solution. ∎

The defining relation for conditional probability may be rewritten as a *product rule*

$$P(AE) = P(E)P(A|E)$$

This rule may be extended to the case of the joint occurrence of three or more events in the following way:

$$
\begin{aligned}
P(A_1A_2A_3) &= \frac{P(A_1)}{P(A_1)} \frac{P(A_1A_2)}{P(A_1A_2)} P(A_1A_2A_3) \\
&= P(A_1) \frac{P(A_1A_2)}{P(A_1)} \frac{P(A_1A_2A_3)}{P(A_1A_2)} \\
&= P(A_1)P(A_2|A_1)P(A_3|A_1A_2)
\end{aligned}
$$

We assume, of course, that $P(A_1) > 0$ and $P(A_1A_2) > 0$, so that the conditional probabilities are defined. By a simple inductive argument, it is easy to extend the result to any finite number n of events. We thus have

(CP1)   *Product rule for conditional probability*
$$P(A_1A_2 \cdots A_n) = P(A_1)P(A_2|A_1)P(A_3|A_1A_2) \cdots P(A_n|A_1A_2 \cdots A_{n-1})$$

As an illustration of the use of the product rule, consider the following example. There are several ways in which the solution might be carried out; the method chosen demonstrates the usefulness of the product rule (CP1).

Example 2-5-2

Suppose a jar contains two white balls and three black balls. The balls are drawn from the jar and placed on the table in the order drawn. What is the probability that they are drawn in the order white, black, black, white, black?

SOLUTION

Let W_i be the event that the ith ball drawn is white
B_i be the event that the ith ball drawn is black

$$(B_i = W_i^c)$$

By (CP1),

$$P(W_1 B_2 B_3 W_4 B_5) = P(W_1)P(B_2|W_1)P(B_3|W_1 B_2)P(W_4|W_1 B_2 B_3)P(B_5|W_1 B_2 B_3 W_4)$$

On the first choice, there are two white balls and three black balls, as shown in Fig. 2-5-2a. We therefore assume $P(W_1) = \frac{2}{5}$. The situation for the second choice, given that the first choice resulted in drawing a white ball (i.e., given W_1), is that there is one white ball and three black balls, as shown in Fig. 2-5-2b. It is natural to assume $P(B_2|W_1) = \frac{3}{4}$. Conditions for the third and fourth choices are shown in Fig. 2-5-2, leading to the assumptions $P(B_3|W_1 B_2) = \frac{2}{3}$ and $P(W_4|W_1 B_2 B_3) = \frac{1}{2}$. The final choice, under the conditions given, is the drawing of a black ball from one black ball. The conditional probability $P(B_5|W_1 B_2 B_3 W_4)$ must be unity. Substitution of these values in the formula above gives the desired probability value of $\frac{1}{10}$. It may be noted that this result is consistent with one based on the classical probability model, where the result is determined by a careful counting of the ways the five balls could be chosen to satisfy the desired conditions. ∎

Use of the product rule makes possible an extension of the basic strategy discussed in Sec. 2-3. There it is pointed out that a standard technique is to express an event as the disjoint union of events whose probabilities are known or which can be obtained from the assumed data. This strategy may be extended in many cases in the following way. The event of interest is expressed as the disjoint union of events, each of which is the intersection of one or more events. The probabilities are calculated by use of the product rule of conditional probability

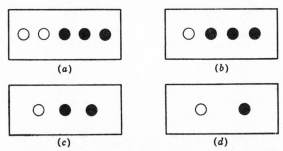

Fig. 2-5-2 Basis for assumption of the values of conditional probabilities in Example 2-5-2. (a) First choice: $P(W_1) = \frac{2}{5}$; (b) second choice, given W_1: $P(B_2|W_1) = \frac{3}{4}$; (c) third choice, given $W_1 B_2$: $P(B_3|W_1 B_2) = \frac{2}{3}$; ($d$) fourth choice, given $W_1 B_2 B_3$: $P(W_4|W_1 B_2 B_3) = \frac{1}{2}$.

and then the additivity rule.   We may illustrate this technique with the following:

Example 2-5-3

A university is seeking to hire two men for positions on its faculty.   One is an experienced teacher who is a leader of research in his field.   The second is a young man just completing his graduate work in the field of the older man's competence.   His interest in accepting the position is conditioned by the decision of the older man, with whom he would like to work.   It is estimated that there is a 50–50 chance that the older man will accept the offer.   It is estimated that if the older man accepts the offer, there is a 90–10 chance that the younger man will do likewise; but if the experienced man declines the offer, the chances are reduced to about 40–60 that the younger man will accept the offer.   What is the probability that the younger man will accept the offer?   What is the probability that both men will accept the offer?

SOLUTION   The problem is amenable to the application of probability theory only if the estimates of the "chances" are taken to be estimates of probabilities. Suppose we let A be the event that the experienced man will accept a position and B be the event that the younger man will accept the offer of a position.   The estimates of chances are taken to be estimates of probabilities, as follows:

$$P(A) = P(A^c) = 0.50$$
$$P(B|A) = 0.90 \quad \text{and} \quad P(B^c|A) = 0.10$$
$$P(B|A^c) = 0.40 \quad \text{and} \quad P(B^c|A^c) = 0.60$$

The problem is to find $P(B)$ and $P(AB)$.

We are concerned with the event B as conditioned by A or A^c.   Now it is plain that $B = BA \uplus BA^c$; that is, the occurrence of event B will be accompanied either by the occurrence of A or by the nonoccurrence of A.   The possibilities are mutually exclusive.   Hence

$$P(B) = P(BA) + P(BA^c)$$

By the rule (CP1), we may write

$$P(B) = P(B|A)P(A) + P(B|A^c)P(A^c)$$
$$= 0.90 \times 0.50 + 0.40 \times 0.50 = 0.65$$

Also

$$P(AB) = P(B|A)P(A) = 0.90 \times 0.50 = 0.45 \quad \blacksquare$$

The specific pattern utilized in the preceding example is frequently encountered.   It may be generalized by applying the product rule (CP1) to each term of the expansion in property (P8), in order to obtain the following important expansion:

(**CP2**)   Let $\mathcal{B} = \{B_i : i \in J\}$ be any countable class of mutually exclusive events, each with positive probability.   If the occurrence of the event A implies the occurrence of one of the B_i (i.e., if $A \subset \underset{i \in J}{\uplus} B_i$), then

$$P(A) = \sum_{i \in J} P(A|B_i)P(B_i).$$

If the class $\mathcal{B}$ is a partition, then the occurrence of any event A implies the occurrence of one of the B_i; in this case, the expansion is always applicable.

Example 2-5-4

Suppose in the problem of drawing the colored balls, posed in Example 2-5-2, it is desired to determine the probability $P(B_3)$ that the third ball drawn is a black one.

SOLUTION   We may easily verify the fact that the class $\{W_1W_2, \ W_1B_2, \ B_1W_2, \ B_1B_2\}$ is a partition.   We may therefore use (CP2) to write

$$P(B_3) = P(B_3|W_1W_2)P(W_1W_2) + P(B_3|W_1B_2)P(W_1B_2) + P(B_3|B_1W_2)P(B_1W_2)$$
$$+ P(B_3|B_1B_2)P(B_1B_2)$$

As before, we assume

$$P(B_3|W_1W_2) = 1 \qquad P(B_3|W_1B_2) = P(B_3|B_1W_2) = \tfrac{2}{3} \qquad P(B_3|B_1B_2) = \tfrac{1}{3}$$

Also,

$$P(W_1W_2) = P(W_1)P(W_2|W_1) = \tfrac{2}{5} \times \tfrac{1}{4} = \tfrac{1}{10}$$
$$P(W_1B_2) = P(W_1)P(B_2|W_1) = \tfrac{2}{5} \times \tfrac{3}{4} = \tfrac{3}{10}$$
$$P(B_1W_2) = P(B_1)P(W_2|B_1) = \tfrac{3}{5} \times \tfrac{1}{2} = \tfrac{3}{10}$$
$$P(B_1B_2) = P(B_1)P(B_2|B_1) = \tfrac{3}{5} \times \tfrac{1}{2} = \tfrac{3}{10}$$

From this, it follows that $P(B_3) = \tfrac{3}{5} = P(B_1)$.   This result, also, is consistent with the usual result based on the classical model. ∎

When the disjoint class $\mathcal{B}$ in the statement of property (CP2) is a partition, the expansion is sometimes referred to as an *average probability*. On any trial, one of the events B_i must occur.   For each such event, the conditional probability $P(\cdot|B_i)$ is a probability measure.   The sum $\sum_{i \in J} P(A|B_i)P(B_i)$ can be considered a probability-weighted average of the various conditional probabilities of the event A.   It is in this sense that the term average probability is used occasionally.   Unfortunately, some writers do not make clear that the probabilities being "averaged" are in fact conditional probabilities.

The next example illustrates an interesting use of the expansion in property (CP2) and develops an important result.

Example 2-5-5   *Random Coding*

In information theory the concept of *random coding* has played an important role in developing certain theoretical expressions for bounds on the probability of error in decoding messages sent over noisy channels.   In this theoretical approach, codes are chosen at random from a large set of possible codes.   The message is encoded and decoded according to the code chosen.   Let E be the event of an error in the random coding scheme, and let C_i be the event that the ith code is chosen.   Note that $\{C_i: i \in J\}$ is a partition, since one and only one code is chosen on each trial.   By (CP2) we have

$$P(E) = \sum_{i \in J} P(E|C_i)P(C_i)$$

$P(E|C_i)$ is the probability (conditional) of making an error if the ith code is used. $P(E)$ is the average error of decoding over all the possible codes (in the sense discussed above). Now let J_a be the index set defined by

$$J_a = \{i: P(E|C_i) \geq aP(E)\}$$

The event $C_a = \underset{i \in J_a}{\uplus} C_i$ is the event of choosing a code such that the conditional probability of an error, given that code, is greater than a times the average probability $P(E)$. We wish to show that $P(C_a) \leq 1/a$; that is, the probability of picking a code which gives a probability of error more than a times the average is no greater than $1/a$.

SOLUTION We note first that $P(C_a) = \sum_{i \in J_a} P(C_i)$. The following string of inequalities is justified by various basic properties and the condition defining J_a.

$$P(E) = \sum_{i \in J} P(E|C_i)P(C_i) \geq \sum_{i \in J_a} P(E|C_i)P(C_i)$$
$$\geq aP(E) \sum_{i \in J_a} P(C_i) = aP(E)P(C_a)$$

Dividing through by $aP(E)$ gives the desired result. ∎

When the disjoint class ℬ is a partition, property (CP2) may be viewed in another manner. If one begins with the conditional probability measures $P(\cdot|B_i)$ and the probabilities $P(B_i)$, these may be combined according to (CP2) to give the total probability measure $P(\cdot)$. It frequently occurs that one makes calculations of the probabilities of various events and then realizes that he has assumed as a condition the occurrence of some specific event E. He has, in effect, restricted his set of allowed outcomes by the condition of the occurrence of E. Thus the probabilities are values of the conditional probability measure $P(\cdot|E)$. This causes no difficulty, since $P(\cdot|E)$ is a probability measure and obeys all the rules for such functions. In fact, as we discuss later in this section, further conditioning events may be introduced into the conditional probabilities, and these may be handled in the same manner as that by which ordinary conditional probabilities are derived from the total probability.

Suppose $P(A|E)$ has been determined, with $0 < P(E) < 1$. How may $P(A)$ be determined? By property (CP2), we could use the partition $\{E, E^c\}$ to get

$$P(A) = P(A|E)P(E) + P(A|E^c)P(E^c)$$
$$= P(A|E)P(E) + P(A|E^c)[1 - P(E)]$$

Thus, to determine $P(A)$, we need $P(E)$ and $P(A|E^c)$, or the equivalent, say, $P(E^c)$ and $P(AE^c)$. It may be helpful to consider these results with the aid of Fig. 2-5-1 and the interpretation of conditional probabilities as fractions of the mass assigned to the conditioning event

E. If it should turn out that $P(E) = 1$, a simpler situation exists.   In
this case we have already noted that $P(\cdot|E) = P(\cdot)$.

It frequently occurs that conditional probabilities are given in one
direction but the conditional probabilities in the other direction are
desired.   That is, we may have the conditional probability $P(A|B)$
available from information about a system, whereas it is the conditional
probability $P(B|A)$ which is desired.   This reversal is possible in the
following general situation, in which we may utilize property (CP2)
to give

(**CP3**) *Bayes' rule*. Let $\mathfrak{B} = \{B_i\colon i \in J\}$ be any countable class of mutually
exclusive events, each with positive probability. Let A be any event with positive
probability such that $A \subset \underset{i \in J}{\biguplus} B_i$. Then

$$P(B_i|A) = \frac{P(A|B_i)P(B_i)}{P(A)} = \frac{P(A|B_i)P(B_i)}{\sum\limits_{j \in J} P(A|B_j)P(B_j)}$$

Note once more that the inclusion relation is automatically satisfied
for any event A if the disjoint class $\mathfrak{B}$ is a partition.   This rule is the
celebrated *Bayes' rule*, which has been the subject of much controversy
in the historical development of probability theory.   Because of mis-
understanding (and misapplication) of this rule, it has often been suspect.
It appears here as a logically derived implication of the probability
model and is thus perfectly valid within the limitations of that model.

As a very simple example of the application of Bayes' rule, we
consider a problem which arises in many situations.   We state the problem in a typical manner, which once more points to the need for distinguishing total probabilities and conditional probabilities.

Example 2-5-6

Two boxes on an electronics service bench contain transistors.   Box A has two good
and two defective transistors.   Box B has three good and two defective units.   One
transistor is selected at random from box A and transferred to box B.   One transistor
is then selected at random from box B; it is tested and found to be good.   What is
the probability that the transistor transferred from box A to box B was good?

SOLUTION AND DISCUSSION   As the question is posed, it is not clear what probability is asked for.   If we let G_1 be the event that the transistor moved from box A
to box B was good and let G_2 be the event that the transistor chosen on the second
selection was good, three possible probabilities could be of interest here: $P(G_1)$,
$P(G_1|G_2)$, or $P(G_1G_2)$.   From the point of view of the mathematical model, all three
may be calculated.   From the point of view of the real-world problem, one must
decide which probability is the one desired.   We shall suppose that the probability
of interest is $P(G_1|G_2)$.   Further, we shall suppose that in making a "random choice"
of a transistor, the selection of any transistor in the box is "equally likely."   With
the aid of the diagrams in Fig. 2-5-3, where defective transistors are shown as dark

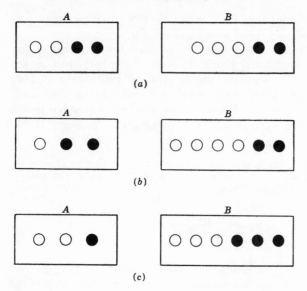

Fig. 2-5-3 Distribution of good transistors (white balls) and defective transistors (dark balls) for Example 2-5-6. (*a*) Original situation; (*b*) situation after transfer of a good transistor; (*c*) situation after transfer of a defective transistor.

balls and good transistors are shown as white balls, we make the following assumptions:

$$P(G_1) = P(G_1{}^c) = \tfrac{1}{2} \qquad P(G_2|G_1) = \tfrac{2}{3} \qquad P(G_2|G_1{}^c) = \tfrac{1}{2}$$

We note that the conditions of the problem lead naturally to the assumption of values for $P(G_2|G_1)$ and $P(G_2|G_1{}^c)$. It is desired to reverse the first of these. This may be done in the following manner, which amounts to applying Bayes' rule.

$$P(G_1G_2) = P(G_2|G_1)P(G_1) = \tfrac{2}{3} \times \tfrac{1}{2} = \tfrac{1}{3}$$
$$P(G_2) = P(G_2|G_1)P(G_1) + P(G_2|G_1{}^c)P(G_1{}^c)$$
$$= \tfrac{2}{3} \times \tfrac{1}{2} + \tfrac{1}{2} \times \tfrac{1}{2} = \tfrac{7}{12}$$
$$P(G_1|G_2) = \frac{P(G_1G_2)}{P(G_2)} = \tfrac{12}{21} \ \blacksquare$$

An example of some interest is provided by the following situation in communication theory.

Example 2-5-7

A noisy communication channel is transmitting a signal which consists of a binary coded message; i.e., the signal is a sequence of 0's and 1's. Noise acts upon one transmitted symbol at a time. A transmitted 0 or a 1 may or may not be perturbed into the opposite symbol in the process of transmission. Physical details of the process may be quite complicated, but in many cases it seems reasonable to suppose that probabilities of error (i.e., of a signal reversal) may be assigned. Let A be the event that a 1 is sent and let B be the event that a 1 is received at a certain time. Suppose that the following conditional probabilities may be assigned:

$P(B^c|A) = p_1$ (the conditional probability of an error, given that a 1 is sent)
$P(B|A^c) = p_2$ (the conditional probability of an error, given that a 0 is sent)

Suppose, moreover, that $P(A) = p$ so that $P(A^c) = 1 - p$. These probabilities are essentially those determined at the sending end. Thus, to evaluate $P(B^c|A)$, one would send 1's and note how often they are perturbed in transmission. The process of reception must work from the receiving end. Thus, when a 1 is received, it is desired to know $P(A|B)$ and $P(A^c|B)$. Similarly, when a 0 is received, it is desired to know $P(A|B^c)$ and $P(A^c|B^c)$. Suppose we wish to know the value of $P(A|B^c)$; that is, we wish to know what is the conditional probability that a 1 is sent, given that a 0 is received.

By Bayes' rule (CP3),

$$P(A|B^c) = \frac{P(B^c|A)P(A)}{P(B^c|A)P(A) + P(B^c|A^c)P(A^c)}$$

$$= \frac{p_1 p}{p_1 p + (1 - p_2)(1 - p)}$$

Similar calculations serve to evaluate the other conditional probabilities. Note that the denominator of the fraction is $P(B^c)$, from which $P(B)$ is easily determined, so that it is relatively simple to invert the other conditional probabilities. ∎

A very similar problem is posed by the question of the reliability of a safety device or of a test for a dangerous condition.

Example 2-5-8

A safety device (or test for a dangerous condition) is designed to have a high conditional probability of operating (or indicating) when a failure (or dangerous condition) occurs and a high conditional probability of not operating (not indicating) when a failure (or dangerous condition) does not occur. Suppose F is the event that there is a failure and W is the event that the safety device works. Put $P(W|F) = p$ and $P(W^c|F^c) = q$. It is desired to make p and q as near unity as possible. Suppose that $P(F) = p_F$, which is normally assumed small. Note that these conditional probabilities are based on given knowledge of the event of a failure. What is really desired is the conditional probability of a failure, given that the safety device has or has not operated.

Again, by Bayes' rule (CP3), we have

$$P(F|W) = \frac{P(W|F)P(F)}{P(W|F)P(F) + P(W|F^c)P(F^c)} = \frac{p p_F}{p p_F + (1 - q)(1 - p_F)}$$

It is instructive to put in some numerical values. If we put $p_F = 0.001$ and $p = q = 0.98$, we obtain

$$P(F|W) = \frac{1}{1 + 20(0.999/0.980)} = \frac{1}{21} \text{ (approximately)}$$

Although it is assumed that the safety device works 98 times in 100, the conditional probability of a failure given that the device operates is only 1 in 21. This seems to be a strange result. Note, however, that $P(FW^c) = (1 - p)p_F = 2 \times 10^{-5}$. The probability of simultaneous failure and nonoperation of the safety device is extremely small. On the other hand, $P(F^cW) = (1 - q)(1 - p_F) = 0.02$, so that the probability of operation of the safety device when there is no failure is approximately $\frac{1}{50}$.

In the case of a test for a dangerous condition, it would be desirable to have $P(F|W)$ as high as possible. It is instructive to consider the conditions required to make $P(F|W)$ larger. The expression above may be rewritten

$$P(F|W) = \frac{1}{1 + (1 - q)(1 - p_F)/p_F p} \approx 1 - \frac{1 - q}{p_F} \frac{1 - p_F}{p}$$

$$\approx 1 - \frac{1 - q}{p_F}$$

The first approximation assumes

$$\frac{1 - q}{p_F} \frac{1 - p_F}{p} \ll 1$$

The second approximation is based on the assumption that $1 - p_F$ and p are both quite close to unity, so that the ratio is also. To make $P(F|W)$ approach unity, we must have $1 - q \ll p_F$; that is, we must have $P(W|F^c) \ll P(F)$. ∎

The effect on the probability of the occurrence of the event A produced by the knowledge that the event E must occur is incorporated into the mathematical model by means of the concept of the conditional probability of A, given E. The latter is derived from the total probability measure by taking the ratio of the probability mass $P(AE)$ to the probability mass $P(E)$. The receipt of additional information that event F has occurred also may be expected to modify further the conditional probability of the occurrence of event A. The new conditional probability, given the occurrence of both E and F, may be derived from the conditional probability given E, in the same manner that the latter is derived from the total probability. As a first step toward examining this fact, let us take note of a product rule closely related to property (CP1) and, in fact, derived therefrom.

(CP4) $P(A_1A_2 \cdots A_n|E) = P(A_1|E)P(A_2|A_1E) \cdots P(A_n|A_1 \cdots A_{n-1}E)$

This may be derived by writing out the expansion (CP1) for the probability $P(EA_1A_2 \cdots A_n)$ and dividing both sides of the equation by $P(E)$.

We may interpret this result with the aid of a hybrid notational scheme for conditional probabilities. The conditional probability $P(\cdot|E)$ is often written $P_E(\cdot)$. This, as we have shown, is a true probability measure [derived, of course, from the total probability measure $P(\cdot)$]. As a special case of (CP4), we may write $P(AF|E) = P(F|E)P(A|EF)$. This may be rewritten

$$P(A|EF) = \frac{P(AF|E)}{P(F|E)} = \frac{P_E(AF)}{P_E(F)} = P_E(A|F)$$

We thus have a sort of "conditional" conditional probability. The form used indicates that $P(A|EF)$ is derived from the conditional probability measure $P(\cdot|E) = P_E(\cdot)$ by the same process that $P(\cdot|E)$ is derived from the total probability measure $P(\cdot)$. "Higher-order" conditional probabilities can be derived in similar fashion. For example,

$$P(A|EFG) = \frac{P(AG|EF)}{P(G|EF)}$$

Using the hybrid notation, we may write

$$P(A_k|A_1 \cdots A_{k-1}E) = P_E(A_k|A_1 \cdots A_{k-1})$$

Also, we may rewrite (CP4) as follows:

(CP1′) $P_E(A_1A_2 \cdots A_n) = P_E(A_1)P_E(A_2|A_1) \cdots P_E(A_n|A_1A_2 \cdots A_{n-1})$

This form emphasizes the fact that in some sense (CP4) is an analog to (CP1). In fact, if $E = S$, this reduces to (CP1).

We may obtain analogs to (CP2) and (CP3). First, we obtain

(CP5)   Let $\mathscr{B} = \{B_i : i \in J\}$ be any finite or countably infinite class of mutually exclusive events, each with positive probability. If the occurrence of the event AE implies the occurrence of one of the B_i (i.e., if $AE \subset \underset{i \in J}{\uplus} B_i$), then

$$P(A|E) = \sum_{i \in J} P(A|B_iE)P(B_i|E).$$

To see this we note from previous properties that

$$P(AE) = \sum_{i \in J} P(AEB_i) = \sum_{i \in J} P(A|B_iE)P(B_iE)$$

Dividing through by $P(E)$ gives the desired result. Utilizing the hybrid notation once more, we may rewrite the expression to give

(CP2′) $P_E(A) = \sum_{i \in J} P_E(A|B_i)P_E(B_i)$

which is an analog to (CP2). A similar argument produces a Bayes' rule (CP3′) for the conditional probability $P_E(\cdot|B_i)$. We are thus able to operate with conditional probabilities with respect to further conditioning events in the same manner that we operate with total probabilities and conditional probabilities derived therefrom.

A variety of special results in particular circumstances may be derived with techniques similar to those utilized above. As a further illustration, we consider an analytical example that is of some interest in connection with the idea of independence, to be introduced in the next section.

Example 2-5-9

Suppose $\{B_i: i \in J\}$ is a finite or countably infinite disjoint class whose union is the event B. Suppose $P(B_i) > 0$ and $P(A|B_i) = p$ for each $i \in J$. Show that $P(A|B) = p$.

SOLUTION   The problem may be understood in terms of the mass picture. $P(A|B_i)$ is the ratio of that part of the mass assigned to the common part of A and B_i to the mass assigned to B_i. If the ratio is the same for each part B_i of B, the ratio for that portion of the mass assigned to A and the union of the B_i to the mass assigned to the whole union must be the same. The result can be obtained analytically by the following argument.

$$P(AB) = \sum_{i \in J} P(AB_i) = \sum_{i \in J} P(A|B_i)P(B_i) = p \sum_{i \in J} P(B_i) = pP(B)$$

Dividing through by $P(B)$ gives the desired result. ∎

2-6   Independence in probability theory

The concept of independence in probability theory, to which we turn attention in this section, has played an indispensable role in the development of the mathematical theory. The notion of independence in the physical world is one that is intuitively grasped but which is not precisely defined. Events are considered to be physically independent when they seem to have no causal relation. In order to develop a satisfactory mathematical counterpart of this vaguely formulated idea, we must identify a precise mathematical condition which seems to correspond to the condition of independence in the real world. Since there is no necessary logical connection between a concept introduced into the mathematical model and a (hopefully) corresponding one in the real world of physical phenomena, it is desirable to distinguish the real-world concept from the mathematical counterpart. It is customary to speak of *stochastic independence* when it is desired to emphasize that the probability concept is intended. Ordinarily, however, we simply speak of independence and know that, in dealing with a probability model, stochastic independence is intended.

Let us consider two events A and B, each with positive probability. We should probably be willing to consider these as independent events if the occurrence of one did not "condition" the probability of the occurrence of the other. Thus we should consider them independent if $P(A|B) = P(A)$ or $P(B|A) = P(B)$. It is easy to show that either of these equalities implies the other. Using the definition of conditional probability, we may express these equalities in the following manner:

$$P(A) = \frac{P(AB)}{P(B)} \qquad P(B) = \frac{P(AB)}{P(A)}$$

The two relationships may be combined into a single product relation

$$P(AB) = P(A)P(B)$$

It would appear that independence may be characterized by this product rule.

Next, let us consider a simple physical situation in which the events under consideration would be judged to be independent in a physical sense.

Example 2-6-1

Suppose two ordinary dice are thrown. In observing the results, we distinguish between the dice. Let A_i be the event that the first die shows face number i ($i = 1$, $2, \ldots, 6$) and B_j the event that the second die shows face number j. On the classical assumption of equally likely outcomes, the events A_i and B_j and the joint event A_iB_j have the following probabilities:

$$P(A_i) = P(B_j) = \tfrac{1}{6} \qquad P(A_iB_j) = \tfrac{1}{36}$$

Hence we have the product relation $P(A_iB_j) = P(A_i)P(B_j)$. ■

Implicit in the assumption of equally likely outcomes is the notion that the outcome of the throw of the first die does not affect the outcome of the throw of the second die, and conversely. For any result of the throw of the first die, the various possibilities of the outcome of the throw of the second die are unaffected, i.e., are "equally likely."

The condition that one outcome does not affect the other may be considered in terms of the relative frequency of occurrence of events in a large number n of trials of an experiment. Suppose n_A is the number of these outcomes which result in the event A, with similar notation for the numbers of occurrences of other events. Under what condition would we be willing to say that events A and B do not affect one another physically? It seems natural to suppose that the occurrence of the event B does not affect the occurrence of the event A if the fraction of the occurrence of A among the occurrences of B is the same as the fraction of the occurrences of A among the nonoccurrences of B. That is, we suppose the occurrence of the event B does not affect the occurrence of the event A if

$$\frac{n_{AB}}{n_A} = \frac{n_{A^cB}}{n_{A^c}} = \frac{n_B}{n}$$

These results may be translated into relative frequencies as follows:

$$\frac{f_{AB}}{f_B} = \frac{f_{AB^c}}{f_{B^c}} = f_A \qquad \text{and} \qquad \frac{f_{AB}}{f_A} = \frac{f_{A^cB}}{f_{A^c}} = f_B$$

For large n it is supposed that the relative frequencies lie close to the corresponding probabilities. Hence, in the case of physical independence, we should expect the probabilities to satisfy the following conditions:

$$P(A|B) = P(A|B^c) = P(A) \quad \text{and} \quad P(B|A) = P(B|A^c) = P(B)$$

These are equivalent to the product rules

$$P(AB) = P(A)P(B)$$
$$P(AB^c) = P(A)P(B^c) \qquad P(A^cB) = P(A^c)P(B)$$

As we shall show in Theorem 2-6A, below, any one of these implies the other two.

On the basis of these arguments, it would seem promising to make the following

Definition 2-6a

Two events A and B are said to be (*stochastically*) *independent* iffi the following *product rule* holds:

$$P(AB) = P(A)P(B)$$

We emphasize that in this mathematical definition stochastic independence is characterized by a *product rule*. The definition is motivated by intuitive notions of physical independence, and we shall show that many of the mathematical consequences of this definition are readily interpreted in terms of ordinary notions of independence in the real world of phenomena. In fact, we may often anticipate mathematical conditions and relations related to stochastic independence by an appeal to intuitive notions based on physical experience. It must be kept firmly in mind, however, that *to establish the independence of two events in the stochastic or probabilistic sense one must show that the product rule holds* (or that some rule proved to be equivalent to it holds).

In the argument above, it may be noted that if A and B are physically independent events, so are A and B^c, A^c and B, and A^c and B^c. The following theorem shows that this pattern is preserved in the mathematical model.

Theorem 2-6A

If any one of the pairs $\{A, B\}$, $\{A, B^c\}$, $\{A^c, B\}$, or $\{A^c, B^c\}$ is an independent pair, then all the pairs are independent pairs.

PROOF   To establish independence we must establish the appropriate product rules.   First we note that

$$P(AB^c) = P(A) - P(AB) = P(A) - P(A)P(B) = P(A)[1 - P(B)]$$
$$= P(A)P(B^c) \qquad \text{iffi the pair } \{A, B\} \text{ is independent}$$

By interchanging the role of A and B in the previous argument, we obtain $P(A^cB) = P(A^c)P(B)$ iffi $\{A, B\}$ is an independent pair.   On replacing B by B^c, we may use the last result to assert $P(A^cB^c) = P(A^c)P(B^c)$ iffi $\{A, B^c\}$ is an independent pair. But this pair is independent iffi $\{A, B\}$ is independent. ∎

The argument leading up to the definition of independence presumes that the events under consideration have positive probabilities. This is because conditional probabilities or probability ratios are used. The formulation of the condition in terms of the product rule allows the removal of this restriction. In fact, it is almost trivially true that the impossible event ∅ and the sure event S are independent of any other events (including themselves). We may formalize this in the following.

Theorem 2-6B

Any event A is independent of the impossible event ∅ and the sure event S.

PROOF

$$P(A∅) = P(∅) = 0 = P(A)P(∅) \quad \text{and} \quad P(AS) = P(A) = P(A)P(S). ∎$$

There is a constant tendency to confuse the idea of mutually exclusive events and the idea of independent events. The following theorem shows that these are quite different concepts.

Theorem 2-6C

If A and B are mutually exclusive events and each has positive probability, they cannot be independent; if they have positive probabilities and are independent, they cannot be mutually exclusive. ∎

PROOF   Under the hypothesis, $P(A)P(B) > 0$. The events can be mutually exclusive only if $P(AB) = 0$. Either the product rule does not hold or the events are not mutually exclusive. ∎

This result, which frequently surprises the beginner, may be understood by realizing that if A and B are mutually exclusive, occurrence of the event A implies the occurrence of B^c and the occurrence of B implies the occurrence of A^c. Thus the occurrence of one of these events very definitely "conditions" the occurrence of the other.

We should note at this point that the independence condition must be introduced into any model problem as an assumption. As we have noted, independence is not precisely defined in the real world of phenomena. And by the very nature of a mathematical model, a condition in the real world does not imply logically a condition in a mathematical model; nor does a mathematical condition imply a real-world condition. The experiential basis for an independence assumption is usually the knowledge that the events to be represented in the mathematical model are physically independent in some appropriate sense. Where statistical examination of the result of a large number of trials is possible, one would check the appropriate product relations for the relative frequencies of occurrences. In introducing independence assumptions into the mathematical model for any problem, one should be guided by two basic considerations.

1. Assumptions should be made in a form most easily checked by physical experience. These are not always the simplest forms mathematically. One of the tasks of the theorist is to find equivalent forms of the assumptions which are more useful for work with the mathematical model.

2. Care should be taken to reduce the assumptions to a practical minimum and to ascertain the consistency of the assumptions made. Superfluous assumptions require unnecessary physical checks and introduce the risk of actual contradictions in assumed conditions. Again, it is the role of the theoretical investigator to determine what assumptions are really required in order that the model problem may be properly stated.

Before very much can be done with the idea of independence, it is necessary to extend the notion to classes of more than two events. Since the independence concept has been introduced in terms of a product rule, it seems natural to consider the possibility of extending it in these terms. As a first step we could consider the case in which the product rule applies to each pair of distinct members of the class. This condition is usually referred to as *pairwise independence*. While there are applications in which this condition is sufficient, experience has shown that a much more general extension is desirable. We shall simply state the more general condition and examine some of its mathematical consequences. The justification of the concept must lie in the nature of these consequences.

Definition 2-6b

A class of events $\alpha = \{A_i : i \in J\}$, where J is a finite or an infinite index set, is said to be an *independent class* iffi the product rule holds for every finite subclass of α.

We may state the defining condition analytically as follows: if J_0 is any finite subset of the index set J, then

$$P(\bigcap_{i \in J_0} A_i) = \prod_{i \in J_0} P(A_i)$$

It should be emphasized that this product rule must hold for any, and hence every, finite subclass. As a simple example to illustrate the independence condition, we consider

Example 2-6-2

The class $\{A, B, C\}$ is independent iffi the following all hold:

$$P(AB) = P(A)P(B) \qquad P(AC) = P(A)P(C) \qquad P(BC) = P(B)P(C)$$
$$P(ABC) = P(A)P(B)P(C) \quad \blacksquare$$

The following example shows that pairwise independence is not sufficient to guarantee independence of the whole class.

Example 2-6-3

Let $\{A_1, A_2, A_3, A_4\}$ be a partition with each $P(A_k) = \frac{1}{4}$.   Put

$$A = A_1 \uplus A_2 \qquad B = A_1 \uplus A_3 \qquad C = A_1 \uplus A_4$$

Then the class $\{A, B, C\}$ has pairwise independence, but is not an independent class, as the following calculations show.

SOLUTION

$$P(AB) = P(A_1) = \frac{1}{4} = P(A)P(B)$$
$$P(AC) = P(A_1) = \frac{1}{4} = P(A)P(C)$$
$$P(BC) = P(A_1) = \frac{1}{4} = P(B)P(C)$$

However,

$$P(ABC) = P(A_1) = \frac{1}{4} \neq P(A)P(B)P(C) \quad \blacksquare$$

On the other hand, validity of the product rule for the entire class does not ensure its holding for subclasses.   The following example illustrates this fact.

Example 2-6-4

Consider four events A, B, C, and D, satisfying the following conditions:

$$AD = BD = \emptyset \qquad C = AB \uplus D \qquad P(A) = P(B) = \frac{1}{4}$$
$$P(AB) = \frac{1}{64} \qquad P(D) = \frac{15}{64}$$

The product rule applies to the class $\{A, B, C\}$, but no two of these events are independent.

SOLUTION   The relationships between the events and their probabilities are indicated in the Venn diagram of Fig. 2-6-1.   It is apparent that

$$P(C) = P(AB) + P(D) = \frac{1}{4} \qquad \text{and} \qquad P(ABC) = P(AB) = \frac{1}{64}$$

We have, therefore, $P(ABC) = P(A)P(B)P(C)$.   On the other hand, no two of the events are independent. $\quad \blacksquare$

It should be noted that independence is a function of the probability measure and is not, except in special cases, a property of the events.

We have seen, in Sec. 2-5, that the product rule for conditional probabilities opens up possibilities for computing the probabilities of compound events.   The product rule for independent events plays a similar role, augmented by the fact that the rule for independent events is inherently simpler than the product rule for conditional probabilities.

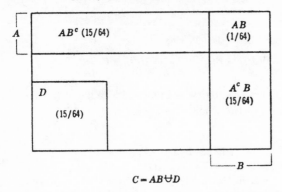

Fig. 2-6-1   Venn diagram for the events in Example 2-6-4.   Probabilities are shown in parentheses.

As a simple example, we return to a consideration of the serial systems in reliability theory discussed briefly in Example 2-3-2.

Example 2-6-5   *Serial Systems in Reliability Theory*

As in the discussion in Example 2-3-2, we let A be the event of a success of the entire system and let A_i be the event of the success of the ith subsystem.   A serial system is characterized by the condition

$$A = \bigcap_{i=1}^{n} A_i$$

That is, the entire system operates successfully iff all subsystems operate successfully.   In many systems, the success or failure of any subsystem is not appreciably affected by the success or failure of any other subsystem.   In such a case, it is natural to assume that the A_i are independent events.   To be precise, we assume that the class $\mathcal{C} = \{A_i : 1 \leq i \leq n\}$ is an independent class.   This really is a model of the fact that the operation of any combination of the subsystems is not dependent upon the operation of any combination of other subsystems.   Under the independence assumption, we have a simple formula for reliability of the whole system:

$$R = P(A) = P(\bigcap_{i=1}^{n} A_i) = \prod_{i=1}^{n} P(A_i) = \prod_{i=1}^{n} R_i$$

The reliability of the entire system is the product of the reliability factors for the various subsystems.   Suppose, for example, there are 10 subsystems, each with reliability $0.99 = 1 - 0.01$.   The reliability of the entire system is $(1 - 0.01)^{10}$.   Since $(1 - x)^n \approx 1 - nx$, for small x, the reliability of the complete system is approximately 0.90. ∎

Example 2-6-6   *Parallel Systems in Reliability Theory*

A second elementary type of system in reliability theory is the so-called parallel system.   Such a system is characterized by the fact that the system operates if any one or more of the subsystems operate.   Utilizing the notation of the previous example, we have

$$A = \bigcup_{i=1}^{n} A_i$$

In the case of independent operation, we assume the class of events $\alpha = \{A_i : 1 \leq i \leq n\}$ to be an independent class.   In order to be able to use the product rule, we may use the rules on complements to obtain the following expression:

$$A = [\bigcap_{i=1}^{n} A_i{}^c]^c$$

Now it seems reasonable to suppose (as we shall show in Theorem 2-6F, below) that the class $\alpha' = \{A_i{}^c : 1 \leq i \leq n\}$ is also an independent class.   In this case, we have

$$P(A) = 1 - P[\bigcap_{i=1}^{n} A_i{}^c] = 1 - \prod_{i=1}^{n} P(A_i{}^c)$$

$$= 1 - \prod_{i=1}^{n} [1 - P(A_i)]$$

Translated into terms of the reliability factors, this becomes

$$R = 1 - \prod_{i=1}^{n} (1 - R_i)$$

Again, we may use some numbers to get a feel for this relation.   Suppose there are two subsystems, each with reliability 0.90.   Then the system reliability is $1 - 0.10^2 = 1 - 10^{-2} = 0.99$. ∎

The figures in the preceding two examples illustrate the general fact that reliability of a serial system is generally smaller than that of any subsystem while the reliability of a parallel system is generally larger than that of any subsystem.

The definition for independent classes is stated for infinite as well as finite classes.   It is important to note that even in the case of infinite classes, only finite subclasses need be considered.   For reference, we state the essential fact in the following

Theorem 2-6D

A class α of events is an independent class iffi every finite subclass consisting of two or more members is an independent class.

This is an obvious consequence of the definition.

We may reformulate the definition of an independent class in a useful manner, which we illustrate by the following analytical example.

Example 2-6-7

Suppose $\{A, B, C, D, E\}$ is an independent class of events. Then the following classes, among others, are independent classes: $\{ABC, E\}$, $\{AC, BE, D\}$, $\{B, C, DE\}$.

What we have done here is to select finite subclasses of the original class in such a way that no two of the subclasses have any member event in common. We have taken the intersections of the members of the various subclasses to form new events. These new events then are collected in a new class. It should be apparent—and the fact may easily be examined in detail—that the product rule holds for each newly formed class of events. On the other hand, if the product rule should hold for every such new class formulated in this manner, it must surely hold for the original class $\{A, B, C, D, E\}$. The pattern indicated here may be generalized into the following

Theorem 2-6E

Suppose $\alpha = \{A_i : i \in I\}$ is any class of events. Let $\{\alpha_j : j \in J\}$ be a family of finite subclasses of α such that no two have any member event A_i in common. Let B_j be the intersection of all the sets in α_j. Put $\mathcal{B} = \{B_j : j \in J\}$. Then α is an independent class iffi every class $\mathcal{B}$ so formed is an independent class.

Theorems 2-6A and 2-6B may be combined and interpreted in the following way. If $\{A, B\}$ is an independent pair, we may replace either or both of the members by $\emptyset$, S, or the complement of the member and still have an independent pair. It would seem reasonable to suppose that this pattern could be extended to larger independent classes. For instance, we should suppose the following assertions to be true:

Example 2-6-8

If $\alpha = \{A, B, C, D\}$ is an independent class, so also are the classes $\{A, B^c, C^c, D^c\}$, $\{A, B^c, C, \emptyset\}$, $\{S, B, \emptyset, D^c\}$, etc.

In order to verify this, it is just about as easy to develop the general theorem as to deal with the special case.

Theorem 2-6F

If $\alpha = \{A_i : i \in J\}$ is an independent class, so also is the class α' obtained by replacing the A_i in any subclass of α by either $\emptyset$, S, or A_i^c. The particular substitution for any given A_i may be made arbitrarily, without reference to the substitution for any other member of the subclass.

PROOF By Theorem 2-6D, it is sufficient to show the validity of the theorem for any arbitrarily chosen (and hence for every) finite subclass. We may argue as follows:

1. We may replace any single A_j in the finite class under consideration by its complement $A_j{}^c$, by $\emptyset$, or by S. By Theorems 2-6A and 2-6B, the product rule still holds if such a replacement is made.

2. Suppose $j_1, j_2, \ldots, j_m$ is any sequence of indices where replacements are to be made. Suppose replacements have been made for all indices up to j_r, and the independence of the modified class is preserved. Then an argument parallel to argument 1 shows that $A_{j_{r+1}}$ may be replaced.

By mathematical induction, all members of the sequence may be replaced. Because of the arbitrariness of the sequence and of the finite subclass considered, the theorem is established. ∎

This theorem allows considerable freedom in dealing with independent classes. It may be convenient to express independence in terms of a class of events; yet, for purposes of calculations, independence is needed in terms of the complements of at least some of the events in the class. We have already encountered such a situation in dealing with parallel reliability systems in Example 2-6-6. In this case we replaced each event in an independent class by its complement. We consider another such case in the next example, which also illustrates a basic strategy in probability calculations.

Example 2-6-9 *Independent Trials*

Suppose three independent tests are made on a physical system. Let A_1 be the event of a satisfactory outcome on the first test, A_2 on the second test, and A_3 on the third test. We suppose these events are independent in the sense that the class $\{A_1, A_2, A_3\}$ is an independent class. Suppose $P(A_i) = p_i$ and $P(A_i{}^c) = 1 - p_i = q_i$. What is the probability of exactly two successful tests?

SOLUTION   If C is the event of exactly two successes in the three tests, then it follows that

$$C = A_1 A_2 A_3{}^c \uplus A_1 A_2{}^c A_3 \uplus A_1{}^c A_2 A_3$$

We have considered all the possible combinations of two successes and one failure. The terms in the union are disjoint, since we cannot have exactly one failure in the ith place and also exactly one failure in the jth place $(i \neq j)$. We have thus expressed the event C as the disjoint union of intersections of events whose probabilities are known. By Theorem 2-6F, the classes $\{A_1, A_2, A_3{}^c\}$, $\{A_1, A_2{}^c, A_3\}$, and $\{A_1{}^c, A_2, A_3\}$ are independent classes for which the product rule holds. Utilizing the additivity property and the product rule, we have

$$P(C) = p_1 p_2 q_3 + p_1 q_2 p_3 + q_1 p_2 p_3 \quad \blacksquare$$

It should be noted that we have used the same basic strategy employed in the section on conditional probability. We have expressed the event whose probability is desired as the disjoint union of intersections of events whose probabilities are known. The product rule for independent events is simpler in character and is therefore easier to use than the product rule for conditional probabilities. We shall employ this strategy frequently in dealing with independent events.

We next consider an example in which we study a slight variation of the game discussed in Example 2-3-1.   The analysis further illustrates the basic strategy of expressing an event as the disjoint union of intersections of independent events whose probabilities are known.   In this case, the number of events to be considered is countably infinite.

Example 2-6-10

Two men spin a roulette wheel alternately, each attempting to get one of several numbers which he has specified.   The first man has probability p_1 of spinning one of his numbers any time he spins.   The second man has probability p_2 of spinning one of his numbers.   The results of the trials are assumed to be independent.   The first man who succeeds in spinning one of his numbers wins the game.   Determine the probability of each man's winning.   If $p_1 < \frac{1}{2}$, show what value p_2 must have to make the probabilities of winning equal.

SOLUTION   Each elementary outcome ξ consists of a sequence of trials, which may not be limited in length.   We may consider each sequence to be of infinite length. The elementary outcomes can be described, however, in terms of the results at any given trial in the sequence.   We let A_i be the set of those outcomes in which the first man turns up one of his numbers on the ith spin.   Since the men spin alternately, it is apparent that $A_i = \emptyset$ (is impossible) for even values of i.   Similarly, we let B_j be the event the second man spins one of his numbers on the jth trial.   It follows that $B_j = \emptyset$ for odd values of j.   The independence condition is taken to mean that the A_i and B_j form an independent class.   We assume $P(A_{2k-1}) = p_1$ and $P(B_{2k}) = p_2$ for $k = 1, 2, \ldots$.   For convenience, we put $q_1 = 1 - p_1$ and $q_2 = 1 - p_2$.   Now we let A be the event the first man wins and B be the event the second man wins. We must find appropriate ways to express A and B, in order that the information provided may be utilized.   A little thought shows that the first man wins if he spins his number on the first trial, or spins it on the third after failure of both on the first and second, etc.   It is obvious that these are mutually exclusive possibilities.   This can be expressed in symbols, as follows:

$$A = A_1 \uplus A_3 A_1^c B_2^c \uplus \cdots \uplus A_{2n+1} A_1^c B_2^c \cdots A_{2n-1}^c B_{2n}^c \uplus \cdots$$

Event B can be expressed in a similar manner, as follows:

$$B = B_2 A_1^c \uplus B_4 A_1^c B_2^c A_3^c \uplus \cdots \uplus B_{2n} A_1^c B_2^c \cdots A_{2n-3}^c B_{2n-2}^c A_{2n-1}^c \uplus \cdots$$

By virtue of Theorem 2-6F and the assumed independence condition, we may apply the product rule to each of the intersection terms.   This gives for event A

$$P(A) = p_1 + p_1(q_2 q_1) + p_1(q_2 q_1)^2 + \cdots + p_1(q_2 q_1)^n + \cdots$$

$$= p_1 \sum_{n=0}^{\infty} (q_2 q_1)^n = \frac{p_1}{1 - q_1 q_2}$$

Here we have made use of the well-known expansion

$$\frac{1}{1 - x} = \sum_{n=0}^{\infty} x^n \qquad \text{for } |x| < 1$$

Similarly

$$P(B) = p_2q_1 + p_2q_1(q_2q_1) + \cdots + p_2q_1(q_2q_1)^n + \cdots$$

$$= p_2q_1 \sum_{n=0}^{\infty} (q_1q_2)^n = \frac{p_2q_1}{1 - q_1q_2}$$

In order to make the probabilities $P(A) = P(B) = \frac{1}{2}$, we must have $p_1 = q_1p_2$ so that

$$p_2 = \frac{p_1}{q_1} = \frac{p_1}{1 - p_1} \qquad 0 \leq p_1 < \frac{1}{2}$$

For $p_1 > \frac{1}{2}$, we should need $p_2 > 1$, which is impossible. For $p_1 = \frac{1}{2}$, we should require $p_2 = 1$, which makes it practically certain that the first man either wins on his first spin or he loses. It is interesting to note that $P(A) + P(B) = 1$, as may be shown by a little algebra. Since $AB = \emptyset$, we must have $S = A \uplus B \uplus A^cB^c$. We cannot say that A^cB^c is empty, for there is at least one sequence for which neither man wins (i.e., there is at least one $\xi \in A^cB^c$). It is true, however, that $P(A^cB^c) = 0$. ∎

This example shows that while we can always assert that $P(\emptyset) = 0$, the fact that $P(E) = 0$ does not imply that $E = \emptyset$. In such a case, however, E would be considered practically impossible, and it would ordinarily be ignored in considering events that could occur.

We consider next a pattern that frequently occurs in the theoretical examination of independence conditions. First we take a simple case.

Example 2-6-11

Suppose event A can occur in one of three mutually exclusive ways and event B can occur in one of two mutually exclusive ways. We have $A = A_1 \uplus A_2 \uplus A_3$ and $B = B_1 \uplus B_2$. If the A_i and B_j are pairwise independent for each possible choice of i and j, then the events A and B are independent.

SOLUTION   We first note that

$$AB = (A_1 \uplus A_2 \uplus A_3)(B_1 \uplus B_2)$$
$$= A_1B_1 \uplus A_1B_2 \uplus A_2B_1 \uplus A_2B_2 \uplus A_3B_1 \uplus A_3B_2$$

Because of the additivity property

$$P(AB) = P(A_1B_1) + P(A_1B_2) + P(A_2B_1) + P(A_2B_2) + P(A_3B_1) + P(A_3B_2)$$

Because of the independence condition

$$P(AB) = P(A_1)P(B_1) + P(A_1)P(B_2) + P(A_2)P(B_1) + P(A_2)P(B_2) + P(A_3)P(B_1) + P(A_3)P(B_2)$$

Simple algebraic manipulations yield

$$P(AB) = [P(A_1) + P(A_2) + P(A_3)][P(B_1) + P(B_2)]$$
$$= P(A)P(B) \qquad \text{(by the additivity property)} ∎$$

This result does not depend upon the small finite number of events A_i and B_j. The general pattern is expressed in the following

Theorem 2-6G

Suppose $\alpha = \{A_i: i \in I\}$ and $\mathcal{B} = \{B_j: j \in J\}$ are finite or count-ably infinite disjoint classes whose members have the property that any A_i is independent of any B_j, that is, that $P(A_iB_j) = P(A_i)P(B_j)$ for any $i \in I$ and $j \in J$.   Then the events

$$A = \biguplus_{i \in I} A_i \text{ and } B = \biguplus_{j \in J} B_j \text{ are independent}$$

PROOF

$$AB = [\biguplus_i A_i] \cap [\biguplus_j B_j] = \biguplus_{i,j} A_iB_j$$
$$P(AB) = \sum_{i,j} P(A_iB_j) = \sum_{i,j} P(A_i)P(B_j) = \sum_i P(A_i) \sum_j P(B_j)$$
$$= P(A)P(B) \ \blacksquare$$

Further generalizations can be made by considering several disjoint classes, say, α, $\mathcal{B}$, $\mathcal{C}$, $\mathcal{D}$, whose unions are A, B, C, D, respectively.   If for any $A_i \in \alpha$, $B_j \in \mathcal{B}$, $C_k \in \mathcal{C}$, $D_h \in \mathcal{D}$ we have $\{A_i, B_j, C_k, D_h\}$ is an independent class, then the class $\{A, B, C, D\}$ is an independent class. Some notational skill but no new ideas are needed to establish such results.

The following example is of some interest in providing an illustration of typical methods in establishing independence.   The conclusion is an immediate consequence of the result in Example 2-5-9, but we present the essentials of the analytical argument.

Example 2-6-12

Suppose $\{B_i: i \in J\}$ is a partition and that $P(A|B_i) = p$ for each $i \in J$.   Then A is independent of each B_i and $P(A) = p$.

SOLUTION

$$P(A) = \sum_{i \in J} P(AB_i) = \sum_{i \in J} P(A|B_i)P(B_i) = p \sum_{i \in J} P(B_i) = p \ \blacksquare$$

We shall return to the topic of independence in Sec. 2-8.   First we shall consider, in the next section, some techniques for systematic handling of compound events which greatly facilitate the handling of both specific examples and general results.

2-7   Some techniques for handling events

In the theory of switching or logic networks, a number of extremely useful techniques have been developed for graphical or symbolic repre-sentation and manipulation of set combinations.   Some of these tech-niques can be exploited to systematize the handling of important classes of problems in applied probability theory. It is the purpose of this section to outline the more pertinent techniques. For a fuller discussion of these topics, the reader is referred to the literature on boolean algebra and switching networks.

Partitions and joint partitions

The development in previous sections has demonstrated the key role in probability theory of the concept of a partition, as introduced in Sec. 2-2. We wish to examine this concept further and to consider some special partitions and techniques for handling them. We begin by noting that the formation of a partition amounts to a classification of the elements into categories, i.e., into the member sets of the partition. Consider, for example, the partition $\alpha = \{A_1, A_2, A_3, A_4\}$ represented on the special Venn diagram of Fig. 2-7-1a. For each element (or elementary outcome) ξ we ask: To which A_i does this element belong? There is a unique answer for each ξ. For example, the particular element ξ_0 shown

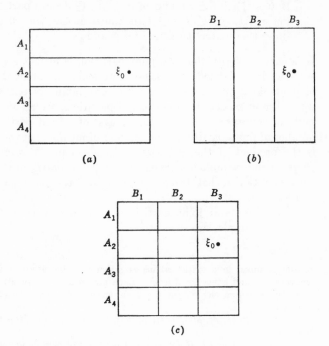

Fig. 2-7-1   Two partitions and a joint partition.

on the diagrams of Fig. 2-7-1 belongs to A_2. When the question has been answered for each element in the basic space S, the elements have in effect been classified into the appropriate categories, namely, the appropriate member events of the partition.

Now suppose the elements of the same basic space are to be classified according to another partition, $\mathfrak{B} = \{B_1, B_2, B_3\}$, as represented on the Venn diagram of Fig. 2-7-1b. Again we may ask of each element ξ: To which member event B_j does ξ belong? For the particular element ξ_0 shown on the diagrams in Fig. 2-7-1, the answer is B_3.

In addition to the two single classifications, one may wish to consider the joint classification in which each element is classified according to both of the criteria. To do this, we ask of each element two questions: To which A_i does it belong? To which B_j does it belong? The unique pair of answers determines one and only one joint event A_iB_j to which the element belongs. The element ξ_0 in Fig. 2-7-1 belongs to the event A_2B_3. The class of all joint events A_iB_j is important enough to be given a special name.

Definition 2-7a

If $\mathcal{Q} = \{A_i : i \in I\}$ and $\mathfrak{B} = \{B_j : j \in J\}$ are partitions, the class $\mathcal{Q}\mathfrak{B} = \{A_iB_j : i \in I, j \in J\}$ is known as the *joint partition* (or sometimes the *product partition*) for $\mathcal{Q}$ and $\mathfrak{B}$.

It is easy to verify that the joint partition is in fact a partition in the sense we have defined such classes. It is also apparent that the concept of joint partition can be extended to any family of partitions. In this work, the index sets for the original partitions are assumed to be either finite or countably infinite. In the case of a joint partition for a finite number of finite partitions, the number of members of the joint partition is the product of the numbers of members in the original partitions. Thus, in the example illustrated in Fig. 2-7-1, partition $\mathcal{Q}$ has four members, partition $\mathfrak{B}$ has three members, and the joint partition $\mathcal{Q}\mathfrak{B}$ has 12 members.

Partitions and joint partitions arise in a variety of ways. The following examples are suggestive.

Example 2-7-1

Persons sampled in a statistical investigation may be classified according to sex, age category, and educational level. Each person in the population to be sampled corresponds to one element in the basic space. We may consider the following partitions.

First partition: $\{M, M^c\}$, where M is the set of males and M^c is the set of females (nonmales).

Second partition: $\{A, B, C\}$, where A is the set of persons under 21 years of age, B is the set of persons 21 to 35, and C is the set of persons over 35.

Third partition: $\{G, H, U\}$, where G is the set of persons whose formal education is limited to grade school or less, H is the set of those persons who went to high school but not to a college or university, and U is the set of those persons with some college or university experience.

The joint partition consists of 8 joint sets (events) of the type MBH, M^cBU, etc. If the author were included in the sampled population, he would belong to the set MCU. ■

Example 2-7-2

Consider as a basic space the set of all pairs of integers, each integer belonging to the set $\{0, 1, \ldots, 9\}$. Two very natural partitions are useful in dealing with such a set. The first partition is determined by the first number in the pair, and the second partition is determined by the second number in the pair. Thus the first partition is

$$\alpha = \{A_i \colon 0 \leq i \leq 9\}$$

where A_i is the set of those pairs for which the first number is i. The second partition is

$$\mathcal{B} = \{B_j \colon 0 \leq j \leq 9\}$$

where B_j is the set of those pairs for which the second number is j. The joint partition $\alpha\mathcal{B} = \{A_iB_j \colon 0 \leq i \leq 9, 0 \leq j \leq 9\}$ has 100 members. Each member event consists of a single pair of integers. For example, $A_3B_0 = \{(3, 0)\}$. ■

Repeated trials

It frequently occurs that a trial in a probability problem is actually a compound trial; each trial is a sequence of component trials. The choice of an element ξ corresponds to making the compound trial. For example, a single choice of ξ may correspond to such a compound operation as

1. Flipping a coin 10 times and noting the particular sequence of 10 heads or tails which turn up

2. Selecting a sequence of 100 decimal digits (each digit being one of the integers $0, 1, \ldots, 9$) and recording these in the order selected

3. Taking a set of readings on five instruments and recording these in a specified order

4. Making a sequence of n trials of a given phenomenon, each component trial of which results in the *success or failure* to achieve a given desired result.

In each of these situations, it is natural to classify the elements according to the result on a given trial in the sequence. For example, in the coin-flipping sequence, we may classify elements (sequences of flips) according to the outcome on the ith trial in the sequence as follows. We let H_i be the event (the set of ξ) for which a head appears on the ith toss in the sequence. If T_i is the event of a tail on the ith trial, we have

$T_i = H_i{}^c$, and $\mathcal{Q}_i = \{T_i, H_i\}$ is a partition.   If we want to classify elements according to the outcome on the first and third trials, we should in fact classify them according to the members of the joint partition $\mathcal{Q}_1\mathcal{Q}_3$. Similar partitions are readily described for the other repeated trial situations.

In general, in dealing with repeated trials, we consider a class of partitions $\{\mathcal{Q}_i : i \in I\}$ where the partition $\mathcal{Q}_i$ is determined by some criterion on the ith component of the sequence.   Events determined by applying jointly the criteria for several component trials, say, trials numbered h, r, and s, are expressible in terms of the members of the joint partition $\mathcal{Q}_h\mathcal{Q}_r\mathcal{Q}_s$.

Example 2-7-3

Suppose each ξ corresponds to a choice of 10 decimal digits.   The event that the first, third, and seventh digit is even may be expressed as follows.   Let $E_k{}^i$ be the event of getting the digit i in the kth position ($i = 0, 1, 2, \ldots , 9$ and $k = 1, 2, \ldots , 10$). The event E_k that the digit in the kth position is even is given by

$$E_k = E_k{}^0 \uplus E_k{}^2 \uplus E_k{}^4 \uplus E_k{}^6 \uplus E_k{}^8$$

This is a union of elements in the kth partition.   The desired event is then $E_1E_3E_7$. ∎

Partitions generated by a class

Consider a class of sets $\mathcal{Q} = \{A_2, A_1, A_0\}$.   We suppose, for purposes of discussion, that each of the sets has a nonempty intersection with each of the others and that none of the sets is contained in any other.   Removal of this restriction simply allows some of the members of the partitions considered below to be empty.   Now consider the class

$$\mathcal{Q}^\circ = \{m_0, m_1, m_2, m_3, m_4, m_5, m_6, m_7\}, \text{ where}$$

$$
\begin{aligned}
m_0 &= A_2{}^cA_1{}^cA_0{}^c \sim 000 & m_4 &= A_2A_1{}^cA_0{}^c \sim 100 \\
m_1 &= A_2{}^cA_1{}^cA_0 \sim 001 & m_5 &= A_2A_1{}^cA_0 \sim 101 \\
m_2 &= A_2{}^cA_1A_0{}^c \sim 010 & m_6 &= A_2A_1A_0{}^c \sim 110 \\
m_3 &= A_2{}^cA_1A_0 \sim 011 & m_7 &= A_2A_1A_0 \sim 111
\end{aligned}
$$

Here we have departed from our usual convention of designating sets or events by capital letters.   We note first that the class $\mathcal{Q}^\circ$ is a partition.   Any two distinct members m_i and m_j are disjoint, since at least one of the factors A_k must appear in uncomplemented form in one member and in complemented form in the other.   For example, m_0m_4 must be empty since it is equal to $A_2{}^cA_1{}^cA_0{}^cA_2A_1{}^cA_0{}^c$, and the intersection $A_2A_2{}^c$ is always empty.   The systematic choice of complemented and uncomplemented factors ensures that each pair $m_im_j = \emptyset$, for $i \neq j$. On the other hand, any element ξ must be in one (and only one) of the m_i. To determine which of the m_i is the correct one, we ask in order: Is it in

A_2? Is it in A_1? Is it in A_0? When the three answers are determined, the event m_i is identified. For example, if the set of answers is "yes, no, yes," respectively, the element in question belongs to

$$A_2 A_1{}^c A_0 = m_5$$

Any such set of answers determines uniquely the set m_i to which an element ξ belongs.

The partition $\mathfrak{a}°$ is determined by the class $\mathfrak{a}$. It should be apparent that the pattern need not be limited to the case in which the class $\mathfrak{a}$ consists of three sets. Hence we may make the general

Definition 2-7b

If $\mathfrak{a} = \{A_{N-1}, A_{N-2}, \ldots, A_1, A_0\}$ is a class of sets, the class $\mathfrak{a}°$ determined by the process described above is called the *partition generated by the class* $\mathfrak{a}$ (or *by the sets* $A_{N-1}, A_{N-2}, \ldots, A_1, A_0$). The sets $A_{N-1}, \ldots, A_0$ are called the *generating sets*. The sets m_i in the generated partition are called *minterms*.

A single set A generates the partition $\{A^c, A\}$. A pair of sets A and B generates the partition $\{A^c B^c, A^c B, A B^c, AB\}$. In general, a class of N distinct sets generates a partition consisting of 2^N members. Some of the members of the generated partition may be empty. If the original class $\mathfrak{a}$ is a partition, the only nonempty terms of the generated partition are the sets of the original class. For three generating sets A, B, C, which themselves form a partition, the nonempty sets in the generated partition are $AB^c C^c$, $A^c BC^c$, and $A^c B^c C$; these are, respectively, the sets A, B, and C.

The partition generated by the class $\{A, B, C\}$ can be expressed as the joint partition $\mathfrak{a}\mathfrak{B}\mathfrak{C}$, where $\mathfrak{a} = \{A^c, A\}$, $\mathfrak{B} = \{B^c, B\}$, and $\mathfrak{C} = \{C^c, C\}$ are the partitions generated by the individual generating sets. It should be apparent that this result extends to the general case.

Minterms

In the example above, the minterms are listed and numbered in a systematic fashion. Let us note the pattern. In the first place, we consider the fact that a given minterm m_i is associated with a set of answers to the questions about any particular element ξ in m_i. We considered above the case that the answers were "yes, no, yes." These answers could be tallied by writing a 0 for each no and a 1 for each yes. This gives for each minterm a specific sequence of 0's and 1's, characteristic of the minterm, if the order of the generating sets is maintained. It is thus natural to make the

Definition 2-7c

The sequence of zeros and ones corresponding to a minterm in the manner just described is called the *binary designator* β_i *for the minterm* m_i.

In our example, the binary designator is 101. Each of the minterms is uniquely determined by its binary designator. Now it is natural to consider the binary designator as a binary number and to designate the minterm by that number. We usually convert the number to its decimal equivalent. Since 101 is the binary equivalent of decimal 5, the minterm $A_2 A_1{}^c A_0 = m_5$. A direct conversion to the binary equivalent is obtained from the product expression by putting a 1 for each uncomplemented variable and a 0 for each complemented variable. The correspondence is indicated for the minterms generated by class $\mathcal{Q}$. It should be apparent that this pattern can be extended to the 2^N minterms generated by the class $\mathcal{Q} = \{A_{N-1}, A_{N-2}, \ldots, A_1, A_0\}$.

Indicator functions

In the next chapter we give considerable attention to functions defined on the basic space S. The concept of a function used is a very simple extension of the ordinary concept of a function of a real variable. We define a *function* $f(\cdot)$ on a set of elements D by assigning to each ξ in D a number $f(\xi)$, known as the *value* of $f(\cdot)$ at ξ. For the present, we consider some very elementary functions whose usefulness will be demonstrated repeatedly throughout the remainder of the book.

Definition 2-7d

If A is any subset of the basic space, S, the *indicator function for A* is the function $I_A(\cdot)$ defined on S by the assignments

$$I_A(\xi) = \begin{cases} 0 & \text{for} \quad \xi \in A^c \\ 1 & \text{for} \quad \xi \in A \end{cases}$$

The term *characteristic function* is also used by some authors.

Part of the importance of the indicator function is a result of the following easily derived properties.

(**IF1**) $\quad I_A \quad \leq I_B \quad$ iffi $\quad A \subset B$

(**IF2**) $\quad I_A \quad \equiv I_B \quad$ iffi $\quad A = B$

(**IF3**) $\quad I_A \quad \equiv 0 \quad$ iffi $\quad A = \emptyset$

(**IF4**) $\quad I_A \quad \equiv 1 \quad$ iffi $\quad A = S$

(**IF5**) $\quad I_{A^c} \quad = 1 - I_A$

(**IF6**) $\quad I_{AB} \quad = I_A I_B = \min(I_A, I_B)$

(**IF7**) $\quad I_{A \cup B} = I_A + I_B - I_{AB} = \max(I_A, I_B)$

The first five properties are obvious consequences of the definition of the indicator function. To prove (IF6) and (IF7) we have to consider each of four possibilities, as follows:

$\xi \in$ both A and B $(\xi \in AB)$

$\xi \in A$ but not B $(\xi \in AB^c)$

$\xi \in B$ but not A $(\xi \in A^cB)$

$\xi \in$ neither A nor B $(\xi \in A^cB^c)$

In each case, evaluation according to the definition and the expressions to be checked shows that the asserted equality holds.

In the special case that the sets A and B are disjoint (mutually exclusive), the rule (IF7) reduces to $I_{A \cup B} = I_A + I_B$. It is apparent that this simple addition rule holds for any disjoint class, as follows:

(IF8)   If $A = \underset{i \in J}{\cup} A_i$ then $I_A = \sum_{i \in J} I_{A_i}$

Boolean functions

The concept of a boolean function arises quite naturally in certain algebraic representations of set combinations. For our purposes, we may make the following

Definition 2-7e

A *boolean function* $F = f(A_{N-1}, A_{N-2}, \ldots, A_1, A_0)$ *of a finite class of sets* is a set obtained by a finite number of applications of the operations intersection, union, and complementation to the members of the class.

For example, two boolean functions of the sets A, B, C are

$$F = f_1(A, B, C) = [A \cup (B \cup C^c)^c]^c$$

and

$$G = f_2(A, B, C) = (A \cup A^cB)^c$$

In the second case, the occurrence of G does not depend upon the occurrence or nonoccurrence of event C. We may consider G a boolean function of A, B, C, with the event C *suppressed*.

If we begin with a finite class of events, a boolean function is a compound event derived from the class by suitable applications of the fundamental set operations intersection, union, and complementation. The simplest boolean functions include AB, $A \cup B$, and A^c, which correspond to single applications of the fundamental set operations. For these functions, we have expressed the indicator function for the compound set as a function of the indicator functions for the component sets or events. A little reflection will show that a similar situation holds for more complicated boolean functions. Once an expression for the compound set F is determined (in terms of members of the finite class and the fundamental set operations), the properties (IF1) through (IF7) provide the basis for a rule for determining the value of $I_F(\cdot)$ from the values of the indicator functions for the component sets. If we consider the minterms in the partition generated by the finite class, we may show that the values of the indicator functions for the component sets cannot change as ξ ranges over any given minterm. For each $i = 0, 1, \ldots, N - 1$, a given minterm will have either A_i or $A_i{}^c$ as a factor. This means that $I_{A_i}(\cdot)$ will have the value one or zero, respectively, for all ξ in that minterm. Since this is true for each of the $I_{A_i}(\cdot)$, it must be true for $I_F(\cdot)$, whose value is determined by the values of the various $I_{A_i}(\cdot)$. These facts are important enough that we state them formally as

Theorem 2-7A

If F is a boolean function of the sets in the class $\mathcal{A} = \{A_{N-1}, A_{N-2}, \ldots, A_1, A_0\}$, the indicator function $I_F(\cdot)$ is a real-valued function of the indicator functions for the members of the class. Moreover, $I_F(\cdot)$ is constant over any given minterm m_i in the partition generated by the class $\mathcal{A}$.

To evaluate the indicator function for a boolean function of a finite class of sets, we need evaluate it for only one element ξ in each of the minterms generated by the class. The values for all ξ are then determined. Let the subclass $\mathcal{A}_F = \{m_i : i \in J_F\}$ be the class such that $I_F(\cdot)$ has the value 1 on each of the minterms m_i in $\mathcal{A}_F$ and has the value zero on all other minterms. Since the m_i are disjoint (mutually exclusive), we can write

$$I_F(\cdot) = \sum_{i \in J_F} I_{m_i}(\cdot)$$

This function determined by the sum has the value 1 whenever ξ belongs to one of the m_i in the class $\mathcal{A}_F$ and has the value zero elsewhere; it is thus the function $I_F(\cdot)$, as asserted. If we consider the set

$$H = \bigcup_{i \in J_F} m_i$$

we may use property (IF8) to assert

$$I_H(\cdot) = \sum_{i \in J_F} I_{m_i}(\cdot)$$

Thus, $I_H(\xi) = I_F(\xi)$ for all ξ. From property (IF2) we conclude that $H = F$. We have thus established the important

Theorem 2-7B   Minterm Expansion Theorem

Any boolean function $F = f(A_{N-1}, A_{N-2}, \ldots, A_1, A_0)$ may be expressed as the disjoint union of an appropriate subclass of the minterms m_i generated by the class $\{A_{N-1}, A_{N-2}, \ldots, A_1, A_0\}$. In symbols,

$$F = \bigcup_{i \in J_F} m_i$$

where J_F is the index set determined by those minterms on which $I_F(\xi) = 1$.

It is convenient to refer to this expansion as follows:

Definition 2-7f

The expansion guaranteed by Theorem 2-7B is called the *minterm expansion* of the boolean function.

The functions used as examples after Definition 2-7e, above, may be expanded to give

$$F = A^cB^cC^c \cup A^cBC^c \cup A^cBC = m_0 \cup m_2 \cup m_3 \qquad J_F = \{0, 2, 3\}$$
$$G = A^cB^cC^c \cup A^cB^cC = m_0 \cup m_1 \qquad\qquad\qquad J_G = \{0, 1\}$$

Note that it is necessary to consider the sets A, B, C in the proper order when numbering the minterms. Direct manipulation of the set relations according to basic rules will verify the expansions for each of the examples. The examples and a discussion of some means for representing boolean functions given in the next subsection further illustrate the essential facts concerning the minte m expansion and demonstrate its importance.

One other fact implied by the above results should be noted. If we are to deal with any boolean function of a given finite class of events, the minterms are the smallest subdivisions of the space which we need to consider. If probability assignments are made to these minterms, the probability of any event expressible as a boolean function of the generating class can be determined.

Representation of boolean functions

A number of closely related means for representing boolean functions have been devised in the literature. These depend, for the most part, on the minterm expansion. Once the minterms have been designated and numbered according to the scheme discussed above, it is sufficient to indicate in some suitable manner the index set J_F.

One convenient way of doing this is to produce a *binary designator* $\beta(F)$ for the event (or boolean function).

Definition 2-7g

The *binary designator $\beta(F)$ for a boolean function F* is a sequence of 0's and 1's determined as follows. A *place* is provided for each minterm m_i in the partition generated by the sets whose combination forms the function. A 1 or a 0 is written in each place, according as the minterm corresponding to that place does or does not appear in the minterm expansion.

For example, with the minterms generated by A, B, C (considered in that order), the designator $\beta(F)$ for the set F described above is 0000 1101. The minterm places are in the order of decreasing numbers from 7 to zero. The designator 0000 1101 indicates that minterms numbered 0, 2, and 3 are in the minterm expansion. In a similar manner, the designator $\beta(G)$ for the set G is 0000 0011. Once this scheme is set up, the binary designator may be interpreted as a number, and each number uniquely determines a

function.   We shall be interested in the binary designator, but shall not
have much need for the numerical interpretation.

Operations of union, intersection, and complementation give rise to
corresponding operations on binary designators for the sets (considered
as boolean functions of a given class of sets).   The binary designator
for the complement of a set is obtained by interchanging 0 and 1 in
each place.   The binary designator for the union of two or more sets is
obtained from the binary designators for the individual sets by putting
in each place the maximum of the numbers in that place.   For the intersection of two or more sets, the number in a given place is the minimum
of the numbers in the corresponding places in the individual sets.   We
may illustrate these rules and show how they are used to obtain the
binary designator for a boolean function by considering the example

$$H = (A \cup B)^c C^c = f_3(A, B, C)$$

$$
\begin{aligned}
\beta(A) &= 1111\ 0000 \\
\beta(B) &= 1100\ 1100 \\
\beta(C) &= 1010\ 1010 \\
\beta(A \cup B) &= 1111\ 1100 \\
\beta[(A \cup B)^c] &= 0000\ 0011 \\
\beta(C^c) &= 0101\ 0101 \\
\beta(H) &= 0000\ 0001
\end{aligned}
$$

This corresponds to the fact that $(A \cup B)^c C^c = A^c B^c C^c$.

Once the rules of operation for the binary designators for boolean
functions are established and the correctness of $\beta(A)$, $\beta(B)$, and $\beta(C)$
is verified, the validity of the minterm expansion theorem for three
events should be evident. $\beta(A)$ must have a 1 in each place corresponding to a minterm having factor A, with similar statements holding for
$\beta(B)$ and $\beta(C)$.   Examination of the table, arranged with $\beta(A)$ in the first
row, $\beta(B)$ in the second, and $\beta(C)$ in the third, shows that the first three
numbers in any place (column), reading from the top down, constitute
the binary designator β_i for that minterm.   Performing the operations on
these binary designators, as indicated by the operations of union, intersection, and complementation, to derive the set H from the sets A, B, and
C, results in the survival of 1's in certain minterm places.   These indicate the terms in the minterm expansion.   The evident generality of this
process indicates the validity of the minterm expansion theorem.

A second popular and useful manner of representing a boolean function is provided by the so-called *minterm map*.   A variety of forms of this
map, or chart, are described and utilized in the literature on switching
circuits.   We shall consider only one, which may properly be designated a
Marquand diagram, since it seems to have been utilized first by Allan
Marquand in 1881, in the study of logic relations.   The minterm map is a

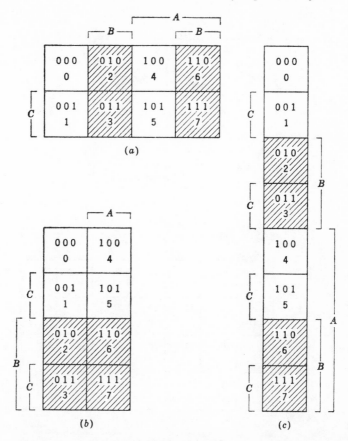

Fig. 2-7-2   Variations of the minterm map for three generating sets,
A, B, C.   Set *B* is shaded.

special form of the Venn diagram for representing sets as regions (sets of
points) in the plane.

Figure 2-7-2 illustrates three variations of the minterm map for
the generating sets *A, B, C*. Each of the sets *A, B*, and *C*, in turn,
divides the map into two complementary parts.   A given set need not be
represented by a single region in the plane, for set membership has nothing
to do with geometrical ideas of continuity.   For each diagram, each
of the small squares represents one of the minterms.   Binary designators
for each of these minterms and the decimal equivalent of the corresponding
minterm number are shown in each square.   For example, minterm m_5
is represented by the designator 101.   Examination of each of the min-

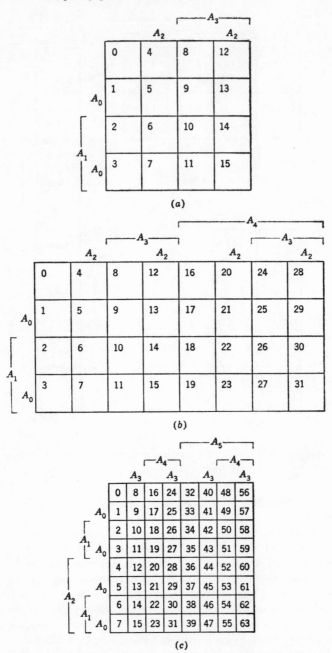

Fig. 2-7-3   Minterm maps for four, five, and six generating sets.

term maps shows that the square so designated is indeed in A, not in B, and in C. A similar examination of each of the squares shows that it represents one of the minterms in the partition generated by A, B, and C. The name *minterm* seems to have been derived from the fact that the set corresponds to one of the *min*imal subdivisions of the diagram. Figure 2-7-3 shows minterm maps for partitions generated by four, five, and six events. The minterm maps are drawn as if each generating set had a nonempty intersection with each of the other generating sets. It may be, in particular examples, that some of the minterms are empty. If m_5 is empty in Fig. 2-7-2, this simply means that there is no element ξ which is in A, not in B, and in C.

Given the minterm map, a boolean function is designated by indicating those minterms which are in the minterm expansion of the function. One convenient way of doing this is to write a 1 in each square representing a minterm in the expansion. For the form of the map in Fig. 2-7-2c, if one also writes a 0 in each square representing a minterm not in the expansion and then turns the map through 90° clockwise, the resulting pattern of 1's and 0's is exactly the same as the binary designator for the function. The minterm-map representation and the binary-designator representation are thus entirely equivalent.

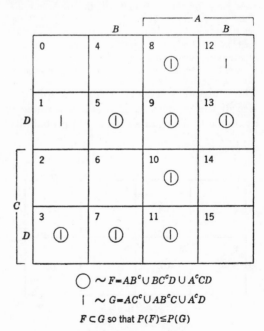

$$\bigcirc \sim F = AB^c \cup BC^cD \cup A^cCD$$

$$\mid \sim G = AC^c \cup AB^cC \cup A^cD$$

$$F \subset G \text{ so that } P(F) \leq P(G)$$

Fig. 2-7-4   Minterm map for Example 2-7-4.

As a simple example of the use of minterm maps, we show how it may be used to check an inclusion relation between two events.

Example 2-7-4

Consider the events $F = AB^c \cup BC^cD \cup A^cCD$ and $G = AC^c \cup AB^cC \cup A^cD$. Show that $P(F) \leq P(G)$.

SOLUTION   The desired result is established if we can show that $F \subset G$. To do this, we employ the minterm-map representation of Fig. 2-7-4. Event F is indicated by circles, and event G is indicated by 1's. The fact that there is a 1 in every minterm square where there is a circle shows that the desired relationship exists. To obtain the minterms in F and G (i.e., to obtain their minterm expansions) we consider the product or intersection terms one by one. For example, AB^c consists of all those minterm squares which are in A but not in B; these are the squares numbered 8, 9, 10, 11. Similarly, BC^cD consists of the squares numbered 5, 13. ∎

A somewhat different use of the minterm map to determine the probability of an event from partial data is illustrated in the next example.

Example 2-7-5

Given that $P(A^cB) = 0.3, P(BC) = 0.4, P(B^c) = 0.4, P(A^cBC) = 0.2$. Find $P(BC^c)$.

SOLUTION   We use the minterm map for three variables in Fig. 2-7-5. Let $p_i = p(m_i)$ be the probability of the ith minterm. Then the desired probability $P(BC^c) = p_2 + p_6$. From $P(A^cBC) = p_3 = 0.2$ and $P(A^cB) = p_2 + p_3 = 0.3$ we get the values for $p_3 = 0.2$ and $p_2 = 0.1$, as shown on the diagram. From $P(BC) = p_3 + p_7 = 0.4$ we get $p_7 = 0.2$, as shown. This gives probabilities for all minterms in event B except m_6. Since $P(B) = 1 - P(B^c) = 0.6$, we have immediately that $p_6 = 0.1$. The desired probability $p_2 + p_6$ is seen to be 0.2. ∎

The probability scheme in this example is not completely determined. Use of the minterm map provides a guide to utilizing the information available to determine the desired probability.

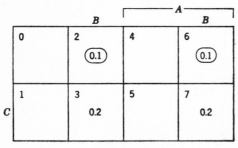

$$P(BC^c) = p_2 + p_6$$

Fig. 2-7-5   Minterm map for Example 2-7-5.

In the preceding two examples, the events are carefully specified. Let us consider a situation in which the problem is posed in physical terms and how it may be expressed in terms of the probability model.

Example 2-7-6

A statistical study of the students at a certain university shows that 55 percent are studying science or engineering (*s-e*); 30 percent of all students are studying *s-e* and have studied both physics and chemistry in high school (referred to as high school science); 20 percent of all students have studied high school science and have one or more parents who are college graduates; 15 percent of the students are not studying *s-e*, have studied high school science, and have no parent who has graduated from college; 55 percent are either not studying *s-e* or have studied high school science and have a college-graduate parent. A student is chosen at random. What is the probability that he is *not* studying science or engineering if it is known that he has studied both physics and chemistry in high school?

SOLUTION   First, we define some events. A choice of ξ amounts to the consideration of a person. Persons are described in terms of three categories: (1) the status with respect to science-engineering, (2) the status with respect to high school science, and (3) the educational experience of the parents. Classification according to these categories determines membership in corresponding sets (or events) as follows:

A = event that student under consideration is studying science-engineering
B = event that student under consideration studied high-school science
C = event that student has at least one college-graduate parent

The problem is interpreted to be the determination of $P(A^c|B)$. We suppose the probabilities of the various events correspond to the fractions of the student population whose situations satisfy the conditions defining the events. Thus the data are interpreted to mean:

$$P(A) = 0.55 \quad P(AB) = 0.30 \quad P(BC) = 0.20$$
$$P(A^c \cup BC) = 0.55 \quad P(A^cBC^c) = 0.15$$

The minterm map of Fig. 2-7-6 is an aid to the organization of the information. The problem is to determine $P(A^c|B) = P(A^cB)/P(B) = (p_2 + p_3)/(p_2 + p_3 + p_6 + p_7)$.

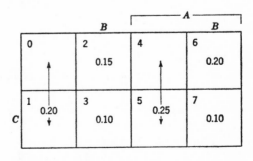

Fig. 2-7-6   Minterm map for Example 2-7-6.

We have

$$P(A) = p_4 + p_5 + p_6 + p_7 = 1 - (p_0 + p_1 + p_2 + p_3) = 0.55$$
$$P(AB) = p_6 + p_7 = 0.30$$
$$P(BC) = p_3 + p_7 = 0.20$$
$$P(A^cBC^c) = p_2 = 0.15$$
$$P(A^c \cup BC) = p_0 + p_1 + p_2 + p_3 + p_7 = 0.55$$

Now p_2 is obtained directly. If p_7 can be obtained, then p_6 and p_3 are easily obtained from the second and third equations. We get

$$p_7 = P(A) + P(A^c \cup BC) - 1 = 0.10$$

From this it follows that

$$p_3 = 0.10 \quad \text{and} \quad p_6 = 0.20$$

The desired result is thus $P(A^c|B) = 0.25/0.55$. ∎

Further examination shows that there is not sufficient information given to determine the probabilities for all boolean functions of A, B, C. The sums $p_0 + p_1$ and $p_4 + p_5$ may be determined to give the values shown on the minterm map. But there is not the information needed to separate the minterm probabilities. This information is not needed for the problem studied above. On the other hand, the information is not sufficient to answer the question: What is the probability that the student chosen has one or more parents who have graduated from college [i.e., what is $P(C)$]? The probability $P(B)$ may be determined easily, however.

2-8   Further results on independent events

The techniques developed in the preceding section make possible a systematic treatment of a large class of problems, particularly when the events under study are boolean functions of independent events whose probabilities are known. We shall consider in this section a number of special problems which are of interest in themselves and which illustrate useful methods of analysis. In addition, we shall obtain several important general results.

First, consider the following assertion:

Example 2-8-1

Suppose $\{A, B, C, D, E\}$ is an independent class of events. Then $F = [A \cup (B \cup C)^c]^c$ and $G = D^c \cup E$ are independent events.

From one point of view, the result might be judged intuitively "obvious." The event F depends upon the events A, B, and C; the event G depends upon the events D and E. The fact that these are members of an independent class and the fact that F and G do not depend upon any common members of this class make it seem plausible that F and G should be

independent.   But when we recall that independence requires the product
relation, the independence of these two boolean functions is not obvious.
Rather than demonstrate independence in this particular case, let us
state and prove a general theorem which includes this case.

Theorem 2-8A

Suppose $\{A_1, A_2, \ldots, A_n, B_1, B_2, \ldots, B_m\}$ is an independent
class, and let

$$F = f(A_1, A_2, \ldots, A_n) \quad \text{and} \quad G = g(B_1, B_2, \ldots, B_m)$$

be boolean functions of the indicated events.   Then F and G are
independent events.

PROOF   Let $\{m_i: 0 \leq i \leq 2^n - 1\}$ be the class of minterms generated by the
class $\{A_1, A_2, \ldots, A_n\}$ and let $\{m_j': 0 \leq j \leq 2^m - 1\}$ be the class of minterms
generated by the class $\{B_1, B_2, \ldots, B_m\}$.   Then, by the minterm expansion theorem
(Theorem 2-7A),

$$F = \underset{i \in J_F}{\uplus} m_i \quad \text{and} \quad G = \underset{j \in J_G}{\uplus} m_j'$$

By Theorems 2-6F and 2-6E, we have

$$P(m_i m_j') = P(m_i)P(m_j') \qquad \text{for each permissible } i \text{ and } j$$

Because the minterm classes are partitions and hence disjoint, we may apply Theorem
2-6G to assert the independence of F and G.   ■

It is easy to extend this result to several functions of events from an
independent class, provided no two of these functions depend on any
common member of the class.

When the problem is to determine the probability of an event which
is a boolean function of a class of independent events whose probabilities
are known, one may use the minterm expansion and exploit the additivity
property and the product relation for independent events.   We consider a
very simple example which points to the general method.   The solution
is developed in an overorganized manner designed to display the general
pattern.

Example 2-8-2

Suppose A, B, C, D are independent events with probabilities

$$P(A) = p_A = 0.4 \quad P(B) = p_B = 0.6 \quad P(C) = p_C = 0.2$$
$$P(D) = p_D = 0.2$$

What is the probability of the event $E = (A \cup BC^c)D \cup A^cCD^c$?

SOLUTION AND DISCUSSION   1. We assume the independence condition, somewhat
loosely stated in a manner all too common in the literature, to mean precisely that
$\{A, B, C, D\}$ is an independent class.

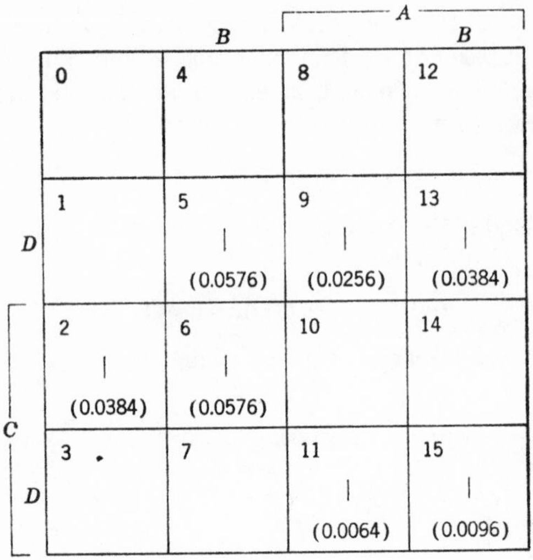

$$E = (A \cup BC^c)D \cup A^cCD^c$$
$$= m_2 \uplus m_5 \uplus m_6 \uplus m_9 \uplus m_{11} \uplus m_{13} \uplus m_{15}$$
$$= m_5 \uplus A^cCD^c \uplus AD$$

Fig. 2-8-1   Minterm map for Example 2-8-2.

2. We reduce the event E to the disjoint union of product terms, by use of the minterm expansion. This expansion for the event E (whose derivation is discussed below) is shown on the minterm map of Fig. 2-8-1 to be

$$E = m_2 \uplus m_5 \uplus m_6 \uplus m_9 \uplus m_{11} \uplus m_{13} \uplus m_{15}$$

An examination of the minterm map shows that the union of the last four minterms gives the simpler intersection term AD, and the union $m_2 \uplus m_6 = A^cCD^c$, so that we may write

$$E = m_5 \uplus A^cCD^c \uplus AD$$

The minterm map facilitates checking the fact that the terms in this expression are disjoint, so that the additivity property of the probability measure may be used. Thus, we may write

$$P(E) = P(m_5) + P(A^cCD^c) + P(AD)$$

Also, we could write $P(E)$ as the sum of the probabilities of the minterms appearing in the minterm expansion.

3. We determine the probabilities of the minterms and other product expressions by using the product rule for independent events. To facilitate writing, we put $P(A^c) = 1 - p_A = q_A$ and do likewise for the other events B^c, C^c, D^c. Then

$$P(AD) = P(A)P(D) = p_Ap_D = 0.08$$
$$P(A^cCD^c) = P(A^c)P(C)P(D^c) = q_Ap_Cq_D = 0.0960$$
$$P(m_5) = P(A^cBC^cD) = q_Ap_Bq_Cp_D = 0.0576$$

Then

$$P(E) = 0.0576 + 0.096 + 0.08 = 0.2336$$

This example exhibits a pattern which is useful in determining the minterm probabilities. Consider the minterm m_2. We may represent it with the binary designator 0010, which indicates that $m_2 = A^cB^cCD^c$. Because of the product rule,

$$P(m_2) = P([0010]) = q_Aq_Bpcq_D = 0.6 \times 0.4 \times 0.8 \times 0.2 = 0.0384$$

A simple pattern may be observed. When the binary designator has a 1 in the A place, use p_A as a factor; when a 0 is in the A place, use q_A as a factor. Proceed similarly with the factors in the other places. The probability of each minterm can be obtained in this fashion. The correspondence between minterm probabilities and minterm binary designators is shown on the minterm map of Fig. 2-8-2. It should be apparent that this correspondence exists for minterms generated by any finite, independent class of events. The numerical values for the various minterms appearing in the minterm expansion for event E are shown on the map in Fig. 2-8-1.

4. The minterm expansion may be obtained in several ways.

a. If the minterm map is used, we note that the minterms represented by the term $(A \cup BC^c)D$ must all lie in D and be in either A or in both B and C^c. Inspection shows that these conditions are satisfied by m_9, m_{11}, m_{13}, m_{15}, m_5 and no other. The minterms for A^cCD^c are m_2 and m_6, as noted above.

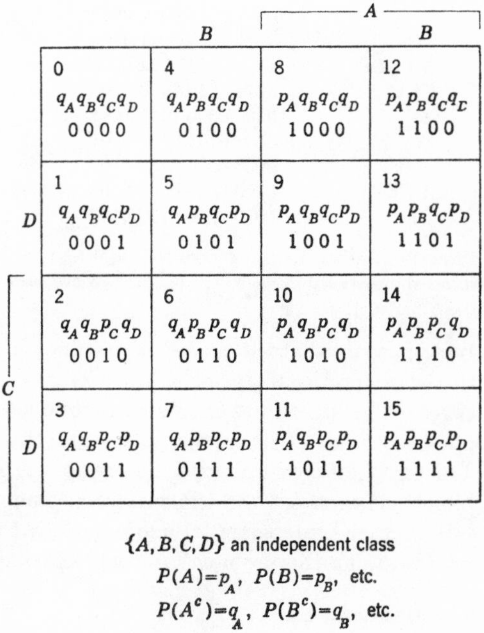

{A, B, C, D} an independent class

$$P(A) = p_A, \quad P(B) = p_B, \text{ etc.}$$
$$P(A^c) = q_A, \quad P(B^c) = q_B, \text{ etc.}$$

Fig. 2-8-2 Minterm map showing correspondence between probabilities and minterm binary designators.

   b. An expansion could be made algebraically.   We first "multiply" the factor D into the parentheses, in keeping with the rules for the algebra of set operations, to give

$$E = AD \cup BC^cD \cup A^cCD^c$$

We should note, with the help of the minterm map, that this is not a disjoint union, since AD and BC^cD share minterm m_{13}.   The product terms have factors missing (A or A^c, B or B^c, etc.).   These may be replaced by the simple expedient of using the theorem $F = FS = F(G \uplus G^c) = FG \uplus FG^c$.   Thus we may expand $BC^cD = ABC^cD \uplus A^cBC^cD$ and $A^cCD^c = A^cBCD^c \uplus A^cB^cCD^c$.   The term AD may be expanded in several successive steps.   The entire expression for the union of the resulting minterms in E will have $m_{13} = ABC^cD$ appearing twice.   One of these identical terms can be dropped, since the union of any set with itself is that set. The result of these operations must be the same as that given by method *a.*

   c. As a third approach, we may utilize the method of the binary designators for boolean functions, as introduced in Sec. 2-7.   When one is familiar with the method and uses paper with guidelines to aid in visual inspection, this can be done quickly and with fewer steps than are given in the following development.

A	1111 1111 0000 0000
B	1111 0000 1111 0000
C	1100 1100 1100 1100
D	1010 1010 1010 1010
C^c	0011 0011 0011 0011
BC^c	0011 0000 0011 0000
$A \cup BC^c$	1111 1111 0011 0000
$(A \cup BC^c)D$	1010 1010 0010 0000
A^c	0000 0000 1111 1111
D^c	0101 0101 0101 0101
A^cD^c	0000 0000 0101 0101
A^cCD^c	0000 0000 0100 0100
E	1010 1010 0110 0100

Inspection shows, counting from the right and beginning with minterm m_0, that minterms numbered 2, 5, 6, 9, 11, 13, and 15 are present, in keeping with the result for the other methods. ∎

It is apparent that the procedure outlined in this example can be extended to the general case of boolean functions of members of an independent class.   The function is expressed in its minterm expansion by some suitable procedure.   The probabilities are calculated by the product rule. This may be systematized by using the correspondence between the binary designators for the minterms and the form of the product or intersection expressions.   Such a systematization could be of interest in programming the problem on a digital computer.

Bernoulli trials

In the next example, we consider an important problem which illustrates the value of a general method of attack.   The condition described in this example is frequently encountered (or assumed) in practice, and several

further examples illustrate some of the possibilities.   Let us identify a
general situation, first studied by James Bernoulli (1654–1705).

Definition 2-8a

By a *sequence of Bernoulli trials* (or a Bernoulli sequence of trials, or
simply Bernoulli trials) we mean an experiment consisting of a
sequence of independent repeated trials with a constant probability
of success.

The defining condition for Bernoulli trials can be given a precise mathematical formulation.   If we let E_i be the event of a success on the ith
trial in the sequence, then the Bernoulli-trial condition is characterized
by the two conditions:

1. $\{E_i: i \in J\}$ is an independent class.
2. $P(E_i) = p$ for all $i \in J$.

Here J is a finite or countably infinite index set, usually of the form
$J = \{i: 1 \leq i \leq n\}$ or $J = \{i: 1 \leq i < \infty\}$.
The fundamental problem for Bernoulli trials is the following.

Example 2-8-3

Consider a sequence of n Bernoulli trials.   What is the probability of exactly r successes in n trials?

SOLUTION AND DISCUSSION   The choice of an element ξ consists of making a
sequence of trials and observing the results.   Let E_i be the event of a success on the
ith trial.   That is, E_i is the set of those ξ (i.e., those sequences of trials) for which the
ith trial in the sequence results in a success.   We assume that the class $\{E_i: 1 \leq i \leq n\}$
is independent and that $P(E_i) = p$ for each i.   Put $1 - p = q$, for convenience of
writing.   Let A_{rn} be the event of exactly r successes in n trials.   We note that the
class $\{A_{rn}: 0 \leq r \leq n\}$ is a partition.   The problem is to find $P(A_{rn})$.   Consider
the minterms generated by the class $\{E_1, E_2, \ldots, E_n\}$, taken in that order.   Any
minterm which has exactly r factors uncomplemented is included in A_{rn}.   We may
then define A_{rn} as the union of minterms in the following manner:

Let $J_r = \{i: \text{binary indicator for } m_i \text{ has } r \text{ ones}\}$.
Then $A_{rn} = \underset{i \in J_r}{\biguplus} m_i$.   It may be noted that $A_{rn}A_{kn} = \emptyset$ for $r \neq k$.
Now, according to the product rule, for each $i \in J_r$, $P(m_i) = p^r q^{n-r}$.   If $n(J_r)$
is the number of i in J_r, the additivity rule implies

$$P(A_{rn}) = n(J_r)p^r q^{n-r}$$

The number $n(J_r)$ is exactly the number of ways r indistinguishable objects (the ones
in this case) may be put into n places (the places in the binary designators for the
minterms in this case).   It is well known (Appendix A) that this number is given by
the expression

$$n(J_r) = C_r^n = \frac{n!}{(n-r)!r!}$$

Thus the solution to the problem is

$$P(A_{rn}) = C_r{}^n p^r (1 - p)^{n-r} \quad \blacksquare$$

It is worth noting that the expression for the probability $P(A_{rn})$ is a term in the *binomial expansion*

$$1 = (p + q)^n = \sum_{r=0}^{n} C_r{}^n p^r q^{n-r}$$

For this reason, $C_r{}^n$ is referred to as the *binomial coefficient*. The coefficient has the property that $C_r{}^n = C_{n-r}^n$. Tables of values of the binomial coefficient are to be found in most handbooks. In addition, many handbooks have tables of values of $C_r{}^n p^r q^{n-r}$, under the name of the *binomial distribution*, and of $\sum_{k=r}^{n} C_k{}^n p^k q^{n-k}$, under the name of the *summed binomial distribution*. The latter expression is the probability of *at least r* successes in n trials.

As an application of the Bernoulli trials, consider the following

Example 2-8-4

Among 10 random decimal digits, what is the probability that there are exactly three zeros?

SOLUTION   Each of the random numbers may be considered the outcome of one of the ten trials making up the experiment. We assume $p = \frac{1}{10}$. The desired probability P_3 is given by

$$P_3 = \frac{10!}{7!3!} \frac{9^7}{10^{10}} = 0.0574 \text{ (from tables)}$$

If it is desired to determine the probability that there are three or fewer zeros, the quantity sought is

$$\sum_{k=0}^{3} C_k{}^{10} (0.1)^k (0.9)^{10-k} = 1 - \sum_{k=4}^{10} C_k{}^{10} (0.1)^k (0.9)^{10-k}$$

which may be calculated from the formulas or determined from tables for the binomial distribution or the summed binomial distribution. From such tables, the value may be found to be $1 - 0.0128 = 0.9872$. $\blacksquare$

The following problem of interest in communication engineering may be treated as a Bernoulli-trial problem although it does not deal with a sequence of trials in time.

Example 2-8-5

A multiplex transmission system has n channels. The channels are used independently, and at any time there is a probability p that any given channel will be in use. What is the probability that exactly k of the channels will be in use? What is the probability that no more than k are in use at any given time?

SOLUTION   If we let E_i be the event that the ith channel is in use at some given time, the conditions of the problem are assumed to mean that the class $\{E_i: 1 \leq i \leq n\}$ is an independent class and that $P(E_i) = p$ for each i. This is the model for Bernoulli trials. Hence

$$P(A_{kn}) = C_k{}^n p^k (1 - p)^{n-k}$$

is the probability that exactly k channels are in use. The probability that no more than k channels are in use is given by

$$\sum_{r=0}^{k} C_r{}^n p^r (1 - p)^{n-r} \quad \blacksquare$$

There are many questions which are closely related to and extensions of the simple Bernoulli-trial problem. As an example, consider the following modification.

Example 2-8-6

Two independent sequences of independent trials are performed; the first sequence consists of n trials, and the second consists of m trials. The probability of a success on any trial in the first sequence is p_1 and on any trial in the second sequence is p_2. What is the probability of a total of r successes in the combined trials?

SOLUTION   First, we define suitable events.

Let E_i = event of a success on ith trial in first sequence
  $E_j^{'}$ = event of a success on jth trial in second sequence
  A_{rn} = event of r successes in n trials in first sequence
  B_{km} = event of k successes in m trials in second sequence
  C_r = event of r success in $n + m$ trials of combined sequences

We assume that $\{E_i, E_j^{'}: 1 \leq i \leq n, 1 \leq j \leq m\}$ is an independent class, with $P(E_i) = p_1$ for each i and $P(E_j^{'}) = p_2$ for each j. Now, according to the development for the case of Bernoulli trials, A_{rn} is the union of minterms generated by the E_i and B_{km} is the union of minterms generated by the $E_j^{'}$. Thus A_{rn} and B_{km} are boolean functions of the E_i and $E_j^{'}$, respectively. By Theorem 2-8A, A_{rn} and B_{km} are independent events for each permissible choice of r and k. Calculations of $P(A_{rn})$ and $P(B_{km})$ are the calculations for Bernoulli trials (i.e., each sequence is a Bernoulli sequence). Because of the independence, $P(A_{rn}B_{km}) = P(A_{rn})P(B_{km})$. The solution to the problem is completed by noting that

$$C_r = A_{rn}B_{0m} \uplus A_{(r-1)n}B_{1m} \uplus \cdots \uplus A_{0n}B_{rm}$$
$$= \biguplus_{k=0}^{r} A_{(r-k)n}B_{km}$$

That is, r successes in the combined sequences are realized by r successes in the first sequence and 0 successes in the second, or $r - 1$ successes in the first sequence and 1 success in the second, etc., to 0 successes in the first sequence and r successes in the second. These various possibilities are mutually exclusive, for one cannot have exactly j successes and also exactly $k \neq j$ successes in the first (or the second) sequence.

Thus

$$P(C_r) = \sum_{k=0}^{r} P(A_{(r-k)n}) P(B_{km})$$

where

$$P(A_{jn}) = C_j{}^n p_1{}^j (1 - p_1)^{n-j}$$

and

$$P(B_{km}) = C_k{}^m p_2{}^k (1 - p_2)^{m-k} \ \blacksquare$$

A variety of extensions and modifications of this problem are found in the literature. Calculations are greatly facilitated by the use of tables of the binomial distribution.

The following Bernoulli-trial problem could be solved by a straight-forward application of the fundamental formula derived in Example 2-8-3, above. However, a special approach is expedient. As a matter of fact, the problem is essentially a particular case of Example 2-6-6, on parallel systems in reliability theory.

Example 2-8-7

The probability p of the malfunction of an electronic circuit on any one operation is about 10^{-5}. What is the probability of at least one malfunction in $n = 10^5$ operations?

SOLUTION   We make the usual Bernoulli-trial assumptions. For the purposes of this problem, a malfunction is a "success." Let F_i be the event of a malfunction on the ith operation, and set $P(F_i) = p = 10^{-5}$. Then $F = \bigcup_{i=1}^{n} F_i$ is the event of at least one malfunction. In order to exploit the independence, we consider $F^c = \bigcap_{i=1}^{n} F_i{}^c$. We then have $P(F^c) = 1 - P(F) = (1 - p)^n$. To evaluate this expression, we take logarithms.

$$\log_{10} P(F^c) = n \log_{10} (1 - p) = n \log_{10} e \log_e (1 - p)$$

For very small p, $\log_e (1 - p) \approx -p$, so that

$$\log_{10} P(F^c) \approx -np \log_{10} e = \log_{10} e^{-np}$$

or

$$P(F^c) \approx e^{-np}$$

For the values assumed, $np = 1$, so that $P(F) \approx 1 - 1/e = 0.63$. $\blacksquare$

In the following example, the Bernoulli-trial situation appears in an interesting way.

Example 2-8-8

A company has n automobiles in a car pool for its sales personnel. Successful operation of the pool is achieved if there is an automobile available each day for each sales-man who needs one. Not all salesmen may need a car on a given day. Experience has shown that the following model, sometimes known in reliability theory as a *partially parallel system*, conforms reasonably well with experience. If we let E_i be the event that the ith automobile in the pool is available for use on a given day, we suppose the class $\{E_i: 1 \leq i \leq n\}$ is an independent class and that $P(E_i) = p$ for each i. Let A_k be the event that k automobiles are available on a given day, and let P_k be the probability that k salesmen will need an automobile on that day. If we let E be the event of successful operation of the pool, then $R = P(E)$ is called the reliability of the pool. The reliability may be determined as follows:

$$P(E|A_k) = P_k \quad \text{and} \quad P(A_k) = Q_k = C_k{}^n p^k (1 - p)^{n-k}$$

From this we obtain

$$R = P(E) = \sum_{k=1}^{n} P(E|A_k)P(A_k) = \sum_{k=1}^{n} P_k Q_k$$

The reliability is determined when the numbers n, p, and P_k for $k = 1, 2, \ldots, n$ are known. ∎

Variations of the partially parallel system are obtained by variations of the distributions of the $P_k = P(E|A_k)$. For example, if it is certain that r salesmen will need automobiles, $P(E|A_k) = 0$ for $k < r$ and $P(E|A_k) = 1$ for $k \geq r$. This means that the system is operative if r or more units (subsystems) are operating. For $r = 1$, this reduces to the ordinary parallel system which is operative if one or more of the subsystems are operative.

The following problem for Bernoulli trials is of interest in the theory of sampling.

Example 2-8-9

Suppose in a Bernoulli sequence of n trials it is known that there are exactly r successes. What is the conditional probability of a success on the ith trial?

SOLUTION AND DISCUSSION   We define E_i and A_{rn} as in Example 2-8-3. Then we seek to evaluate

$$P(E_i|A_{rn}) = \frac{P(E_i A_{rn})}{P(A_{rn})}$$

Now $E_i A_{rn} = E_i A'_{(r-1)(n-1)}$, where $A'_{(r-1)(n-1)}$ is the event of $r - 1$ successes in the $n - 1$ trials remaining when the ith trial is deleted from the original sequence. The independence conditions assumed and Theorem 2-8A assure us that E_i and $A'_{(r-1)(n-1)}$ are independent events. The sequence of $n - 1$ trials in the deleted sequence is also a Bernoulli sequence. Hence

$$P(A'_{(r-1)(n-1)}) = C_{r-1}^{n-1} p^{r-1}(1 - p)^{n-r}$$

since $(n - 1) - (r - 1) = n - r$. Thus

$$
\begin{aligned}
P(E_i|A_{rn}) &= \frac{P(E_i)P(A'_{(r-1)(n-1)})}{P(A_{rn})} = \frac{p C_{r-1}^{n-1} p^{r-1}(1 - p)^{n-r}}{C_r{}^n p^r (1 - p)^{n-r}} \\
&= \frac{C_{r-1}^{n-1}}{C_r{}^n} = \frac{(n - 1)!}{(n - r)!(r - 1)!} \cdot \frac{(n - r)!r!}{n!} \\
&= \frac{r}{n}
\end{aligned}
$$

Note that this result is the same for each $i = 1, 2, \ldots, n$; moreover, it does not depend on the value of p. ∎

This result corresponds to the intuitive notion that if one knows that there are r successes among n trials, the conditional probability of obtaining a success on any trial is the ratio of the number of possibilities of success to the total number of possible outcomes.

We now extend the preceding result to a situation that often arises in sampling theory.

Example 2-8-10

A "random sample" of size n is taken from an "infinite" population and put into a box. These units have probability p of satisfying a specified condition. One of the units in the box is chosen from the sample of size n. We consider two questions: (1) What is the conditional probability that the unit chosen from the box will satisfy the criterion, given that there are r such units among the n in the box? (2) What is the total probability of choosing a unit which satisfies the criterion?

SOLUTION   Here we have a compound process. The choice of an elementary outcome corresponds to (1) taking a sample of size n and (2) making a choice from the n units in the sample. We shall find it expedient to define events in terms of the outcomes of each of these processes. The process of taking a sample of size n is a process of repeated trials. If the sample is random and from an infinite population (or from a finite population with replacement), it is assumed that the outcomes of the trials are independent and have the same probability. If the event of choosing a unit satisfying the specified criterion is considered a "success," the process is thus equivalent to a Bernoulli sequence of length n with $P(E_i) = p$, where E_i is defined as in Example 2-8-3. We define A_{rn} as before. In addition, we must represent the process of the second sampling from the box in a suitable fashion. Each elementary outcome ξ can be described in terms of which of the n units that are chosen in the first sampling operation is also chosen on the second sampling. We define C_i to be that set of outcomes for which the unit chosen from the box on the second operation of sampling from the box is the ith unit chosen in the original operation of sampling from the original population. The condition of making the choice from the box is assumed to be such that, for each $j = 1, 2, \ldots, n$, the class $\{C_j, E_i: 1 \leq i \leq n\}$ is an independent class. We note that the class $\{C_j: 1 \leq j \leq n\}$ is a partition, since one of these events must occur but only one can occur. We let E be the event that the unit chosen from the box on the final selection satisfies the specified criterion. Now E occurs iff for some i there is a success on the ith trial in the original sampling operation and the ith unit in this sample is selected from the box. Thus, for some i, the element ξ must belong to both E_i and C_i. Recalling the fact that the C_i form a partition, we have

$$E = \biguplus_i E_i C_i.$$

The problem is thus to determine (1) $P(E|A_{rn})$ and (2) $P(E)$. Under the assumed conditions, we have from Example 2-8-9

$$P(E_i|A_{rn}) = \frac{r}{n}$$

Because of the assumed independence,

$$P(E_i C_i A_{rn}) = P(E_i A_{rn})P(C_i)$$

so that

$$P(EA_{rn}) = \sum_i P(E_i C_i A_{rn}) = \sum_i P(E_i A_{rn})P(C_i)$$

and

$$P(E|A_{rn}) = \sum_i P(E_i|A_{rn})P(C_i) = \sum_i \frac{r}{n} P(C_i)$$

$$= \frac{r}{n}$$

since the sum of the probabilities of the C_i is unity. This answers the first question. To get $P(E)$, we consider

$$P(E) = \sum_{r=0}^{n} P(E|A_{rn})P(A_{rn}) = \frac{1}{n} \sum_{r=0}^{n} r C_r{}^n p^r (1 - p)^{n-r}$$

$$= p$$

Evaluation of the sum may be carried out by the following classical algebraic manipulations:

$$\sum_{r=0}^{n} r \frac{n!}{(n-r)!r!} p^r (1-p)^{n-r} = np \sum_{(r-1)=0}^{(r-1)=(n-1)} \frac{(n-1)!}{(n-r)!(r-1)!} p^{r-1}(1-p)^{n-r}$$

The second expression is obtained by factoring out np and dropping the $r = 0$ term, which is zero. If we put $k = r - 1$ and $m = n - 1$, this expression may be written

$$np \sum_{k=0}^{m} C_k{}^m p^k (1-p)^{m-k} = np$$

Upon division by n, we have the result stated previously. ∎

It is interesting to note that the results obtained above do not depend upon how the probabilities $P(C_i)$ are distributed. It is only necessary that each of these events be independent of the events E_i. This is a natural and necessary assumption of any process of sampling that can be considered random. This means that the choices from the box do not have to be equally likely. All sorts of schemes for selection from the box (which would modify the distributions of the probabilities of the various choices) should give the same result. In fact, as a limiting case, we could let $P(C_j) = 1$ for some particular j and let all other $P(C_i) = 0$. This would be the deterministic case of choosing a specified unit from the box.

Conditional independence

We have noted that independence is a function of the probability measure, rather than a characteristic of the events (except in the special cases of the sure event and the impossible event). We have also seen that a conditional probability measure $P_E(\cdot)$ is a probability measure. This suggests the possible usefulness of the following concept.

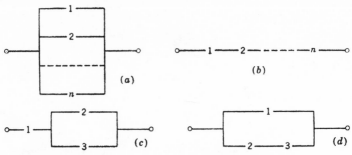

Fig. 2-8-3   Switching-circuit configurations for Example 2-8-11

Definition 2-8a

The events A, B are said to be *conditionally independent, given E,* $(P(E) > 0)$ iffi $P(AB|E) = P(A|E)P(B|E)$.

The defining product rule may be written $P_E(AB) = P_E(A)P_E(B)$, emphasizing that conditional independence is independence with respect to the conditional probability measure $P_E(\cdot) = P(\cdot|E)$.

This pattern of conditioning appears in many applications traditionally labelled independent, although conditional independence does not imply ordinary independence, nor does the latter imply conditional independence. The following example, discussed in considerable detail in Lloyd and Lipow [1962], provides an interesting illustration of such a situation.

Example 2-8-11   *Safety-switch Circuits*

A system has one or more safety switches designed to open when a certain dangerous condition (e.g., an improper temperature) exists. The operation of any switch is affected only by the dangerous condition and not by the operation or nonoperation of any other switches. It is desired to determine the probability of failure of the safety-switch circuit to operate properly for various possible circuit configurations. We suppose there are n switches in the circuit.

Let A_i = event ith switch opens

A = event composite control circuit is open

D = event dangerous condition exists

F = event circuit fails to operate properly

We note first that the circuit fails to operate properly if it opens when the dangerous condition does not exist or if it fails to open when the dangerous condition exists. The two possibilities are mutually exclusive. In symbols we have

$$F = AD^c \uplus A^cD$$

so that

$$P(F) = P(A|D^c)P(D^c) + P(A^c|D)P(D)$$

The event A may be expressed as a boolean function of the A_i. The particular function depends upon the circuit configuration. We consider four circuit configurations, as shown in Fig. 2-8-3.

1. n switches in parallel. The circuit is open iffi all the n switches are open.

$$A = A_1A_2 \cdots A_n \qquad A^c = A_1{}^c \cup A_2{}^c \cup \cdots \cup A_n{}^c$$

2. n switches in series. The circuit is open if any one or more of the switches are open.

$$A = A_1 \cup A_2 \cup \cdots \cup A_n \qquad A^c = A_1{}^c A_2{}^c \cdots A_n{}^c$$

3. Switch 1 in series with the parallel combination of switches 2 and 3. The circuit is open if switch 1 or both switches 2 and 3 are open.

$$A = A_1 \cup A_2 A_3 = A_1 \uplus A_1{}^c A_2 A_3$$
$$A^c = A_1{}^c (A_2{}^c \cup A_3{}^c)$$

4. Switch 1 in parallel with the series combination of switches 2 and 3. The circuit is open if switch 1 and either switch 2 or 3 are open.

$$A = A_1 (A_2 \cup A_3) \qquad A^c = A_1{}^c \uplus A_1 A_2{}^c A_3{}^c$$

In order to calculate the conditional probabilities, we use (CP4) and the assumed conditional independence, given D and given D^c.

1. For the first case we have

$$P(A|D^c) = P(A_1|D^c)P(A_2|A_1 D^c) \cdots P(A_n|A_{n-1} \cdots A_1 D^c)$$

Since we assume the probabilities are conditioned only by D or D^c and not by the events A_i, this expression is assumed to reduce to

$$P(A|D^c) = P(A_1|D^c)P(A_2|D^c) \cdots P(A_n|D^c)$$

We may argue similarly for $P(A|D)$, from which it follows that

$$P(A^c|D) = 1 - P(A|D) = 1 - P(A_1|D)P(A_2|D) \cdots P(A_n|D)$$

2. For the switches in series, we use the intersection expression for the complement and the partial independence to obtain

$$P(A^c|D) = P(A_1{}^c|D)P(A_2{}^c|D) \cdots P(A_n{}^c|D)$$
$$P(A|D^c) = 1 - P(A_1{}^c|D^c)P(A_2{}^c|D^c) \cdots P(A_n{}^c|D^c)$$

3. For the third case, we use the second expression for A to obtain

$$P(A|D^c) = P(A_1|D^c) + P(A_1{}^c|D^c)P(A_2|D^c)P(A_3|D^c)$$
$$P(A^c|D) = 1 - [P(A_1|D) + P(A_1{}^c|D)P(A_2|D)P(A_3|D)]$$

4. For the fourth case, we use the expression given for A^c.

$$P(A^c|D) = P(A_1{}^c|D) + P(A_1|D)P(A_2{}^c|D)P(A_3{}^c|D)$$
$$P(A|D^c) = 1 - [P(A_1{}^c|D^c) + P(A_1|D^c)P(A_2{}^c|D^c)P(A_3{}^c|D^c)]$$

In the simplest case, we may suppose all switches have the same probabilities (conditional) of operating. Suppose we let

$$P(D) = p \qquad P(A_i|D^c) = \alpha \qquad P(A_i{}^c|D) = \beta$$

We then have the following expressions for $P(F)$ in the various cases:

(1) $\quad P(F) = (1 - p)\alpha^n + p[1 - (1 - \beta)^n]$

(2) $\quad P(F) = (1 - p)[1 - (1 - \alpha)^n] + p\beta^n$

(3) $\quad P(F) = (1 - p)[\alpha + (1 - \alpha)\alpha^2] + p[1 - (1 - \beta) - \beta(1 - \beta)^2]$

(4) $\quad P(F) = (1 - p)[1 - (1 - \alpha) - \alpha(1 - \alpha)^2] + p[\beta + (1 - \beta)\beta^2]$

These expressions may be simplified and put in various forms. If $n = 3$ in the first two cases, various possible configurations for three switches may be compared for different values of the parameters. ∎

A rather detailed analysis for various cases is given in Lloyd and Lipow [1962, pp. 243–248]. For our purposes it suffices to set up the model and utilize probability theory to obtain the basic formulas.

2-9   Some comments on strategy

In this chapter we have laid the foundations of an analytical theory of probability. In doing so, we have distinguished sharply between the real world of phenomena and the abstract, idealized world of the mathematical model. It has been emphasized that the utility of the mathematical theory for solving real-world problems lies in the ability to do three things:

1. *Translate* the real-world problem into a corresponding model problem expressed in terms of the precise concepts and relationships of the mathematical model.

2. *Manipulate* the mathematical model to obtain a solution to the model problem. This is frequently aided and guided by an appeal to certain auxiliary models which facilitate the visualizing and remembering of significant aspects of the mathematical model.

3. *Interpret* the results by correlating model phenomena with appropriate real-world phenomena.

The content of the mathematical theory consists of working with the mathematical model. Much of the effort of an introductory study such as this must be devoted to the purely mathematical task of studying the model for its properties, patterns, and possibilities. Success in application of the theory to real-world problems depends upon skill in translation and interpretation, as described above. If the investigator has sufficient skill in translating his real-world problem into the mathematical model, he ordinarily experiences no real difficulty in making the correlation in the other direction, i.e., from the mathematical system to the real world.

Formulation of problems

A first step in successful formulation of the mathematical problem is usually the identification and definition of the appropriate events. One must consider both those events about which probability information exists or can be assumed reasonably and those events whose probabilities are to be determined. In addition, certain auxiliary events may be required in order that the appropriate calculations may be carried out.

Sometimes the required events can be defined without any real thought of the basic space or of elementary outcomes. In other situations it is necessary to be quite clear about what constitutes a trial (choice of an element ξ) and to use this understanding as a guide to the specification of events. In most cases it is not really necessary to describe completely the structure of the basic space. The basic space serves, however, as the *basic organizing concept*, which aids in clear thinking about events and their occurrences.

When appropriate events are defined, careful attention must be given to the probability information, which is available or which must be assumed. Logical relationships must be noted, and special conditions such as stochastic independence carefully identified. There is, in fact, considerable interplay between the process of defining events and the process of identifying the relationships which are assumed to exist.

Mathematical solution

The development of this chapter has indicated that there are certain elements of strategy which apply to large classes of problems. Not all these elements of strategy are used in every problem. But the deliberate exploitation of these often leads to a systematic method of solution, not only for individual problems but for important classes of problems. A good grasp of strategies and general techniques, based on sound theoretical perspective, is valuable both for insight and as an aid in finding specific strategies for special situations. Among the general patterns that should be specifically noted are the following:

1. Exploit the additivity property of probability by expressing events whose probability is desired as the disjoint union of events whose probabilities are known.

2. Extend the first strategy by use of the product rules for conditional probability and for independent events. Expand the event whose probability is desired as the disjoint union of intersections of other events about which appropriate probability information exists. In particular, minterm maps and minterm expansions may be useful in deriving the appropriate expressions.

3. Since probability information is frequently in the form of conditional probabilities, properties of conditional probability should be utilized. In particular, Bayes' rule is frequently useful in "reversing" conditional probabilities.

Guides to analysis

Once a model has demonstrated its usefulness by virtue of its correspondence with significant patterns in the real world of phenomena, it serves a more important role than as a mere language for expressing real-world problems.   It serves as a rational basis for posing questions to the real world.   Success in analysis may well turn on the ability to put the correct questions.   If the problem is to be translated into the model, the questions put to the real-world situation must be meaningful in terms of the mathematical counterpart.   Much of the confusion in probability applications stems from the fact that assertions and questions about the real-world situation have no counterparts (and hence no meaning) in terms of the model.   It may be that the model as it exists does not really provide the concepts and apparatus needed for the problem under study.   In such a case, one must seek to find ways of extending the model in an appropriate fashion.   More often, the model is quite applicable if the problem is analyzed in a fashion appropriate to its character.

The model, as a logical, mathematical structure, should be studied theoretically to discover what assumptions and conditions are really needed to specify a problem.   Unnecessary assumptions can sometimes be eliminated by careful attention to the theoretical model.   Alternative forms of the assumptions may be discovered which make for simpler formulation of the physical problem and for easier checking of the experimental validity of the assumptions.

Considerations such as these point to the need and desirability of a more careful study of the mathematical model itself than has ordinarily been the practice among scientists and engineers.

Problems

2-1. Let the universal set S be the set of all positive integers, and consider the following subsets.

$A = \{\xi: \xi \leq 12\}$
$B = \{\xi: \xi < 8\}$
$C = \{\xi: \xi \text{ is even}\}$
$D = \{\xi: \xi \text{ is a multiple of } 3\}$
$E = \{\xi: \xi \text{ is a multiple of } 4\}$

Express in terms of A, B, C, D, and E (and possibly their complements) the following sets:

(a) $\{1, 3, 5, 7\}$ ANSWER: BC^c
(b) $\{3, 6, 9\}$
(c) $\{8, 10\}$ ANSWER: AB^cCD^c
(d) The even integers greater than 12.
(e) The positive integers which are multiples of 6.
(f) The integers which are even and less than or equal to 6 or which are odd and greater than 12.

2-2. A deck has 10 cards with numbers 0 through 9; i.e., each number is on one and only one of the cards.   A simple game is played in which four successive draws are made. At each draw the number on the card chosen is recorded, and the card is replaced in the deck before the next draw. An elementary outcome thus consists of a set of four decimal digits, recorded in the order of their choice.

(a) How many elementary outcomes are there?

(b) Let A be the event a 2 appears in the first or third places. Express event A in terms of the events B and C where

B = event that a 2 appears in first place
C = event that a 2 appears in third place

(c) How many elementary outcomes comprise the event A (i.e., how many ways can the event A occur)?   ANSWER: 1,900

2-3. How many ordered sets of initials can be formed if every person has the following number of given names plus a surname:

(a) Exactly two?   ANSWER: 26^3
(b) At most two, i.e., one or two?   ANSWER: $26^2 + 26^3$
(c) At most three?   ANSWER: $26^2 + 26^3 + 26^4$

2-4. Each domino is marked by two numbers. The pieces are symmetrical, so that the number pair is not ordered. How many different pieces can be made using the numbers $1, 2, \ldots, n$? (*Hint*: Each domino except the "doubles" has two distinct numbers.)   ANSWER: $C_2{}^n + n = C_2{}^{n+1}$

2-5. Consider 10-word telegrams in English, no word of which has more than eight letters.

Let A = set of all such telegrams ending with word "love"
    B = set of all such telegrams beginning with word "will" and ending with word "love"

Which set has the greater probability? Why?

2-6. Suppose an experiment is repeated a large number of times. Let A and B be two events, and let $f(A)$ and $f(B)$ be the fraction of times that these events occur (i.e., these are the relative frequencies of occurrences of events A and B). Show that

(a) $f(A \cup B) = f(AB^c) + f(A^cB) + f(AB)$
(b) $f(A^c) = 1 - f(A)$
(c) $f(A) = f(AB) + f(AB^c)$

2-7. Find the (classical) probability that among three random digits (each random digit being one of the numbers $0, 1, 2, \ldots, 9$) there occur

(a) All three alike.   ANSWER: 0.01
(b) No two alike.   ANSWER: 0.72
(c) Exactly two alike.   ANSWER: 0.27

2-8. What is the probability that among k random digits the digit 0 appears:

(a) Exactly three times?   ANSWER: $C_3{}^k 9^{k-3}/10^k$

(b) Three times or less?   ANSWER: $\displaystyle\sum_{r=0}^{3} C_r{}^k 9^{k-r}/10^k$

2-9. A box contains 90 good and 10 defective screws. If 10 screws are used, what is the probability that none are defective?   ANSWER: $C_{10}{}^{90}/C_{10}{}^{100}$

2-10. A shipment of n diodes is received. Of these, b are defective. From the shipment, k are chosen.

(a) What is the (classical) probability that all k chosen are good?

(b) What is the probability that one or more of the k chosen are defective?

2-11. Four men order dinner at a restaurant and then decide to go to the bar for a drink. The waiter places their orders on the table at random. What are the probabilities that exactly four, three, two, one, and zero of them find what they ordered in the places they had chosen? ANSWER: $\frac{1}{24}$, 0, $\frac{1}{4}$, $\frac{1}{3}$, $\frac{3}{8}$

2-12. Determine $P(A \cup B \cup C \cup D)$ in terms of the probabilities of the separate events and suitable intersections.

2-13. Use fundamental properties of probability to show that

(a) $P(AB) \leq P(A) \leq P(A \cup B) \leq P(A) + P(B)$

for any two events A and B

(b) $P(\bigcap_{i \in J} A_i) \leq P(A_k) \leq P(\bigcup_{i \in J} A_i) \leq \sum_{i \in J} P(A_i)$ for any $k \in J$

2-14. Given that $P(A) = \frac{1}{3}$, $P(B) = \frac{1}{2}$, and $P(A|B) = \frac{1}{2}$. Find $P(B|A)$.

2-15. Three dice are rolled. Given that no two faces are the same, what is the probability (conditional) that one of the faces is a four? Justify your answer in terms of reasonable assumptions and properties of conditional probability.

ANSWER: $P = \frac{1}{2}$.

2-16. A die is loaded in such a way that the probability of a given number turning up is proportional to that number (e.g., a 6 is twice as probable as a 3). What is the probability of a 3, given that an odd number is rolled? ANSWER: $\frac{1}{3}$

2-17. Consider two boxes containing black and white balls. Box 1 has initially 2 white and 3 black balls. Box 2 has initially 2 white and 2 black balls.

(a) Draw a ball at random from box 1 and transfer it to box 2 (without determining which color ball was transferred).

(b) Draw at random a ball from the second box (after the transfer of the ball of unknown color).

What is the probability of drawing a white ball from box 2 on step (b)? [*Suggestion:* Let A be the event of transferring a white ball and B be the event of choosing a white ball from the second box. Use the property (CP2)]. ANSWER: $P(B) = \frac{12}{25}$

2-18. Let the probability that the weather on one day is of the same kind (rain or no rain) as the previous day be p. Let P_1 be the probability of rain on the first day of the year. What is the probability P_n of rain on the nth day?

ANSWER: $P_n = \frac{1}{2}[1 + (2P_1 - 1)(2p - 1)^{n-1}]$

2-19. Suppose $0 < P(E) < 1$ and $0 < P(F) < 1$. Let the relation $*$ indicate $<$, $=$, or $>$, and let the relation $*^{-1}$ indicate $>$, $=$, or $<$, respectively. Use the definition of conditional probability to show that

(a) $P(E|F) * P(E)$ iffi $P(F|E) * P(F)$ iffi $P(EF) * P(E)P(F)$

(b) $P(E|F) * P(E)$ iffi $P(E|F^c) *^{-1} P(E)$

(c) $P(E|F) * P(E)$ iffi $P(E^c|F^c) * P(E^c)$

2-20. On an attempt to land an unmanned rocket on the moon, the probability of a successful landing is 0.4. The probability (conditional) that the monitoring systems will give the correct information concerning the landing is 0.9 in either case.

A shot is made, and a successful landing is indicated by the monitoring system. What is the probability (conditional) of a successful landing?   ANSWER: 0.86

2-21. There are two good transistors and two defective ones in a box. These are tested one by one. What is the probability that the second defective unit to be tested is the second unit tested? The third tested? The fourth tested?

ANSWERS: $\frac{1}{6}$, $\frac{1}{3}$, $\frac{1}{2}$

2-22. Two production lines turn out the same type of manufactured item. In a given time, line 1 turns out n_1 units, of which $n_1 p_1$ are defective; in the same time, line 2 turns out n_2 units, of which $n_2 p_2$ are defective. Suppose a unit is selected at random from the combined lot produced by the two lines. Let D be the event of a defective unit, A be the event that the unit was produced on line 1, and B be the event that the unit was produced on line 2. Determine $P(A|D)$ and $P(B|D)$, and show that if $n_1 p_1 > n_2 p_2$, we must have $P(A|D) > P(B|D)$.

2-23. A carnival man hides a pea under one of three nut shells. By a series of complicated movements, he attempts to confuse the bystander so that he no longer knows which of the shells hides the pea. The bystander attempts to guess which shell covers the pea. If he follows the sleight of hand, he will guess correctly. If he cannot detect the proper shell throughout the series of movements, he will pick a shell at random. There is a probability of 0.10 that he will know the correct shell. The game is played, and the bystander determines the correct shell. What is the probability (conditional) that he has detected the correct shell and not merely guessed the answer randomly?   ANSWER: $\frac{1}{4}$

2-24. On a multiple-choice examination there are five possible answers for each question, only one of which is correct. If the student knows the correct answer, he will determine this answer with conditional probability 1. If he does not know the right answer, he guesses, so that his probability (conditional) is $\frac{1}{5}$. A good student may be expected to know 90 percent of the answers. If a good student has the correct answer to a given problem, what is the probability (conditional) that he was guessing?

2-25. The following is a generalization of Probs. 2-23 and 2-24. Let A be the event of a correct choice from among n alternatives. Let B be the event of knowing prior to the selection which of the alternatives is correct. Suppose $P(B) = p$, with $0 < p < 1$.

   (a) Determine $P(B|A)$.
   (b) Show $P(B|A) \geq P(B)$, and show $P(B|A)$ increases with increasing n for fixed p.

2-26. Each of three boxes has two drawers. Box 1 has one gold coin in each drawer; box 2 has one silver coin in each drawer; box 3 has one gold coin in one drawer and a silver coin in the other. A box is chosen at random; one of the two drawers is chosen at random; it has a gold coin. What is the probability (conditional) that the other drawer has a silver coin?   ANSWER: $\frac{1}{3}$

2-27. *Polya's urn model of a contagious disease.* An urn contains initially b black and r red balls. A ball is drawn and then replaced along with c additional balls of the same color. The process is repeated. Let B_k be the event the kth ball drawn is black and R_k the event the kth ball drawn is red. Suppose on any drawing it is equally likely that any ball in the urn will be drawn. Determine

   (a) $P(B_2|R_1)$ (b) $P(R_1R_2)$ (c) $P(B_1R_2B_3)$ (d) $P(R_1|B_2)$

2-28. Two boxes contain large numbers of pieces of hard candy in three flavors: lemon, butterscotch, and mint.   The fractions are as follows:

Box 1: 0.30 lemon   0.40 butterscotch   0.30 mint
Box 2: 0.40 lemon   0.50 butterscotch   0.10 mint

A box is chosen at random, and two pieces of candy are chosen at random therefrom. Assume the number of pieces is large enough so that choice of the first piece does not appreciably affect the probabilities of choice of the second piece.

(a) What is the probability that two butterscotch pieces are chosen?

ANSWER: 0.205

(b) What is the probability that the second piece chosen is mint-flavored?

ANSWER: 0.20

2-29. In logic, two statements p and q are said to be logically independent iffi each of the four truth-table combinations FF, FT, TF, TT (where F means the proposition is false and T means it is true) is possible.   If we let A be the event p is true and B be the event q is true, this requirement of logical independence is equivalent to the requirement that none of the events in the class $\{A^cB^c, A^cB, AB^c, AB\}$ is impossible.   Show that there is the possibility of a probability measure $P(\cdot)$ such that A and B are stochastically independent events iffi p and q are logically independent statements, with $0 < P(A) < 1$ and $0 < P(B) < 1$.

2-30. Let $\{B_i: i \in J\}$ be a finite or countably infinite disjoint class of events whose union is the event B.   Suppose for each $i \in J$ the pair of events A, B_i is an independent pair.   Are the events A and B independent?   If so, prove that they are; if not, give a counterexample.

2-31. Consider a class $\{A, B, C\}$ of events.   Suppose it is known that A, B is an independent pair and that B, C is an independent pair.   Does it follow that A, C is an independent pair?   Justify your answer.

2-32. An integer expressed in the decimal system requires five digits.   What is the probability of guessing the number?   State reasonable assumptions.

ANSWER: $\frac{1}{9} \times 10^{-4}$

2-33. Consider two partitions $\mathfrak{A} = \{A_1, A_2, A_3, A_4\}$ and $\mathfrak{B} = \{B_1, B_2, B_3, B_4\}$. These partitions are said to be independent iffi $P(A_iB_j) = P(A_i)P(B_j)$ for each pair A_i, B_j.   The following probabilities are given:

$$P(A_1) = 0.1 \qquad P(B_1) = 0.2$$
$$P(A_3) = 0.3 \qquad P(B_2) = 0.2$$
$$P(A_4) = 0.4 \qquad P(B_3) = 0.1$$

The following is a partial table of joint probabilities $P(A_iB_j)$:

	$i = 1$	$i = 2$	$i = 3$	$i = 4$
$j = 1$	0.02		0.06	0.08
$j = 2$	0.02			0.08
$j = 3$		0.01	0.03	
$j = 4$	0.04		0.15	0.20

(a) Determine $P(A_2)$ and $P(B_4)$, and complete the table of joint probabilities.

(b) Determine whether or not the partitions are independent.

ANSWER: Not independent.

2-34. A machine can fail if any one of *five* independent parts fails. If the probability of each one of them failing in a year is $\frac{1}{10}$, what is the probability of a machine failure within the year? Assume that physical independence of the parts makes the events stochastically independent. ANSWER: $1 - (\frac{9}{10})^5$

2-35. A man seeks advice regarding one of two possible courses of action from three advisers, who arrive at their recommendations independently. He follows the recommendation of the majority. The probabilities that the individual advisers are wrong are 0.1, 0.05, and 0.05, respectively. What is the probability that the man takes incorrect advice? ANSWER: 0.012

2-36. Assume that the probability that a family chosen at random in a certain locality will have one child is $\frac{1}{4}$; two children, $\frac{1}{4}$; three children, $\frac{1}{8}$ (each family has at least one child). One such family is known to have three or fewer children, all of whom are boys. What is the probability that there are exactly two children? Assume that the probability that any given child will be a boy is $\frac{1}{2}$, independently of the number and sexes of other children in the family. ANSWER: $\frac{4}{13}$

2-37. Three men flip the same coin in succession. They continue to flip until a head appears. The man who obtains the first head wins the game.

(*a*) Suppose the coin is "honest" and that the outcomes of successive throws are stochastically independent. Calculate the probabilities for each man's winning. [*Suggestion:* For each man, list the sequences of heads and tails for which he can win. Use the relation $\sum_{k=1}^{\infty} x^k = x/(1-x)$ for $|x| < 1$.]

ANSWER: $P_1 = p/(1-q^3)$, $P_2 = qp/(1-q^3)$, $P_3 = q^2p/(1-q^3)$

(*b*) Show that even if the coin is biased so that the probability p of a head is different from $\frac{1}{2}$ (but not equal to zero or one), the first man has a probability greater than $\frac{1}{3}$ of winning.

2-38. Show that the product of the partition generated by the class $\{A, B, C\}$ with that generated by the class $\{D, E\}$ is the partition generated by the class $\{A, B, C, D, E\}$.

2-39. Rearrange the variables in the minterm map of Fig. 2-7-3*b* so that the minterms are numbered sequentially in rows from left to right, with minterm numbered zero in its same position in the upper left-hand corner of the diagram.

2-40. Consider the partition generated by the sets A, B, C, D, E, and represent the minterms on a map arranged as in Fig. 2-7-3*b*.

(*a*) Show that every product term, for example, B^cCDE^c, obtained by suppressing the variable A from a minterm, must give a pattern consisting of a pair of minterm blocks on the same row, separated by three minterm blocks.

(*b*) What is the pattern produced by suppressing the variable E from a minterm?

(*c*) What is the pattern produced by suppressing both A and E from a minterm?

2-41. A set of four cards is used to determine any integer from 0 through 15. Each card has eight numbers on the front, and each card is assigned a "value," or "weight." Cards A, B, C, and D have values 8, 4, 2, and 1, respectively. These may be used in a simple game as follows. A person is asked to pick a number. He is shown each card in turn and asked, "Is your number on this card?" The values of the cards for which the answer is "yes" are added to obtain the specified number.

(*a*) Determine the numbers which should be on each of the cards, i.e., those which are examined before answering the question. (*Suggestion:* Use a minterm map.)

(*b*) Suppose this game were extended to determine numbers—say, the age of the subject—not exceeding 60. How many cards are needed, and how many numbers are on each? How many if the numbers to be determined do not exceed 80?

2-42. One hundred students are questioned about their course of study and their plans for entering graduate school. The results of the questionnaire are:

There are 55 men students.
There are 23 engineering students, one of whom is a woman.
There are 75 students studying a foreign language, including all women students.
28 men and 9 women plan to enter graduate studies.
Among those who take engineering, no women students and 12 men students plan to enter graduate school.
Of those studying foreign languages, 9 women students and 10 engineering students plan to enter graduate school.
Of those nonengineers who do not take a foreign language, only one is planning to go to graduate school.

(*a*) How many students planning to enter graduate school are studying a foreign language?   ANSWER: 34

(*b*) How many men students are either planning to enter graduate school or are studying a foreign language, but not both?   ANSWER: $3 + 5 = 8$

2-43. Obtain minterm expansions of the following boolean functions, using the method of binary designators.

(*a*) $AB^c \uplus A^cB = A \oplus B$
(*b*) $AB^c \cup (A \cup \check{B})C^c$
(*c*) $AB(C \cup D) \cup ACD^c$
(*d*) $AB \cup AC \cup BD$

2-44. Obtain the minterm expansion of the following boolean functions, using minterm maps.

(*a*) $AB(C \cup D^c)^c$
(*b*) $A^c \cup B(CD)^c$
(*c*) $AB \cup AC \cup (B \cup D)^c$

2-45. Show that $P(A^cC \cup AB^c) \leq P(A) + P(C)$.

2-46. Show that $P[AB \cup (A \cup B)^c \cup C^c] \leq P(A \cup B^c \cup C^c)$.

2-47. Show that $P(A^cB^cC^c \cup ABC) \leq P(A) + P(B^c) + P(C^c)$.

2-48. Define the symbol $\oplus$ to mean the set combination $A \oplus B = AB^c \uplus A^cB$ (Prob. 2-43). Show $P(AB \oplus A_0B_0) \leq P(A \oplus A_0) + P(B \oplus B_0)$.

2-49. Show $|P(A) - P(B)| \leq P(A \oplus B)$.

2-50. Given $P(A) = 0.6$, $P(A \cup B^c) = 0.8$, $P(A \cup B \cup C) = 0.9$, and $P(A^cC) = 0.15$. Determine $P(BC|A^c)$.   ANSWER: $\frac{1}{8}$

2-51. A lot of n manufactured units is to be accepted or rejected on the basis of testing a small number of units from the lot. The lot has r defective units; the units to be tested are selected at random without replacement. Three decision rules are to be tested: (1) Test two units. If both are good, accept the lot; otherwise reject the lot. (2) Test two units. If both are bad, reject the lot; otherwise accept the lot. (3) Test two units. If both are good, accept the lot; if both are bad, reject the lot;

if one is good and one is bad, select and test a third unit; accept or reject the lot according as this third unit is good or bad. Let G_i be the event the ith unit tested is good. Let A, B, and C be the events leading to acceptance according to the rules 1 through 3, respectively.

(a) Express A, B, and C in terms of the G_i.

(b) Determine $P(A)$, $P(B)$, and $P(C)$, using reasonable assumptions on the conditional probabilities.

(c) Suppose $n = 100$. Determine numerical values of these probabilities for $r = 5, 10, 25$. Compare the decision rules in these cases from the viewpoint of the buyer.

2-52. Given the following probabilities:

$$P(C) = 0.40 \qquad P(AB^c) = 0.25$$
$$P(A^cB) = 0.30 \qquad P(A^cC^c) = 0.40$$
$$P(A^cC) = 0.15 \qquad P(ABC) = 0.10$$
$$P(A^cBC^c) = 0.20$$

Determine which, if any, of the events A, B, and C are pairwise independent.

ANSWER: B and C are independent.

2-53. Suppose $\{A, B, C, D, E\}$ is an independent class of events. Consider the boolean functions

$$F = f(A, B, C) = AB \cup A^c(B \cup C^c)^c$$
$$G = g(D, E) = D \cup D^cE$$

Use Theorems 2-6F and 2-6G to show that the events F and G are independent. (*Suggestion:* Express F and G as the union of minterms.)

2-54. A drawer has a large number of resistors, having one of the values 1,000, 1,500, and 2,500 ohms. One-third of them are 1,000 ohms; one-half of them are 1,500 ohms; one-sixth of them are 2,500 ohms. Assume that the probabilities are not appreciably changed by removing a small number of resistors. (This is the statistical assumption of "infinite population," or equivalently, of "sampling with replacement.") What is the probability of obtaining exactly *one* 1,000-ohm resistor if three are taken from the drawer? ANSWER: $\frac{4}{9}$

2-55. Five random digits are chosen. What is the probability that *at least three* (i.e., three or more) of them will be ones? Use tables to get numerical answers.

ANSWER: $\displaystyle\sum_{r=3}^{5} C_r{}^5(\frac{1}{10})^r(\frac{9}{10})^{5-r}$

2-56. A man closes an electronic counter circuit provided with an a-c input. Assume that it is equally likely that he will close on the positive half cycle and on the negative half cycle. If he closes the switch on the positive half cycle, the counter adds one unit. If the switch is closed on the negative half cycle, the counter subtracts one unit. What are the possible totals at the end of 10 closures, and what is the probability of each of them? The counter begins with zero total.

ANSWER: $n = 2k - 10$, $0 \leq k \leq 10$; $p_n = C_k{}^{10}2^{-10}$

2-57. A game is played with dice as follows. A player rolls a pair of dice five times. He scores a "hit" on any given throw in the sequence if he throws either a 7 or an 11. He wins if he scores an *even* number of hits (zero counted as an even number) in the sequence of five throws.   Make the usual assumptions on the throwing of fair dice.

(a) What is the probability p of scoring a hit on any given throw?

ANSWER: $p = \frac{2}{3}$

(b) What is the probability that the dice thrower wins?   It is sufficient to write out a suitable formula such that the routine of numerical calculation is clearly indicated.   ANSWER: $(1 - p)^5 + C_2{}^5 p^2 (1 - p)^3 + C_4{}^5 p^4 (1 - p)$

2-58. A man claims he can pick the winning horse in a race 90 percent of the time. In order to test his claim, he picks horses to win in 10 races.   We may assume the results to be Bernoulli trials.

(a) What is the probability of his picking at least 9 of the 10 races if his probability p of a correct choice on any race is 0.90?   ANSWER: 0.736

(b) Suppose the man is in effect guessing and that there are five horses in each race from which he chooses at random.   What is the probability that he will pick as many as five races correctly?   ANSWER: 0.033

Selected references

Events and sets

GOLDBERG [1960], chap. 2, sec. 2.   Cited at the end of our Chap. 1.   This elementary work, designed for the freshman level, provides a careful, lucid exposition of the elements of probability theory in the case of a finite basic space.

PARZEN [1960]: "Modern Probability Theory and Its Applications," chap. 1, sec. 4. This modern treatment is characterized by sound mathematics (at about the level of the present book) and a wealth of examples and problems.

Basic probability model

BRUNK [1964]: "An Introduction to Mathematical Statistics."   Although devoted primarily to statistical applications, this book provides, in part 1 (some 94 pages), a clear, succinct, modern introduction to basic probability theory.

FISZ [1963]: "Probability Theory and Mathematical Statistics," 3d ed. (transl. from the Polish).   A detailed treatment of the fundamental probability model as well as a fundamental text on mathematical statistics.   The work is characterized by sound mathematics and detailed treatments of important probability distributions and methods, with considerable introduction to the relevant literature.

GOLDBERG [1960]: cited above.   See chap. 2.

KOLMOGOROV [1933, 1956]: "Foundations of the Theory of Probability."   Written for the mathematician, this work is listed here because of its historical interest.   The axiomatic formulation of the probability model as well as the mathematical development—much of it beyond the scope of the present book—has had a definitive influence on mathematical work in probability theory.

McCORD AND MORONEY [1964]: "Introduction to Probability Theory," chaps. 1 through 4. A careful development of the axiomatic model from a viewpoint and at a level similar to that of the present work.   Distinguishes carefully between mathematical statements and intuitive statements.   Sound and readable.

PARZEN [1960]: cited above.   See chaps. 1 through 3.

Probability as mass

MUNROE [1953]: "Introduction to Measure and Integration."   This work, written for the mathematician, provides a rigorous basis for many of the facts stated without proof in the present work.   Has a brief statement on measure as mass on p. 72.

chapter 3

Random variables and probability distributions

The probability model described in Chap. 2 provides the basis for an extremely useful mathematical theory. In that model, we represent elementary outcomes of some fundamental trial or experiment by elements ξ in a basic space S. There is one element in this space for each possible outcome. When the trial is performed, the outcome is examined with regard to one or more properties. Occurrence or nonoccurrence of any event is determined by noting whether or not the appropriate conditions characterizing that event are met by the properties associated with the particular outcome observed. In many situations, the property to be observed is the value of a number or the values of a set of numbers associated with the outcome. In other cases, the property to be observed is not directly expressed in numerical form, but may be represented in a very natural and useful manner by a number or a set of numbers. For example, in the Bernoulli trials a success can be represented by the number 1 and a failure can be represented by the number 0. In a sampling procedure, the results may be categorized and the various categories may be represented by appropriate numbers. If a sample of size n is taken, the result presents itself as a set of n numbers.

In this chapter, we wish to extend the mathematical model by introducing the concept of a random variable and certain associated analytical means of describing the distribution of values of random variables.   Utilization of these new tools makes possible a very powerful analytical formulation of the theory of probability, capable of handling problems of great complexity.   In fact, the utility of modern probability theory is to a large degree due to the introduction of the concept of random variables and its generalization in the concept of random processes.

3-1   Random variables and events

The idea of numerical-valued phenomena subject to statistical or "chance" fluctuations is familiar in many fields.   One throws a pair of dice and counts the total number of spots which appear on the faces which turn up.   This number will vary from throw to throw.   It is a number characteristic of one or more of the possible outcomes of the experiment of throwing the dice.   One studies the ages at death of the population in a given geographical area.   This age will be different for each individual.   The selection of a given individual from the population corresponds to the selection of a possible outcome.   The age at death is a number characteristic of this person and hence of this possible outcome. One counts the rate at which automobiles pass a given check point during a given traffic period each day.   The number resulting from this count may be expected to vary from day to day, although one might expect some "pattern" in the statistical behavior of these numbers.   It is precisely for this type of phenomena that the concept of random variable is introduced into probability theory.

Random variables as functions

How can phenomena of the type illustrated above be represented mathematically in the probability model?   A partial answer is readily apparent. We have noted in the examples above that the outcome of each trial or experiment is characterized by a number (or set of numbers).   In the case of the single number associated with the outcome, we have a simple correspondence:

To each element ξ corresponds a number

This is precisely the concept of a *function*, which is fundamental in modern mathematics.   Let us examine the possibility that the proper mathematical model for numerical-valued phenomena in probability theory is the concept of a function.   First, let us be sure that the concept of a function is familiar in the sense in which we shall need it.

For our purpose, we shall be interested in the idea of a function as a *mapping*, or *transformation*.   The function provides a mapping from a set of elements D, known as the *domain* of the function, to another set of

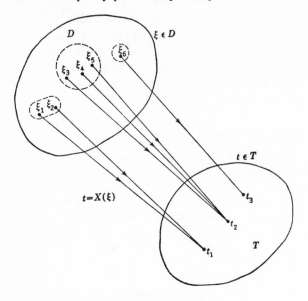

Fig. 3-1-1   Mapping produced by the function $X(\cdot)$.

elements T, known as the *range* of the function.   Each element of D is
mapped into one and only one element of T.   On the other hand, many
elements in D may be mapped into a single element in T.   The situation
is represented schematically in Fig. 3-1-1, in a manner which we shall
employ frequently in subsequent developments.   The function $X(\cdot)$ produces a correspondence between elements of D and those of T in such a
way that each element in D is paired with one and only one element in T.
If ξ is an element in D and t is the element in T which corresponds to it,
we write

$$X(\xi) = t$$

The element t is called the *value* of $X(\cdot)$ at ξ.   With reference to the
mapping concept, elements ξ of the domain are sometimes called *object
points*, and the elements t which correspond are called *image points*.   In
this section, we suppose the domain D to be the whole basic space S and
the values to be real numbers, so that the values t are elements of the
real line R; that is, T is a subset of R.   We shall ordinarily designate
functions with capital letters near the end of the alphabet: $X(\cdot)$, $Y(\cdot)$,
$Z(\cdot)$, etc.   Exceptions should be clear from the context.   Sometimes, to
simplify notation, we write X, Y, Z, etc., when there is no danger of
confusing this designation for a function with the letter designation for a
set or event.

Consider a very simple example. Suppose in a sampling process a 1 is recorded each time the event E occurs and a 0 is recorded otherwise. This natural tallying procedure is equivalent to defining the indicator function introduced in Sec. 2-7.

Example 3-1-1   *The Indicator Function for an Event*

If E is any event (or in fact any subset of S) the indicator function $I_E(\cdot)$ for E is defined by

$$I_E(\xi) = \begin{cases} 0 & \text{if } \xi \notin E \\ 1 & \text{if } \xi \in E \end{cases}$$

The name *indicator function* (the term *characteristic function* is also used) is very suggestive. Whenever the event E occurs, i.e., whenever the element ξ chosen belongs to E, the function takes on the value 1 and so indicates the occurrence of E. If the event E does not occur on a trial, the function has the value zero. The mapping produced by this function is extremely simple, as shown in Fig. 3-1-2. Every element ξ in the set E is mapped into the point 1 on the real line; every element in the set E^c is mapped into the point 0 on the real line.

The indicator function can be used in the construction of more complicated functions for representing numerical-valued random phenomena.

Example 3-1-2

Suppose $\{A, B, C\}$ is a class of mutually exclusive events, one of which is sure to occur. If event A occurs, the number observed is 0; if the event B occurs, the num-

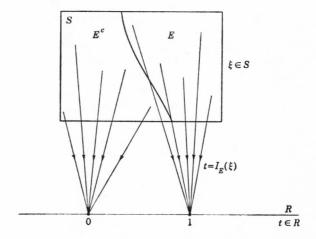

Fig. 3-1-2   Schematic representation of the mapping from the basic space S to the real line R produced by the indicator function $I_E(\cdot)$.

ber observed is 1; and if the event C occurs, the number 3 is observed. We wish to describe a function (random variable) $X(\cdot)$ which assigns to each elementary outcome ξ the appropriate value. The desired function may be written in terms of indicator functions for the events in the partition as follows:

$$X(\cdot) = 0 \cdot I_A(\cdot) + 1 \cdot I_B(\cdot) + 3 \cdot I_C(\cdot)$$

If we choose a ξ, one and only one of the indicator functions will have the value unity; the others will have the value zero. This indicator function is multiplied by the coefficient needed to give the function the required value. Thus, if $\xi \in C, I_C(\xi) = 1$, $3I_C(\xi) = 3$, and therefore $X(\xi) = 3$, as required. ∎

This function produces a mapping similar to that in Fig. 3-1-2. In this case, every element in the set A is mapped into the point 0 on the real line R; every element in the set B is mapped into the point 1; and every element in the set C is mapped into the point 3 on the real line. We shall give considerable attention to functions of the type illustrated in this example.

Not all random variables produce mappings as simple as this. Many random variables have a continuum of values, so that the image points are spread along the real line rather than concentrated at discrete points. We shall consider later how these mappings may be represented in a manner that facilitates visualization and makes possible the analytical handling of random variables.

Random variables and events

In dealing with random variables, we need to be able to make probability statements. It is not sufficient to know what values the function may take on (i.e., what the range of the function is). It is also necessary to be able to speak of the probability that a given value will be assumed or that the value of the random variable will be among a given set of values, say, in a certain interval on the real line. Thus we are interested in the probability that the throw of the dice will result in the value seven. Or we may be interested in the probability that the number of cars passing the check point in a 1-minute interval may exceed 50. In the case of Bernoulli trials, we may be interested in the probability that the number of successes in 10 attempts (which is certainly a random variable) should exceed 7.

Probabilities are defined for events, which are sets of elements ξ on the basic space. How can sets of values of the function be related to sets of elements ξ on the domain space S? We may return to the characterization of a function as a mapping, as diagramed in Fig. 3-1-1, to note a second kind of correspondence produced by a function. Referring to this figure, we note that the value t_1 is paired with ξ_1 and with ξ_2. The value t_2 is paired with each of the elements ξ_3, ξ_4, and ξ_5. Similarly, t_3 is paired with ξ_6.   We may think of these as the correspondences

element t_1 and set $\{\xi_1, \xi_2\}$
element t_2 and set $\{\xi_3, \xi_4, \xi_5\}$
element t_3 and set $\{\xi_6\}$

Thus, to each *element t*, there corresponds a *set* of elements ξ in S, called the *inverse image* of t.   If t is not in the range set T, the inverse image of t is the empty set $\emptyset$.   It is customary to designate the inverse image of t by $X^{-1}(t)$.   The relation between elements t and ξ sets $X^{-1}(t)$ is of the nature of an abstract function having ξ sets as values.   This relation is called the *inverse function*, or *inverse mapping induced by the function*, $X(\cdot)$.   It should be noted, however, that many authors reserve the term inverse function for those functions $X(\cdot)$ whose mappings are one-to-one, so that $X^{-1}(\cdot)$ is also a point function.   The inverse mapping $X^{-1}(\cdot)$ is conceived as carrying each element t in the space containing T into a subset $X^{-1}(t)$ of the domain S.   If we suppose that in Fig. 3-1-1 all the elements ξ_i which are mapped into t_1, t_2, or t_3 are shown, then

$$X^{-1}(t_1) = \{\xi_1, \xi_2\}, X^{-1}(t_2) = \{\xi_3, \xi_4, \xi_5\}, \text{ and } X^{-1}(t_3) = \{\xi_6\}$$

In a similar way, to each t set M there corresponds a ξ set $X^{-1}(M)$, called the *inverse image* of M, which is the set of all those ξ which are mapped into the set M by the function $X(\cdot)$.   Thus, in symbols,

$$X^{-1}(M) = \{\xi : X(\xi) \in M\}$$

For example, in Fig. 3-1-1, if

$$M = \{t_1, t_2, t_3\}, X^{-1}(M) = \{\xi_1, \xi_2, \xi_3, \xi_4, \xi_5, \xi_6\}$$

This inverse relation provides an inverse mapping which produces a correspondence between t sets (subsets of the space containing T) and ξ sets (subsets of S).   The term *inverse function*, or *inverse mapping*, is also applied to this set-to-set relation, and the symbolic designation $X^{-1}(\cdot)$ is used, as in the previous case.   We have, in fact, a function whose domain is an appropriate class of subsets of the space containing T and whose range is a corresponding class of subsets of S.   The set-to-set relation $X^{-1}(\cdot)$ may be considered to include the element-to-set relation previously designated $X^{-1}(\cdot)$ by virtue of the fact that $X^{-1}(\{t\})$ (set-to-set relation) is the same ξ set as $X^{-1}(t)$ (element-to-set relation).   The notation $\{t\}$ indicates a single-element set whose only element is t.

It is an important fact that the inverse mapping $X^{-1}(\cdot)$ preserves set operations in the following sense.

Theorem 3-1A

Suppose $X(\cdot)$ is a mapping carrying elements ξ of the domain S into elements t in the range T. Then

1. The inverse image of the union (intersection) of a class of the t sets is the union (intersection) of the class of inverse images of the separate t sets.

2. The inverse image of the complement of a t set is the complement of the inverse image of the t set.

3. If a class of t sets is a disjoint class, the class of the inverse images is a disjoint class.

4. The relation of inclusion is preserved by the inverse mapping.

The validity of these statements is easily established by a careful consideration of the definitions. The statements may be illustrated by the schematic representation in Fig. 3-1-3. The t set M_1 is the set $\{t_1, t_2\}$; its inverse image is the ξ set $X^{-1}(M_1) = \{\xi_1, \xi_2, \xi_3\}$. The set M_2 is the set $\{t_3, t_4\}$; its inverse image $X^{-1}(M_2) = \{\xi_4, \xi_5, \xi_6, \xi_7, \xi_8\}$. The inverse

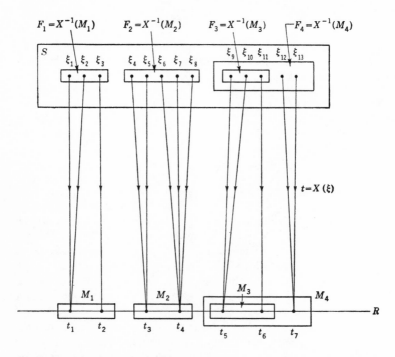

Fig. 3-1-3 Diagrammatic representation of the preservation of set operations and inclusion by the inverse mapping $X^{-1}(\cdot)$.

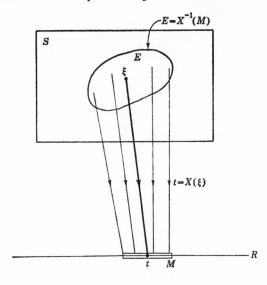

$$\xi \in E \ \text{iffi} \ X(\xi) \in M$$

Fig. 3-1-4　A t set M and its inverse image $E = X^{-1}(M)$.

image of any point not in M_1 is not in $X^{-1}(M_1)$. The inverse image of those t which are in the union of M_1 and M_2 is in the union of the inverse images of these two sets. In the diagram shown, there are no points in the intersection of M_1 and M_2, and hence none in the intersection of the inverse images. On the other hand, M_3 is a subset of M_4, which implies that $X^{-1}(M_3)$ is contained in $X^{-1}(M_4)$. Although this is a specific example, it portrays accurately the role of the inverse function in preserving set operations and inclusion.

The condition that the value $X(\xi)$ belongs to a specific set M is equivalent to the condition that ξ belongs to a corresponding subset $X^{-1}(M)$ of the basic space. If M belongs to a suitable class of t sets, $X^{-1}(M)$ can be expected to be an event, to which probability is assigned. This may be visualized with the help of Fig. 3-1-4. The points of the set $E = X^{-1}(M)$ are all mapped into the set M. This means that an element ξ belongs to the set (event) E iffi $X(\xi) = t$ is one of the real numbers in the set M. In the language of events, the event E occurs iffi the value of the random variable $X(\cdot)$ belongs to the set of numbers M.

For the indicator function in Example 3-1-1, the event E occurs iffi the value of the random variable $I_E(\cdot)$ is 1. For the somewhat more complicated function in Example 3-1-2, the event A occurs iffi the value of the random variable $X(\cdot)$ has the value 0; the event B occurs iffi the value is 1; etc. The event that $X(\xi) \le 2$, which is the event that $X(\xi) \in (-\infty, 2]$, is the event $A \uplus B$.

Measurability

We have indicated, in passing, that the t set M must belong to such a class that the inverse image $X^{-1}(M)$ can be expected to be an event. To deal with this problem fully would be to deal with the problem of *measurability*, discussed briefly in Appendix C. This means that we require certain properties of those functions which are to be considered random variables. Fortunately, not much attention need be given to this topic for most practical applications.

The essential facts are these. The class of t sets needed is the class called the *Borel sets* on the real line, designated $\mathfrak{B}_R$. The properties of this class are discussed briefly in Appendix B. The class is sufficiently general to include all those sets which normally arise in practice. Thus, if M is in $\mathfrak{B}_R$, it is assumed that $X^{-1}(M)$ is in $\mathcal{E}$, the class of events. The following theorem is developed in Appendix C.

Theorem 3-1B

If the function $X(\cdot)$ is such that $\{\xi: X(\xi) \le t\}$ is an event for each real t, then $X^{-1}(M)$ is an event for each Borel set M.

We are now in a position to give a precise definition of the mathematical concept of a random variable in a way that is mathematically satisfactory and which serves as an adequate representation of numerical-valued phenomena in the real world.

Definition 3-1a

A real-valued function $X(\cdot)$ from the basic space S to the real line R is called a (real-valued) *random variable* iffi, for each real t, it is true that $\{\xi: X(\xi) \le t\}$ is an event.

A random variable is thus a special case of what is known in mathematics as a *measurable function*. Because the random variable is in fact a function, it is perhaps unfortunate that the term variable is used rather than function (or some other more suitable term). The usage is firmly fixed in the literature, and once the nature of the concept is firmly grasped, the nomenclature need cause no difficulty.

As we have defined it above, the concept of a random variable is limited to the real-valued case. The concept may be extended to complex-valued functions and vector-valued functions. Vector-valued random variables are introduced in Sec. 3-5. Some attention is given to complex-valued random variables in later chapters and in Appendix E.

In this and the next three sections we limit consideration to the real-valued case.

We may illustrate the measurability idea by considering some simple random variables. First we reconsider the indicator function described in Example 3-1-1.

Example 3-1-3

Let $I_E(\cdot)$ be the indicator function for the event E. Then

$$\{\xi: I_E(\xi) = 1\} = E$$
$$\{\xi: I_E(\xi) = 0\} = E^c$$

$$\{\xi: I_E(\xi) \in M\} = \begin{cases} \emptyset & \text{if } 0 \notin M, 1 \notin M \\ E & \text{if } 0 \notin M, 1 \in M \\ E^c & \text{if } 0 \in M, 1 \notin M \\ S & \text{if } 0 \in M, 1 \in M \end{cases}$$

$$\{\xi: I_E(\xi) \le t\} = \begin{cases} \emptyset & \text{if } t < 0 \\ E^c & \text{if } 0 \le t < 1 \\ S & \text{if } 1 \le t \end{cases}$$

The set $I_E^{-1}(M)$ is thus an event for any t set M, so that the function $I_E(\cdot)$ meets all the requirements for a random variable. ∎

As a second illustration of the measurability condition, we may consider the random variable which gives the number of successes in a sequence of n repeated trials. This number will ordinarily vary from sequence to sequence. Since each sequence corresponds to one elementary outcome, we have a function of elementary outcomes which we would expect to be a random variable in the technical sense of the definition.

Example 3-1-4

Let $X(\cdot)$ be the number of successes in a sequence of n repeated trials (the Bernoulli trials form a special case).

SOLUTION AND DISCUSSION As in the Bernoulli case, let E_i be the event of a success on the ith trial in the sequence, and let A_{rn} be the event of exactly r successes in n trials. The choice of ξ is the mathematical representation of making a sequence of trials. The value $I_E(\xi)$ is unity iff there is a success on the ith trial in the sequence corresponding to ξ. The sum of the indicator functions thus counts the number of successes in the first n trials of the sequence. We may therefore write

$$X(\cdot) = \sum_{i=1}^{n} I_{E_i}(\cdot)$$

In the Bernoulli case, the class $\{E_i: 1 \le i \le n\}$ is an independent class, with $P(E_i) = p$ for each i. The expression for the random variable need not be limited to this case, however; it holds for any sequence of repeated trials if the E_i are defined as indicated.

The random variable may be put into another form which is both instructive and useful. We recall that $\{A_{rn}: 0 \le r \le n\}$ is a partition. This means that for any ξ, one and only one of the indicator functions $I_{A_{rn}}(\cdot)$ is different from zero. The random variable $X(\cdot)$ has the value r when $I_{A_{rn}}(\xi) = 1$. We may thus write

$$X(\cdot) = \sum_{r=0}^{n} r I_{A_{rn}}(\cdot)$$

The range of $X(\cdot)$ is $\{0, 1, 2, \ldots, n\}$. The set of ξ such that $X(\xi) \le t$ may be expressed as follows:

$$\{\xi : X(\xi) \le t\} = \overset{[t]}{\underset{r=0}{\biguplus}} A_{rn}$$

where $[t]$ is the largest integer which is no greater than t, and $A_{rn} = \emptyset$ for r greater than n. For each real t the set $\{\xi : X(\xi) \le t\}$ is the union of events and is thus an event, so that we are justified in referring to $X(\cdot)$ as a random variable. ∎

The next example gives further evidence of why it is important that the class $\mathcal{E}$ of events be closed under complements, countable unions, and countable intersections. Even very common events defined by a random variable utilize these operations.

Example 3-1-5

Suppose $X(\cdot)$ is any random variable. Put $E_t = \{\xi : X(\xi) \le t\}$. Consider several sets defined by $X(\cdot)$.

SOLUTION

1. $\{\xi : X(\xi) \in (a, b]\} = E_b E_a^c$, which is an event, since $X(\xi)$ belongs to $(a, b]$ iffi $X(\xi) \le b$ and $X(\xi) > a$; that is,

$$\{\xi : X(\xi) \in (a, b]\} = \{\xi : X(\xi) \le b\} \cap \{\xi : X(\xi) > a\}$$

Now

$$\{\xi : X(\xi) > a\} = \{\xi : X(\xi) \le a\}^c = E_a^c$$

2. Noting that $t \ge a$ iffi $t > a - 1/n$ for every positive integer n, we have

$$\begin{aligned}
\{\xi : X(\xi) \in [a, b]\} &= \{\xi : a \le X(\xi) \le b\} \\
&= \bigcap_{n=1}^{\infty} \left\{\xi : a - \frac{1}{n} < X(\xi) \le b\right\} \\
&= \bigcap_{n=1}^{\infty} E_b E_{a-1/n}^c
\end{aligned}$$

which is an event

3. $\{\xi : X(\xi) = a\} = \{\xi ; X(\xi) \in [a, a]\} = \bigcap_{n=1}^{\infty} E_a E_{a-1/n}^c$

which is an event.

4. $\{\xi : X(\xi) = -\infty\} = \bigcap_{n=1}^{\infty} E_{-n}$ and $\{\xi : X(\xi) = \infty\} = \bigcap_{n=1}^{\infty} E_n^c$ which are events.

These are usually assumed to have probability zero. ∎

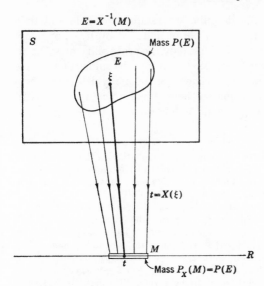

Fig. 3-2-1   Probability measure $P_X(\cdot)$ induced on the real line R by the random variable $X(\cdot)$.

Definition 3-2a

The probability measure $P_X(\cdot)$ defined on the class of Borel sets $\mathfrak{B}_R$ by

$$P_X(M) = P[X^{-1}(M)] \qquad \text{for each } M \in \mathfrak{B}_R$$

is called the *probability measure induced by the random variable* $X(\cdot)$.

The quantity $P_X(M)$ is the probability mass associated with the Borel set M on the real line R by the mapping $t = X(\xi)$.

At this point we may consider some very simple random variables and the mass distributions induced by them on the real line. More complicated examples are deferred until the appropriate analytical means for description are introduced in subsequent sections.

Example 3-2-1

Consider again the discrete-valued random variable $X(\cdot) = 0 \cdot I_A(\cdot) + 1 \cdot I_B(\cdot) + 3 \cdot I_C(\cdot)$ discussed in Example 3-1-2. Since $\{A, B, C\}$ is assumed to be a partition, we must have $A = \{\xi\colon X(\xi) = 0\}$, $B = \{\xi\colon X(\xi) = 1\}$, and $C = \{\xi\colon X(\xi) = 3\}$. The entire probability mass is mapped onto the points 0, 1, 3 on the real line. The mass associated with the point 0 is $P(A)$; the mass associated with the point 1 is $P(B)$; and that associated with 3 is $P(C)$. This mass distribution determines uniquely the mass associated with any set of points on the real line. Now the mass associated with any t set M is precisely the probability that $X(\cdot)$ takes on a value in M. ■

In the first example, we fully describe the random variable and then determine the probability mass distribution induced by it. For many purposes, as we shall see in later sections, only the probability mass distribution on the real line is of interest; that is, it may be sufficient to know that the random variable under consideration is one which produces a given probability mass distribution on the real line. One type of distribution often assumed is that given in the following

Example 3-2-2   *Uniform Distribution*

The random variable $Y(\cdot)$ is said to have a uniform distribution on the interval $[a, b]$ iffi the probability mass distribution induced on the real line is uniform over that interval and there is no other probability mass. The mass density in the interval must be $1/(b - a)$, since the total probability mass is unity. Suppose M is any t set. The probability $X \in M$ is equal to the mass of that part of the interval $[a, b]$ which is common to the set M. Suppose for simplicity the interval $[a, b]$ is the unit interval $[0, 1]$. The mass density in this interval is unity. The mass density outside this interval is zero. The probability that $Y \geq 0.6$ is the probability mass in the interval $[0.6, 1]$, which is 0.4. It is interesting to note that for this random variable with a continuum of possible values, $P(Y = t)$ for any given t is zero since the mass in the common part of the set $\{t\}$ and the unit interval is zero. The event $\{\xi: Y(\xi) = t\}$ is not impossible, however, if t is in the unit interval. ■

We have suggested that while the probability mass distribution is sufficient for answering probability questions about the values taken on by a random variable, the probability mass distribution induced on the real line does not determine uniquely the random variable. The following simple illustration will make this plain.

Example 3-2-3

Consider the partition $\{A, A^c\}$, and suppose $P(A) = P(A^c) = \frac{1}{2}$. Then the random variables $I_A(\cdot)$ and $I_{A^c}(\cdot)$ induce the same probability mass distribution on the real line. Each random variable maps a mass of $\frac{1}{2}$ onto the point $t = 0$ and onto the point $t = 1$. The random variables are quite different, for they are equal for no value of ξ. ■

We shall consider, in later sections, probability mass distributions on the plane and on higher-dimensional spaces induced by vector-valued random variables, that is, by several real-valued random variables considered jointly. Also, we shall consider how the mass distributions induced on the line, in the plane, etc., may be described analytically in a way to make various analytical operations feasible. For the present, we emphasize the mental image of the mass distribution and the intuitive notion of the transfer of mass produced by the mapping. A clear grasp of this concrete representation of the analytical procedure of defining the induced probability measure serves as an important aid to thinking about an essentially abstract process.

We have extended our mathematical model and the associated auxiliary model in an important way, which we may summarize as follows:

Real world	Mathematical model	Auxiliary model
Empirical relative frequencies	Mathematical probabilities of events	Probability masses
Observed numerical-valued phenomena	Random variables as functions and induced probabilities	Mappings and transfer of mass

We shall need to introduce analytical means of describing the distribution of induced probabilities so that we can develop analytical procedures. Before doing that, we consider in some detail the important case of discrete random variables.

3-3   Discrete random variables

Random variables which have a finite or at most a countably infinite set of possible values play a very important role in practice. Many numerical phenomena are inherently discrete-valued. Because of approximations, random variables which in principle have a continuum of possible values may be reduced effectively to discrete-valued variables. This is particularly true for values determined experimentally. Measurements are inherently discrete, since the limits of accuracy of instruments place a limit on the number of distinct values which can be determined experimentally.

Much of the treatment of such random variables in the literature is hampered by the lack of suitable notation and a precise analytical formulation of problems and relationships. In this section, we consider the task of formal representation and characterization of random variables whose range is a discrete set of numbers. Also, we show that any random variable may be expressed as the limit of a sequence of random variables, each of which has a finite set of values. First we make a

Definition 3-3a

A random variable $X(\cdot)$ whose range T consists of a finite set of values is called a *simple random variable*. If the range T consists of a countably infinite set of distinct values, the function is referred to as an *elementary random variable*. The term *discrete random variable* is used to indicate the fact that the random variable is either simple or elementary.

The terms simple and elementary, as applied to random variables, are commonly employed in mathematical works on probability theory and on measure theory.   In the latter, the reference is generally to measurable or integrable functions rather than to random variables.

Canonical form

We shall discuss primarily the case of simple random variables.   Extensions to the more general case are usually obvious.   Suppose the range is $T = \{t_i : i \in J\}$.   Here J is taken to be finite, and the t_i are supposed to be distinct.   Obvious modifications may be made when several different t_i have the same value.   We let $A_i = \{\xi : X(\xi) = t_i\}$; it is apparent that the class $\{A_i : i \in J\}$ is a partition.   For convenience of writing, we put

$$P(A_i) = P(X = t_i) = p_i$$

The p_i must satisfy the conditions

$$p_i \geq 0 \qquad \text{and} \qquad \sum_{i \in J} p_i = 1$$

If $I_{A_i}(\cdot)$ is the indicator function for an event A_i (cf. Example 3-1-1), it follows that one and only one of these indicator functions differs from zero for any ξ, since ξ must belong to one and only one member of the partition.   If ξ belongs to A_i, the random variable must have the value t_i. Thus we may write the following analytical expression:

$$X(\cdot) = \sum_{i \in J} t_i I_{A_i}(\cdot)$$

On the other hand, if $T = \{t_i : i \in J\}$ is a set of distinct constants and $\{A_i : i \in J\}$ is a partition, the function

$$X(\cdot) = \sum_{i \in J} t_i I_{A_i}(\cdot)$$

is a discrete random variable whose range is T.   Also, it follows that $\{\xi : X(\xi) = t_i\} = A_i$.   This analytical form is so fundamental that it is convenient to give it a name, as follows:

Definition 3-3b

The discrete random variable $X(\cdot)$ is said to be in *canonical form* iffi it is written

$$X(\cdot) = \sum_{i \in J} t_i I_{A_i}(\cdot)$$

where $T = \{t_i : i \in J\}$ is a set of distinct constants and $\{A_i : i \in J\}$ is a partition.

Usually, in writing the expression for a specific random variable, the term with a zero coefficient (if 0 is in the range) is omitted.   It is therefore convenient to make the following

Definition 3-3c

The discrete random variable $X(\cdot)$ is said to be in *reduced canonical form* iffi it is written

$$X(\cdot) = \sum_{i \in J} t_i I_{B_i}(\cdot)$$

where $T' = \{t_i : i \in J\}$ is a set of distinct, nonzero constants and $\{B_i : i \in J\}$ is a disjoint class.

If 0 is not in the range T of a discrete random variable, the canonical form and the reduced canonical form are identical. The canonical form may be obtained from the reduced canonical form in an obvious manner by using the fact that

$$A_0 = \{\xi : X(\xi) = 0\} = [\bigcup_{i \in J} B_i]^c = \bigcap_{i \in J} B_i{}^c$$

Example 3-3-1

For any event A, the indicator function $I_A(\cdot)$ is a simple random variable (cf. Example 3-1-1). The expression $X(\cdot) = I_A(\cdot)$ is in reduced canonical form; the expression $X(\cdot) = 0 \cdot I_{A^c}(\cdot) + 1 \cdot I_A(\cdot)$ is in canonical form. ∎

Example 3-3-2

In a sequence of trials (cf. Example 3-1-4), let E_i be the event of a success on the ith trial and A_{rn} be the event of exactly r successes in n trials.

$$X(\cdot) = \sum_{i=1}^{n} I_{E_i}(\cdot) \qquad \text{is } not \text{ in canonical form, since the } E_i \text{ are not disjoint}$$

$$X(\cdot) = \sum_{r=0}^{n} r I_{A_{rn}}(\cdot) \qquad \text{is in canonical form}$$

$$X(\cdot) = \sum_{r=1}^{n} r I_{A_{rn}}(\cdot) \qquad \text{is in reduced canonical form}$$

$$A_{0n} = \bigcap_{r=1}^{n} A_{rn}{}^c \quad ∎$$

For every ξ the appropriate indicator function has the value unity, and this indicator function has the coefficient needed to give the correct value for the function.

Mass distribution

A simple random variable $X(\cdot)$ provides a mapping that transfers all the mass in set A_i to the single point t_i on the real line. The situation is diagramed in Fig. 3-3-1, in a manner that is sometimes helpful in the visualization of arguments and results in analytical developments. The various members of the partition are indicated by regions on the Venn diagram. Although these sets can be quite complicated in character,

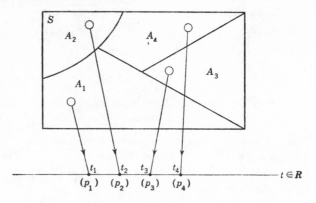

Fig. 3-3-1   Mapping provided by a simple random variable.

the basic space has no geometrical properties of continuity or distance, so that the essential fact of the nonoverlapping character of the A_i can be represented in the schematic manner shown.   Since position in the space S really has no meaning, the points representing elements may be located on the Venn diagram so that the proper set of points and its associated mass is localized and separated as shown.   A representative point from each of the A_i is shown as mapped into t_i.   The actual situation is that every point of A_i is mapped into t_i.   The resultant probability mass distribution on the real line R is a point mass distribution, with mass p_1 at t_1, p_2 at t_2, etc.   The total unit mass is concentrated on these points, and none is distributed continuously elsewhere on the real line.

Step function

It is sometimes helpful conceptually to view the simple random variable as a step function on the basic space S.   If the partition $\{A_i : i \in J\}$ is represented in Fig. 3-3-2 as in Fig. 3-3-1, the function may be represented as a surface over the Venn diagram; the height of the surface at any point is equal to the value of the function.   The resulting surface has the step character shown in Fig. 3-3-2.

Measurability

In the case of a discrete, real-valued random variable, the sigma field $\mathcal{E}(X)$ determined by $X(\cdot)$ is easily characterized.   As before, we let $T = \{t_i : i \in J\}$ be the range of $X(\cdot)$ and for each $i \in J$ we put $A_i = \{\xi : X(\xi) = t_i\} = X^{-1}(\{t_i\})$.   As we have noted previously, the A_i form a

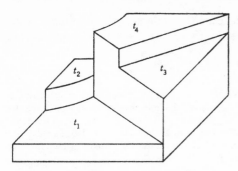

Fig. 3-3-2 A simple random variable viewed as a step function on the basic space. Ordinate heights are values of the function at various ξ.

partition, which we denote by $\mathcal{P}(X)$ and call the *partition determined by the discrete function* $X(\cdot)$.

Since $\{t_k\}$ is a Borel set on the real line, if $X(\cdot)$ is a random variable then $A_i = X^{-1}(\{t_k\})$ is an event.   On the other hand, suppose that M is any Borel set on the real line.   We let $J_M = \{i \in J : t_i \in M\}$.   We then have

$$X^{-1}(M) = \{\xi : X(\xi) \in M\} = \bigcup_{i \in J_M} A_i$$

If each A_i is an event, then the inverse image of any Borel set is an event. We have thus established

Theorem 3-3A

A discrete, real-valued function $X(\cdot)$ defined on S is a random variable iff each $A_i = \{\xi : X(\xi) = t_i\}$ is an event.

We have also characterized the class $\mathcal{E}(X)$ of inverse images of Borel sets by showing that it is the class of all countable unions of the A_i.   The empty set $\emptyset$ appears as the degenerate union of no members of the class. $\mathcal{E}(X)$ is the sigma field generated by the partition $\mathcal{P}(X)$.

It frequently occurs in practice that a simple random variable is specified by giving its range T and the set of corresponding probabilities.   No basic space or probability measure thereon is specified.   If only a single random variable is to be considered, a very simple discrete probability space suffices.   We simply introduce an element ξ_i for each value t_i and assign to the elementary events $A_i = \{\xi_i\}$ the probabilities $P(A_i) = p_i$.   A discrete probability space is thus defined.   We consider in a later section the discrete space that is naturally associated with several random variables.

Reduction to canonical form

It occurs frequently in practice that the most natural way to specify a simple random variable is to express it as a linear combination of indicator functions.   As an example, we refer once more to the random variable whose value is the number of successes in n repeated trials (cf. Examples

3-1-4 and 3-3-2).   Let us consider another simple example which may be used to illustrate a general approach to the problem of reduction to canonical form.

Example 3-3-3

Suppose $X(\cdot) = 3I_A(\cdot) - 5I_B(\cdot) + 4I_C(\cdot)$.   Consider the partition generated by the sets A, B, and C.   It is easy to see that each of the indicator functions must have the same value throughout any minterm.   Hence the function $X(\cdot)$ must have the same value throughout any minterm.   The function is thus uniquely determined if its value is determined for one point in each of the minterms.   This can be done systematically in tabular form, as follows:

	I_A	I_B	I_C	$3I_A$	$-5I_B$	$4I_C$	X
m_0	0	0	0	0	0	0	0
m_1	0	0	1	0	0	4	4
m_2	0	1	0	0	-5	0	-5
m_3	0	1	1	0	-5	4	-1
m_4	1	0	0	3	0	0	3
m_5	1	0	1	3	0	4	7
m_6	1	1	0	3	-5	0	-2
m_7	1	1	1	3	-5	4	2

Thus $T = \{-5, -2, -1, 0, 2, 3, 4, 7\}$.   In each case, the value is taken on over only one minterm.   Should any of these minterms be empty, the indicator function for that minterm is identically zero, so that in effect the value over that minterm is removed from the range T.   ■

One situation that comes up frequently is that the union $A \cup B \cup C$ is the whole space S.   In that case, m_0 must always be empty.   Since the value assigned to m_0 is always 0, this may remove the value 0 from the range.   It may be noted that the first three columns of numbers in the table above are superfluous.   They serve only to indicate a pattern; if this pattern is familiar, they may be omitted.   In fact, we can use a simpler form of the table, analogous to that developed for the binary designator for a boolean function (Sec. 2-7).   A further example will illustrate this and some other aspects of the problem.

Example 3-3-4

Consider

$$Y(\cdot) = 4I_A(\cdot) - 2I_B(\cdot) + 2I_C(\cdot) + I_D(\cdot)$$

and suppose minterms m_4, m_5, and m_{11} in the partition generated by sets A, B, C, and D are empty.   We arrange the values on the various minterms in rows, with the values for m_0 at the right and the values for m_{15} at the left.   The minterms which are empty are indicated by the symbol $\emptyset$.

$4I_A$	4 4 4 4	$\emptyset$444	0	0$\emptyset\emptyset$	0000
$-2I_B$	-2 -2 -2 -2	$\emptyset$000	-2	$-2\emptyset\emptyset$	0000
$2I_C$	2 2 0 0	$\emptyset$200	2	2$\emptyset\emptyset$	2200
I_D	1 0 1 0	$\emptyset$010	1	0$\emptyset\emptyset$	1010
Y	5 4 3 2	$\emptyset$654	1	0$\emptyset\emptyset$	3210

For this random variable, the range $T = \{0, 1, 2, 3, 4, 5, 6\}$.   If we let A_i be the set on which the value i is taken, i in T, we may write each A_i as the union of the appropriate minterms.   Thus

$$A_0 = m_0 \uplus m_6 \qquad A_3 = m_3 \uplus m_{13}$$
$$A_1 = m_1 \uplus m_7 \qquad A_4 = m_8 \uplus m_{14}$$
$$A_2 = m_2 \uplus m_{12} \qquad A_5 = m_9 \uplus m_{15}$$
$$A_6 = m_{10}$$

In canonical form,

$$Y(\cdot) = 0I_{A_0}(\cdot) + I_{A_1}(\cdot) + 2I_{A_2}(\cdot) + 3I_{A_3}(\cdot) + 4I_{A_4}(\cdot) + 5I_{A_5}(\cdot) + 6I_{A_6}(\cdot)$$

The first term would normally be omitted, to give the reduced canonical form.   ∎

The generality of this process should be evident from the examples.   Any random variable expressed as a linear combination of a finite number of indicator functions may be handled in this manner.   The same process could also be carried out on a minterm map, in an obvious way.

Random variables as limits

Simple random variables are important not only because they frequently appear in practice, but also because any random variable can be expressed as the limit of a sequence of random variables.   This fact is quite important in developing the theory of integration and in the application of that theory to the concept of mathematical expectation, in the next two chapters.

First, we note that a random variable can always be expressed as the difference between two nonnegative random variables.   We use the simple device of defining

$$X_+(\xi) = \begin{cases} X(\xi) & \text{for } X(\xi) \geq 0 \\ 0 & \text{for } X(\xi) < 0 \end{cases} = \max\,[X(\xi),\,0]$$

$$X_-(\xi) = \begin{cases} -X(\xi) & \text{for } X(\xi) \leq 0 \\ 0 & \text{for } X(\xi) > 0 \end{cases} = X_+(\xi) - X(\xi)$$

Then both $X_+(\cdot)$ and $X_-(\cdot)$ are nonnegative random variables and $X(\cdot) = X_+(\cdot) - X_-(\cdot)$.   We shall limit the discussion below to nonnegative random variables, and then extend the results to the general case by using this decomposition.

Second, we note two basic facts about the inverse mapping $X^{-1}(\cdot)$:

1. A partition of the space R produces a partition of S.
2. A partition of any t set M produces a partition of the inverse image $X^{-1}(M)$.

These facts follow directly from the elementary properties noted in Theorem 3-1A.

Suppose $X(\cdot)$ is a *nonnegative* random variable. For simplicity of exposition, we suppose $X(\cdot)$ is bounded above by the integer N [that is, $0 \leq X(\xi) \leq N$]. Extension of the results to unbounded random variables involves some complications in details but no essential complications in ideas. The range T is a subset of $R(N) = \{t: 0 \leq t \leq N\}$. We make a sequence of partitions of $R(N)$ as follows (see Fig. 3-3-3 for graphical representations):

nth partition: divide the interval at points $t_{in} = i/2^n$,

$$0 \leq i \leq N2^n$$

$(n + 1)$st partition: divide the interval at points $t_{i,n+1} = i/2^{n+1}$,

$$0 \leq i \leq N2^{n+1}$$

Each partition divides the interval $R(N)$ into half-open intervals. Consider a typical interval M in the nth partition, as illustrated in Fig. 3-3-3a. If a is the point of the subdivision at the left end of the interval and if $\delta = (\tfrac{1}{2})^n$ is the length of the interval, then $M = [a, a + \delta)$. In the $(n + 1)$st subdivision, the interval M is divided into two nonoverlapping, half-open intervals; this situation is represented by

$$M = M' \uplus M'' = [a, a + \delta/2) \uplus [a + \delta/2, a + \delta)$$

The situation is diagramed in Fig. 3-3-3b. We let

$$E = X^{-1}(M) \qquad E' = X^{-1}(M') \qquad E'' = X^{-1}(M'')$$

As noted above,

$$E = E' \uplus E''$$

To form $X_n(\cdot)$: Map each point ξ of E into the point $t = a$; do this for each interval in the nth partition.

To form $X_{n+1}(\cdot)$: Map each point ξ of E' into the point $t = a$; map each point ξ of E'' into the point $t = a + \delta/2$; do this for each interval in the $(n + 1)$st partition.

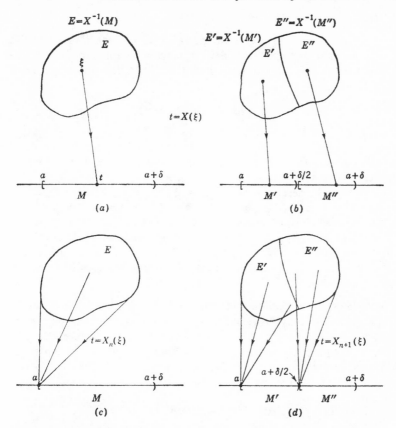

Fig. 3-3-3   Mappings and partitions for a sequence of simple functions $X_n(\cdot)$ approximating a nonnegative random variable $X(\cdot)$.   (a) Typical interval M in the nth partition; (b) subdivision of M in the $(n+1)$st partition; (c) mapping for $X_n(\cdot)$; (d) mapping for $X_{n+1}(\cdot)$.

From this construction, it is apparent that, for every ξ and any n, the image point $X_n(\xi)$ lies at or to the left of the image point $X_{n+1}(\xi)$, which in turn lies at or to the left of the image point $X(\xi)$.   This means that B

$$X_n(\xi) \leq X_{n+1}(\xi) \leq X(\xi)$$
$$X(\xi) - X_n(\xi) < \delta = (\tfrac{1}{2})^n$$

The function $X_n(\cdot)$ is obviously a simple function, with values $t_{in} = i/2^n$, $0 \leq i \leq N2^n$.

The sequence $\{X_n(\cdot) \colon 1 \leq n < \infty\}$ is thus a nondecreasing sequence of simple functions whose limit for any ξ is $X(\xi)$.   Because of the last inequality, above, the convergence is uniform in ξ (that is, the approximation does not depend on ξ).   We have thus established

Theorem 3-3.B

A bounded, nonnegative random variable $X(\cdot)$ can be represented as the limit of a nondecreasing sequence of simple functions. The convergence of this sequence is uniform in ξ over the whole basic space S.

We make several comments, by way of extension:

1. The hypothesis of boundedness can be removed. The proof involves some slight technical complications in the argument, but no essentially new ideas. For unbounded random variables, the uniformity of convergence does not hold.

2. Use of the decomposition $X(\cdot) = X_{+}(\cdot) - X_{-}(\cdot)$ makes it possible to extend the results to the nonnegative case in an obvious way.

3. The sequence of simple functions is not unique. Any sequence of partitions of $R(N)$ into nonoverlapping intervals could be used, provided each partition divides the intervals of the previous partition in such a way that the maximum length of the interval goes to zero with increasing n.

Many properties of random variables and the results of operations on random variables may be visualized and interpreted by considering the case for simple random variables. Properties which are preserved in the limit become properties of the general random variable. This is particularly true of the averaging process provided by the abstract integrals introduced and studied in Chaps. 4 and 5.

3-4   Probability distribution functions

The discussion in Sec. 3-2, in which a random variable is viewed as producing (or inducing) a probability mass distribution on the real line, suggests a natural analytical description of the probability characteristics of a random variable. In the case of discrete random variables, the characterization is simple. Probability mass is concentrated at points in the range of the random variable. This situation is easily stated and visualized. In the general case of continuous or mixed continuous and discrete distributions of probability mass, analytical and graphical means of representation are needed. The basic tool for such representation is presented in the following

Definition 3-4a

For any real-valued random variable $X(\cdot)$, we define the *distribution function* $F_X(\cdot)$ by the expression

$$F_X(t) = P(\{\xi\colon X(\xi) \leq t\}) = P(X \leq t)$$

for each real t.

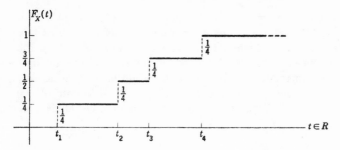

Fig. 3-4-1   Probability distribution function for a random variable with equally likely values.

In terms of the probability mass distribution induced on the real line by the random variable $X(\cdot)$, *the value $F_X(t)$ is the amount of that mass on the real line R which is at or to the left of the point t.*   In terms of the induced probability measure $P_X(\cdot)$, we have

$$F_X(t) \; = \; P_X(M_t) \; = \; P[X^{-1}(M_t)]$$

where $M_t = (-\infty, t]$ is the semi-infinite interval $-\infty < x \leq t$.

It should be noted that some authors define the distribution function slightly differently, using the value $P(X < t)$.   Some care should be taken in reading a work to determine which of the two possibilities is employed by the author.

Discrete random variables

For a simple random variable $X(\cdot)$ with range $T = \{t_i \colon i \in J\}$, the distribution function $F_X(\cdot)$ has a very simple character.   Probability mass p_i is located at $t = t_i$ for each $i \in J$.   The value $F_X(t)$ is the value of all the probability mass at or to the left of the point t.   This value does not change for $t_i < t < t_{i+1}$.   When one of the points t_j is reached, the value of the distribution function must jump by an amount p_j.   We thus have a *step function* with jumps of the amount p_i at $t = t_i$.   The function is defined at the jump point as the value at the top of the jump (this is the value the function approaches as the jump point is approached from the right).   Figure 3-4-1 shows the distribution function $F_X(\cdot)$ for a random variable having four values, each taken on with probability ¼.   Analytical expressions may be written with the aid of the *unit step function*.   We define the function $u_+(\cdot)$ as follows:

$$u_+(x) \; = \; \begin{cases} 0 & \text{if } x < 0 \\ 1 & \text{if } x \geq 0 \end{cases}$$

We then have $u_+(t - a) = 0$ for $t < a$ and $u_+(t - a) = 1$ for $t \geq a$. The graph of the function so defined is zero to the left of the point $t = a$ and "steps" to one at $t = a$, where it remains for all t to the right of the point a.   When the subscript $+$ is omitted, we indicate by $u(\cdot)$ the unit step but leave the question of the value at the origin open; that is, $u(0)$ may be undefined or may be defined arbitrarily.   With the aid of the function $u_+(\cdot)$, we may define $F_X(\cdot)$ for the simple random variable above by the expression

$$F_X(t) = \sum_{i \in J} p_i u_+(t - t_i)$$

For the particular case shown in Fig. 3-4-1, $F_X(t)$ is given by

$$F_X(t) = \tfrac{1}{4}u_+(t - t_1) + \tfrac{1}{4}u_+(t - t_2) + \tfrac{1}{4}u_+(t - t_3) + \tfrac{1}{4}u_+(t - t_4)$$

Each step function $u_+(t - t_i)$ assumes the value 1 at the proper value of t to provide the jump in value by the amount p_i.   This constant value p_i is added to the expression for the function for all t at or to the right of the point $t = t_i$.   Thus the expression given does indeed equal all the probability mass at or to the left of the point t, for every t.

Although the argument and examples above are for simple random variables with only a finite number of possible values in the range, it should be apparent that everything carries over to elementary random variables having a countable infinity of possible values.   If the values t_i should approach some limit point $t = a$, the values of p_i must go to zero (since we have an infinity of positive quantities whose sum is 1) and the step character of the function may be difficult to visualize.

Absolutely continuous random variables

In many cases of interest, the probability mass is distributed smoothly and continuously on the real line in such a way that a mass density function (a lineal density function) may be defined.   In this case, as we shall discuss below, the distribution function $F_X(\cdot)$ is continuous and a density function $f_X(\cdot)$ is defined in such a way that the probability mass in any interval $M = [a, b] = \{t : a \leq t \leq b\}$ is given by

$$P_X(M) = P(X \in M) = \int_a^b f_X(t)\, dt$$

In this case we also have

$$F_X(t) = P(X \leq t) = \int_{-\infty}^{t} f_X(u)\, du$$

and

$$f_X(t) = \frac{d}{dt} F_X(t) \qquad \text{wherever } f_X(\cdot) \text{ is continuous}$$

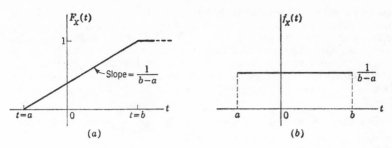

Fig. 3-4-2   (a) Probability distribution and (b) density functions for a uniformly distributed random variable.

Example 3-4-1   *Uniform Distribution*

If the probability mass is distributed uniformly over some interval $a \leq t \leq b$, the random variable $X(\cdot)$ inducing this distribution is said to be *uniformly distributed* over the interval $[a, b]$.   Both the density and the distribution functions are defined. They are probably most simply described graphically, as in Fig. 3-4-2.   The density function $f_X(\cdot)$ has a constant, positive value over the interval $[a, b]$ and is zero elsewhere.   The value of the constant is equal to $1/(b - a)$, the reciprocal of the length of the interval, since the total mass is unity.   The distribution function $F_X(\cdot)$ has a graph which rises linearly from the value 0 at $t = a$ to the value 1 at $t = b$.   It has value 0 below $t = a$ and value 1 for t greater than b.   Explicit formulas can be written with the aid of the step function, as follows:

$$F_X(t) = \frac{u(t - a)(t - a) - u(t - b)(t - b)}{b - a}$$

$$f_X(t) = \frac{u(t - a) - u(t - b)}{b - a} \quad \blacksquare$$

Conditions under which the density function exists are discussed later in this section.   Ordinarily, if the probability mass is smoothly distributed with no mass concentrations, we are safe in supposing that the mass distribution may be described by either the distribution function $F_X(\cdot)$ or by the probability density function $f_X(\cdot)$, which is the derivative of the distribution function.   When the density function exists, its graph provides a powerful aid in visualizing the manner in which probability mass is spread on the real line.   This distribution serves to indicate the probability that values of the random variable will be found in any given set of numbers.

Example 3-4-2   *Triangular Distribution*

Consider the density function $f_X(\cdot)$ defined graphically in Fig. 3-4-3.   The probability mass assigned to the set M designated on the figure is determined by obtaining the area of the region under the density curve and over the set M; this region is indicated by the shading in the figure.   It is apparent that no probability mass lies outside the interval $[a, c]$, so that the probability of finding a value of the random variable $X(\cdot)$ outside this interval is zero.   The probability mass concentration is greater near the point $t = b$, where the density curve is highest.   The total area under the curve must be unity, since it is assumed that all probability mass is accounted for. The value of the distribution function $F_X(t)$ for any t is the area of the region under the curve and to the left of the point t on the real line. The slope of the distribution function at any t must equal the value of the density function at that point. The distribution function must have zero slope for $t < a$, its slope must increase linearly from $t = a$ to $t = b$, then decrease linearly from $t = b$ to $t = c$, where the curve levels off to hold a constant value (which must be unity). The curve for $F_X(\cdot)$ is shown in Fig. 3-4-3b. ∎

One of the most important and most frequently encountered distributions in many applications is the following.

Example 3-4-3 *The Normal (or Gaussian) Distribution*

A continuous random variable $X(\cdot)$ is said to be normally distributed if it has the probability density function determined by

$$f_X(t) = \frac{1}{\sigma\sqrt{2\pi}} \exp\left[-\frac{1}{2}\left(\frac{t-\mu}{\sigma}\right)^2 \right]$$

where σ is a positive constant and μ is a constant. The factor $1/(\sigma\sqrt{2\pi})$ is needed to make the area under the density curve unity. The distribution function can most easily be expressed by the integral formula

$$F_X(t) = \int_{-\infty}^{t} f_X(x)\, dx = \frac{1}{\sigma\sqrt{2\pi}} \int_{-\infty}^{t} \exp\left[-\frac{1}{2}\left(\frac{x-\mu}{\sigma}\right)^2 \right] dx \quad ∎$$

The graph of the density function $f_X(\cdot)$ is the well-known bell-shaped curve shown in Fig. 3-4-4. Its integral, the distribution function $F_X(\cdot)$,

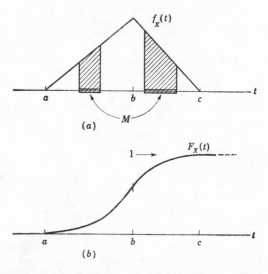

Fig. 3-4-3 (a) Density function and (b) distribution function for a continuous random variable with a triangular distribution.

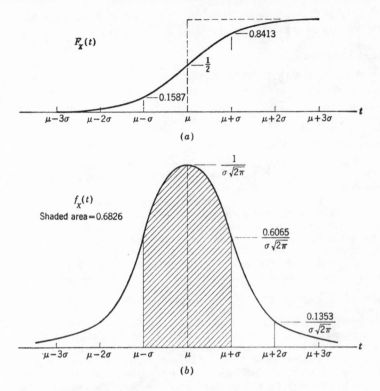

Fig. 3-4-4   Graphs for (*a*) the normal distribution function $F_X(\cdot)$ and (*b*) the normal density function $f_X(\cdot)$ for a normally distributed random variable.

is the familiar S-shaped curve also shown in that figure. Symmetry about the point $t = \mu$ and the dependence of the width and height upon the parameter σ may be noted. The significance of these parameters and some applications of the normal distribution are given in Chap. 5.

The importance of the normal distribution rests in large part on the so-called central limit theorem, discussed briefly in Chap. 6. Under very general conditions, the sum (or average) of a large number of independent random variables is a random variable which is approximately normally distributed. Suppose a physical quantity (such as the nonsystematic, random component of the error of reading of some instrument) is the resultant of a large number of "randomly varying" contributing components whose physical connection is so "loose" that these components are "independent" in an appropriate sense; the central limit theorem leads one to expect that the physical quantity should be represented as a normally distributed random variable. That this is, in fact, the case in many situations is amply demonstrated by experience.

Mixed distributions

It sometimes occurs that the probability mass distribution is neither entirely concentrated at discrete points nor smoothly distributed in a continuous manner. The following simple example will indicate the possibility of a mixed case.

Example 3-4-4

An idealization of the following game leads to a random variable with mixed distribution. Flip a coin. If a head appears, spin a wheel having a uniform, continuous scale from 0 to 10. If a tail appears, throw a die. Record the resulting number in either case. Let $X(\cdot)$ be the random variable whose value represents the recorded number. If we idealize the wheel to suppose that any number in the range is possible and that the wheel is balanced and smooth-acting, we may make the following idealization of the distribution of the numbers. Let H be the event that a head appears and T the event that a tail appears. Then the conditional probability that $P(X \leq t|H)$ corresponds to a uniform distribution over the interval [0, 10]. Likewise, $P(X \leq t|T)$ corresponds to equal probability of the values 1, 2, . . . , 6. We then have

$$F_X(t) = P(X \leq t) = P(X \leq t|H)P(H) + P(X \leq t|T)P(T)$$

We have one-half the probability mass distributed uniformly over the interval [0, 10] and one-half the probability mass concentrated in amounts of $\frac{1}{12}$ at each of the six points $t = 1, 2, . . . , 6$. The distribution function has the graph shown in Fig. 3-4-5. The continuous portion of the mass distribution corresponds to a continuous and increasing component of the distribution function. The discrete portion of the mass distribution produces steps in the distribution function. The summed height of the steps corresponds to the total mass concentrated at the various discrete points. The remaining rise of the distribution function is due to the continuously distributed portion of the probability mass. ∎

Study of the foregoing example should make it evident that for a mixed distribution it is possible to separate the two parts of the probability mass. Each can be represented by a distribution function. The complete probability distribution function is the sum of the two, as follows:

$$F_X(\cdot) = F_{Xc}(\cdot) + F_{Xd}(\cdot)$$

where $F_{Xc}(\cdot)$ is the distribution function for the continuous part, and $F_{Xd}(\cdot)$ is a step function which determines the distribution for the discrete part of the probability mass (Fig. 3-4-5b). In most cases, the function $F_{Xc}(\cdot)$ is differentiable and its derivative function $f_{Xc}(\cdot)$ is a mass density function for the continuous portion of the probability mass. This may be utilized in the same manner that the density function for the continuous case is used to determine the probability mass in any given interval or other set of points on the real axis.

distribution function is the value approached from the right.   That is,

(**F5**) $F_X(\cdot)$ is continuous from the right.

PROOF   Consider the sets $B_n = \{\xi: a < X \le a + 1/n\}$.   These form a decreasing sequence whose limit is empty.   Hence $\lim_{n \to \infty} P(B_n) = 0$.   Now by (F3) we may write $F_X(a + 1/n) = F_X(a) + P(B_n)$, from which it follows that $\lim_{n \to \infty} F_X(a + 1/n) = F_X(a)$. ∎

It should be noted that if, as some authors do, we had defined the distribution function to be the probability mass to the left of the point t [i.e., had defined $F_X(t) = P(X < t)$], the function would be continuous from the left.   We shall hold to the convention adopted in our definition.

The continuity condition

The discussion above of random variables with absolutely continuous distributions is somewhat intuitive in character and simply refers to the possibility of describing a probability mass distribution with a density function.   We turn now to a brief discussion of the mathematical situation which ordinarily exists when the probability mass is distributed smoothly on the real line.

In order to speak with some precision, we must utilize the concept of the *Lebesgue measure* (Sec. 4-5), which generalizes the notion of the "length" of a set of points on the real line.   This measure is a completely additive set function on a class of sets which is somewhat more general than (but which includes) the class of Borel sets.   This set function assigns to each interval a "measure" equal to its length.   In the case that the probability mass is distributed smoothly and uniformly on the unit interval [0, 1], the Lebesgue measure of any Borel subset of this interval is equal to the probability mass assigned to that set.

When the probability mass distribution induced by a real-valued random variable $X(\cdot)$ has no point mass concentrations, the following condition usually prevails: the probability mass assigned to any point set of length (Lebesgue measure) zero must be zero.   This condition is of sufficient importance to justify identifying it with the following terminology.

Definition 3-4b

If the probability measure $P_X(\cdot)$ induced by the real-valued random variable $X(\cdot)$ is such that it assigns zero probability to any point set of Lebesgue measure (generalized length) zero, the probability measure and the probability distribution are said to be *absolutely continuous* (with respect to Lebesgue measure).   In this case, the random variable is also said to be *absolutely continuous*.

If $P_X(\cdot)$ is absolutely continuous, there can be no point mass concentrations. Examples have been constructed to show that the converse is not true. The nature of these examples is such, however, that one would expect to find them rarely, if ever, in practice. As a practical rule, one is quite safe in assuming that the absence of point mass concentrations ensures absolute continuity of the random variable and its induced probability mass distribution.

In the case of absolutely continuous probability distributions, a general theorem known as the Radon-Nikodym theorem ensures the existence of the function named in the following

Definition 3-4c

If the probability measure $P_X(\cdot)$ induced by the real-valued random variable $X(\cdot)$ is absolutely continuous, the function $f_X(\cdot)$ defined on the real line such that

$$\int_M f_X(u)\ du = P_X(M) \qquad \text{for each Borel set } M$$

is called the *probability density function* for $X(\cdot)$.

Several important properties of $f_X(\cdot)$ may be stated as follows:

(f1) $f_X(\cdot) \geq 0$ almost everywhere

(f2) $\displaystyle\int_{-\infty}^{\infty} f_X(u)\ du = 1$

(f3) $\displaystyle\int_{-\infty}^{t} f_X(u)\ du = F_X(t)$

The term "almost everywhere" in the statement of property (f1) means that the condition holds for all points on the real line except possibly a set of Lebesgue measure (length) zero, and hence on a set to which has been assigned zero probability. Practically speaking, it allows the function to be undefined or to be defined arbitrarily at isolated points. Such peculiarities do not affect the value of the integrals.

According to property (F4), the condition of a smooth mass distribution (i.e., no probability mass concentrations at a point) requires that $F_X(\cdot)$ be continuous. From properties of integrals, we also have

(f4) $f_X(t) = \dfrac{d}{dt} F_X(t)$ almost everywhere [in particular, at points of continuity of $f_X(\cdot)$]

Comment on notation

A variety of notational conventions are used in the literature. Many writers use $p(\cdot)$ [or $p_X(\cdot)$] to denote the density function. Others, how-

ever, use the symbol $p(\cdot)$ to indicate the function defined by

$$p(x) = P(X = x)$$

We have designated by a subscript the random variable whose distribution is described, for example, $F_X(\cdot)$, $f_X(\cdot)$, etc. Many authors use a simpler, but in many ways confusing and inadequate, scheme; they designate the random variable by the letter chosen for the argument of the function, for example, $F(x)$, $f(y)$, etc. The form adopted in this book is generally superior, for we are free to designate any specific values of the argument of the function by any convenient specified letter, for example, $F_X(a) = P(X \leq a)$, $F_Y(b) = P(Y \leq b)$, etc. Compare this with the situation that arises when the two distribution functions are represented by $F(x)$ and $F(y)$. There is hopeless confusion if we write $P(X \leq a) = F(a)$ and $P(Y \leq b) = F(b)$, and there is unnecessary complication if we must resort to some scheme such as

$$P(X \leq a) = F(x)\Big|_{x=a} \qquad \text{and} \qquad P(Y \leq b) = F(y)\Big|_{y=b}$$

Some further special distributions

We conclude this section with a brief consideration of several important distributions.

Example 3-4-5　*Binomial Distribution*

The random variable $X(\cdot)$ which gives the number of successes in a sequence of n Bernoulli trials (Example 3-1-4) is said to have the binomial distribution. It has range $\{i: 0 \leq i \leq n\}$ and probabilities $p_i = P(X = i) = C_i^n p_i^i (1 - p)^{n-i}$. Reasons for this terminology are given in the discussion following Example 2-8-3. ∎

Example 3-4-6　*Negative Binomial Distribution*

Consider sequences of Bernoulli trials of unlimited length, with probability p of success and $q = 1 - p$ of failure at any trial in the sequence. Let $X_m(\cdot)$ be the random variable whose value for any sequence is the number of trials required for exactly m successes $(m = 1, 2, \ldots)$. Note that if the trials occur at a fixed rate in time, this number represents the waiting time until the mth success. We must have $P(X_m = n) = C_{m-1}^{n-1} p^m q^{n-m}$ for $n \geq m$. To see this, we note that $X_m(\xi) = n$ provided there are exactly $m - 1$ successes in the first $n - 1$ trials, with probability $C_{m-1}^{n-1} p^{m-1} q^{n-m}$ and a success on the nth trial with probability p. Use of the product rule for independent events gives the indicated result. This set of probabilities may be put into another form by considering $P(X_m = m + k)$, $0 \leq k < \infty$. Use of expressions and relationships for the binomial coefficient lead to the result

$$P(X_m = m + k) = P(X_m - m = k) = C_{k}^{-m} p^m (-q)^k \qquad 0 \leq k < \infty$$

where

$$C_{k}^{-m} = \frac{(-m)(-m-1)\cdots(-m-k+1)}{k!}$$

It may be shown that $C_{k-m}p^m(-q)^k$ is a term in the binomial expansion of $p^m(1-q)^{-m}$, from which the term *negative binomial distribution* is derived. For the special case $m=1$, $P(X_1 = k+1) = pq^k$, $0 \le k < \infty$. In this case we have the *geometric* distribution. ■

Example 3-4-7 Poisson Distribution

A random variable which may take on the value 0 or any positive integer k with probability $p_k = P(X = k) = e^{-\mu}\mu^k/k!$, where μ is a positive constant, is said to have the *Poisson distribution*. Under certain natural conditions (see the development in Sec. 7-3), a random variable $N(\cdot, t)$ which counts the number of occurrences of a given phenomenon in a time interval of length t has the distribution

$$P(k, t) = P[N(\cdot, t) = k] = e^{-\lambda t}\frac{(\lambda t)^k}{k!}$$

where λ is a positive constant and $t \ge 0$. This counting process is commonly assumed in studies of traffic patterns, utilization of telephone switchboards, estimates of the number of system failures in a given period, and many other applications. ■

Example 3-4-8 The Gamma Distribution

Suppose

$$P(k, t) = e^{-\lambda t}\frac{(\lambda t)^k}{k!} \qquad t > 0$$

is the probability that exactly k automobiles pass a given point in a time interval of length t. Let $X_n(\cdot)$ be the random variable whose value is the time elapsed (the so-called "waiting time") until exactly n automobiles pass the point. We wish to determine the density function for $X_n(\cdot)$. It is obvious that $X_n(\cdot)$ cannot take on negative values. Now the waiting time for n automobiles to pass will be less than or equal to t iff the number of cars passing in time t is n or greater. Thus

$$F_{X_n}(t) = P(X_n \le t) = \sum_{k=n}^{\infty} P(k, t) = \sum_{k=n}^{\infty} e^{-\lambda t}\frac{(\lambda t)^k}{k!}$$

This series expression may be differentiated with respect to t on a term-by-term basis to obtain the density function. When this is done, all terms in the resulting infinite sums cancel out except one, to give

$$f_{X_n}(t) = n(t) = \frac{\lambda^n t^{n-1}}{(n-1)!}e^{-\lambda t}$$

Now the factorial is related to the gamma function $\Gamma(\cdot)$ by the relation $(n-1)! = \Gamma(n)$. The density function is therefore a special case of the density function for the so-called *gamma distribution*, which is given by

$$f_X(t) = n(t) = \frac{t^\alpha e^{-t/\beta}}{\beta^{\alpha+1}\Gamma(\alpha+1)}$$

Putting $\beta = 1/\lambda$ and $\alpha = n - 1$ in this expression gives the waiting-time density function. We note further that for $n = 1$ the waiting-time density function becomes the density function for the *exponential distribution*, which plays an important role in reliability theory. The waiting time, in this case, is the waiting time to failure (i.e., to the first occurrence of the phenomenon of failure). ■

3-5   Families of random variables and vector-valued random variables

In the previous sections, we have considered single random variables having real values.   In this section, we consider families of two or more random variables.   For the most part, we shall deal with two random variables, since the extension of the ideas discussed to larger numbers is simple in principle, but complicated by the mechanics of writing out the details.

Joint mapping

Suppose we are interested in two real-valued random variables $X(\cdot)$ and $Y(\cdot)$.   This means that, corresponding to each elementary outcome ξ, we are interested in two numbers, $t = X(\xi)$ and $u = Y(\xi)$.   Now, we may look upon this pair of numbers in two ways: we may consider them separately, and hence deal with each random variable without reference to the other, or we may consider the pair of numbers as the coordinates of a vector or of a point in the plane.   These two points of view may be considered in terms of the mappings produced by the random variables (see Fig. 3-5-1 for a diagrammatic representation).

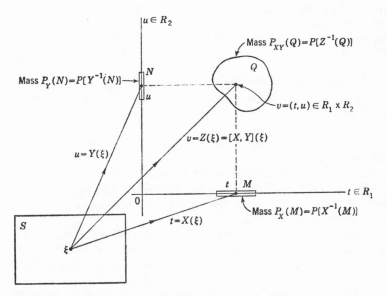

Fig. 3-5-1   The mappings produced by two random variables $X(\cdot)$ and $Y(\cdot)$, considered separately and jointly.

Two mappings from the basic space to the real line:

$X(\cdot)$ maps S into R_1, a copy of the real line; $t \in R_1$.

$Y(\cdot)$ maps S into R_2, a copy of the real line; $u \in R_2$.

$X(\cdot)$ induces probability measure $P_X(\cdot)$ on the Borel sets of R_1.

$Y(\cdot)$ induces probability measure $P_Y(\cdot)$ on the Borel sets of R_2.

$\{\xi : X \in M\} = X^{-1}(M)$ is an event for each Borel set $M \subset R_1$.

$\{\xi : Y \in N\} = Y^{-1}(N)$ is an event for each Borel set $N \subset R_2$.

Joint events of the following kinds are defined:

$\{\xi : X \in M \text{ and } Y \in N\} = X^{-1}(M)Y^{-1}(N)$

$\{\xi : X \in M \text{ or } Y \in N\} = X^{-1}(M) \cup Y^{-1}(N)$

When we consider the pair of values $[X(\xi), Y(\xi)]$, we may view this, in standard fashion, as represented by a point on the plane $R_1 \times R_2$ or as a two-dimensional vector. To each ξ corresponds a point on the plane or, equivalently, a two-dimensional vector. We are led naturally to consider a *vector-valued function* $Z(\cdot)$ defined by

$$Z(\xi) = [X, Y](\xi) = [X(\xi), Y(\xi)]$$

Thus $Z(\cdot) = [X, Y](\cdot)$ maps S into the plane $R_1 \times R_2$. The random variables $X_1(\cdot)$ and $X_2(\cdot)$ are called the *coordinate random variables* for the vector-valued function $Z(\cdot)$. It should be apparent how these ideas can be extended to any number of dimensions. That is, $Z(\cdot)$ could have any number of coordinate random variables. To any family of n real-valued random variables $\{X_i(\cdot): 1 \leq i \leq n\}$ there corresponds a vector-valued function $Z(\cdot) = [X_1, X_2, \ldots, X_n](\cdot)$ whose i*th* coordinate random variable is $X_i(\cdot)$.

Measurability of joint mapping

We have been careful not to use the term random variable for the vector-valued function $Z(\cdot)$ introduced above. As in the case of a real-valued random variable, it is necessary to consider the question of measurability. We carry out the discussion for the case $n = 2$, since we may use graphical representations such as provided in Fig. 3-5-1 as aids to understanding. An examination of the development will show that the reasoning can be extended to any finite class of coordinate random variables. With sufficient care, many of the results may be extended to an infinite sequence of coordinate random variables.

We begin by recalling that if M and N are any two sets of real numbers, the set $M \times N$ is the set of all pairs of real numbers t, u such that $t \in M$ and $u \in N$. Now $Z(\xi) = [X(\xi), Y(\xi)] \in M \times N$ iffi both $X(\xi) \in M$ and $Y(\xi) \in N$. This may be expressed by $\{\xi : Z(\xi) \in M \times N\} = \{\xi : X(\xi) \in M\}$ $\cap \{\xi : Y(\xi) \in N\}$, or more succinctly by the inverse mapping relationship $Z^{-1}(M \times N) = X^{-1}(M) \cap Y^{-1}(N)$. Now if M and N are Borel sets on the real line, we must have $Z^{-1}(M \times N)$ is the intersection of two events and hence an event.

The class $\mathfrak{B}$ of Borel sets on the plane is the sigma field generated by the class $\mathcal{I}$ of all rectangle sets of the form $M_t \times N_u$, where M_t and N_u are the semi-infinite intervals $(-\infty, t]$ and $(-\infty, u]$, respectively. The class $\mathfrak{B}$ is also the sigma field generated by the class $\mathfrak{R}$ of all rectangles of the form $A = M \times N$, where M and N are Borel sets on the real line. Theorem 3-1B may be extended easily to the multidimensional case. Thus, we may assert that $Z^{-1}(Q)$ is an event for each Borel set Q on the plane iffi $X(\cdot)$ and $Y(\cdot)$ are random variables.

In analogy to the one-dimensional case, we consider the *sigma field* $\mathcal{E}(Z) = Z^{-1}(\mathfrak{B})$ *determined by* $Z(\cdot)$. This is the class of all inverse images of Borel sets on the plane. We also consider the sigma field $\mathcal{E}(X, Y)$ generated by all sets of the form $X^{-1}(M_t) \cap Y^{-1}(N_u)$. This is the same as the sigma field generated by all sets of the form $X^{-1}(M) \cap Y^{-1}(N)$, where M and N are any Borel sets on the real line. Thus, $\mathcal{E}(X, Y)$ is the smallest sigma field with respect to which both $X(\cdot)$ and $Y(\cdot)$ are measurable. It is known that $\mathcal{E}(Z) = \mathcal{E}(X, Y)$. As a result, we may restate the assertion of the last sentence of the previous paragraph as follows: $\mathcal{E}(Z) \subset \mathcal{E}$ iffi both $\mathcal{E}(X) \subset \mathcal{E}$ and $\mathcal{E}(Y) \subset \mathcal{E}$.

Induced probabilities

In direct analogy to the single-variable case, we may determine a probability mass distribution on the Borel sets on the plane by the assignment $P_{XY}(Q) = P[Z^{-1}(Q)]$. Properties (P1) and (P2) are immediately evident. Since the inverse image of a disjoint union of Borel sets is a disjoint union of the corresponding inverse images of the Borel sets, the countable additivity property for $P_{XY}(\cdot)$ follows from property (P3) for the original measure $P(\cdot)$.

In dealing with a pair $X(\cdot)$, $Y(\cdot)$ of random variables, we thus have three induced probability measures, $P_X(\cdot)$, $P_Y(\cdot)$, and $P_{XY}(\cdot)$. The three mappings and the probability mass distributions induced are represented on the single diagram in Fig. 3-5-1, in a manner that will be helpful in later discussions. The mass distribution on R_1, represented by $P_X(\cdot)$, is to be viewed as a line distribution. A mass of unit weight is distributed along the line. Similarly, the mass distribution on R_2, represented by $P_Y(\cdot)$, consists of unit mass distributed along that line. The mass distribution on the plane $R_1 \times R_2$, represented by $P_{XY}(\cdot)$, spreads unit mass over the plane in the manner indicated. It is reasonably apparent that the mass distributions on the coordinate axes and the distribution on the plane are related. We consider the nature of this relation.

The two points of view regarding the mappings have been presented separately. They are connected in a fundamental way, however, as shown by the following facts. We state the facts analytically, but they may be viewed geometrically with the aid of Fig. 3-5-2.

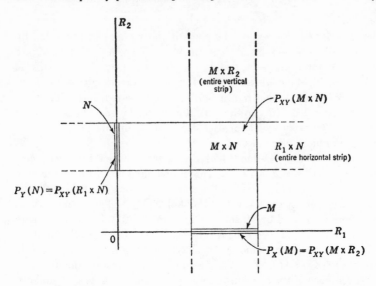

Fig. 3-5-2   Relation of joint mass distribution on the plane and marginal mass distribution on the coordinate axes.

If M is a Borel set in R_1 and N is a Borel set in R_2, the set $M \times N = \{(t, u) : t \in M \text{ and } u \in N\}$ is a Borel set in $R_1 \times R_2$ (it is known as a *rectangle set*—see Appendix B).   We have

$$Z^{-1}(M \times N) = \{\xi : X \in M \text{ and } Y \in N\} = X^{-1}(M)Y^{-1}(N)$$

Thus, the probability distribution on the plane satisfies the relation

$$P_{XY}(M \times N) = P(Z \in M \times N) = P(X \in M, Y \in N)$$
$$= P[X^{-1}(M)Y^{-1}(N)]$$

If N is the whole real line R_2, $Y^{-1}(N) = S$, so that

$$P_{XY}(M \times R_2) = P[X^{-1}(M)] = P_X(M)$$

Similarly, if M is the whole real line R_1, $P_{XY}(R_1 \times N) = P_Y(N)$.

These facts can be given a simple interpretation with the aid of Fig. 3-5-2. Sets M and N are shown as intervals on their respective axes.   The set $M \times N$ is the rectangle so indicated; a little thought shows that this is indeed the set of those $v = (t, u)$ such that t is in M and u is in N.   The set $M \times R_2$ is the set of those v such that the t coordinate is in M but the u coordinate can have any value.   This is the strip of all those points v which project onto the set M.   Now the joint mapping may spread probability mass throughout $M \times R_2$. According to the results above, the single mapping produced by $X(\cdot)$ puts exactly the same mass on the set M. Thus, for any set M, the mass $P_X(M)$ is the total mass on the plane lying above or below M. A similar interpretation holds for the mass distribution on the R_2 axis. For these geometrical reasons, we make the following

Definition 3-5a

The probability measure $P_{XY}(\cdot)$ defined on the Borel sets in the plane is called the *joint probability measure* induced by the joint mapping $(t, u) = Z(\xi) = [X, Y](\xi)$. The probability mass distribution is called the *joint distribution*. The probability measures $P_{XY}(\cdot \times R_2) = P_X(\cdot)$ and $P_{XY}(R_1 \times \cdot) = P_Y(\cdot)$ are called the *marginal probability measures* induced by $X(\cdot)$ and $Y(\cdot)$, respectively. The corresponding probability mass distributions are called the *marginal distributions*.

These mass-distribution ideas and their relationship are important for visualizing the analytical representation of probability distribution in the next section.

Discrete random variables

In the discrete case, the joint mapping produces a point mass distribution on the plane; the marginal mass distributions are also point mass distributions on the coordinate axes. Consider the pair of simple random variables $X(\cdot)$ and $Y(\cdot)$, expressed in canonical form as follows:

$$X(\cdot) = \sum_{i=1}^{n} t_i I_{A_i}(\cdot) \qquad \text{and} \qquad Y(\cdot) = \sum_{j=1}^{m} u_j I_{B_j}(\cdot)$$

The joint mapping must produce point mass $P(A_i B_j)$ at each point (t_i, u_j) on the plane, where i and j run through the permissible values. The total mass on points having t-coordinate t_i must be $P(A_i)$; this is the mass located at the point $t = t_i$ on the line R_1 in the marginal distribution. Similarly, the other marginal distribution has mass $P(B_j)$ at the point $u = u_j$ on R_2. This must be the sum of the masses on all points in the joint distribution which have u-coordinate u_j. If we put

$$P(A_i B_j) = P(X = t_i, Y = u_j) = p(i, j)$$
$$P(A_i) = P(X = t_i) = p(i, *)$$

and

$$P(B_j) = P(Y = u_j) = p(*, j)$$

we must have

$$\sum_{i,j} p(i, j) = \sum_{i} p(i, *) = \sum_{j} p(*, j) = 1$$

$$\sum_{j} p(i, j) = p(i, *) \qquad \sum_{i} p(i, j) = p(*, j)$$

This set of relations is illustrated in Fig. 3-5-3. Probability mass values for the joint distribution are shown on the plane. Probability mass values for the marginal distributions are shown along the coordinate axes, in parentheses.

It is apparent that a pair of simple random variables, considered jointly, is characterized by (1) a set of pairs of values, (2) a joint, or product, partition, and (3) probabilities of the joint events in the partition. As in the case of a single real-valued random variable, the pair of simple real-valued random variables is often specified in practice by the set of values and the set of joint probabilities. If these are the only random variables to be considered in the investigation, it is customary to introduce a discrete space with one point for each possible pair of values. This could be done in any of several ways. A common way of doing this is to let the points (t_i, u_j) on the plane $R_1 \times R_2$ serve as both object and image points. Thus this set of points serves as the basic space, and the joint mapping maps each point into itself. The marginal mappings $X(\cdot)$ and $Y(\cdot)$ map each of these points into the associated coordinate point on the coordinate axis R_1 or R_2. In general, if we set up a basic space with element ξ_{ij} carried into the point (t_i, u_j) by the joint mapping, we have $X(\xi_{ij}) = t_i$ and $Y(\xi_{ij}) = u_j$. In the case where ξ_{ij} is the point (t_i, u_j),

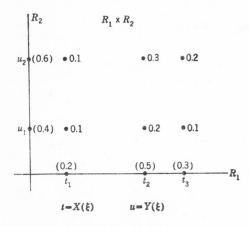

Fig. 3-5-3   Mass distributions for a pair of simple random variables.

as described above, the random variables $X(\cdot)$ and $Y(\cdot)$ are the *coordinate functions* for the space. It should be apparent that other equivalent schemes are possible. A variety of these are found in the literature, sometimes without careful description or definition. It is usually necessary to identify an appropriate model when attempting to follow an exposition in the literature.

3-6   Joint distribution functions

The analytical description of the manner in which the probability mass for random variables is distributed can be extended readily to two or more functions. To simplify writing, we consider almost exclusively the case of two random variables. Extension to larger families is immediate.

If $X(\cdot)$ and $Y(\cdot)$ are two random variables, the joint mapping produced by the pair induces a joint mass distribution on the plane $R_1 \times R_2$, as well as marginal mass distributions on the coordinate lines R_1 and R_2. These may be described by a joint distribution function and marginal distribution functions, defined as follows:

Definition 3-6a

The function $F_{XY}(\cdot, \cdot)$ defined by

$$F_{XY}(t, u) = P(X \leq t, Y \leq u)$$

is called the *joint distribution function* for $X(\cdot)$ and $Y(\cdot)$. The special cases $F_{XY}(\cdot, \infty)$ and $F_{XY}(\infty, \cdot)$ are called the *marginal distribution functions* for $X(\cdot)$ and for $Y(\cdot)$, respectively.

We note, first, that the marginal distribution functions, as functions of a single variable, are just the ordinary distribution functions for the single variables, since

$$F_{XY}(t, \infty) = P(X \leq t, Y \leq \infty) = P(X \leq t) = F_X(t)$$
and
$$F_{XY}(\infty, u) = P(X \leq \infty, Y \leq u) = P(Y \leq u) = F_Y(u)$$

The significance and the properties of the distribution functions can be visualized geometrically. The following analytical relationships indicate what the geometrical situation is. If we let $M_t = (-\infty, t]$ and $N_u = (-\infty, u]$

$$F_{XY}(t, u) = P(X \leq t, Y \leq u) = P_{XY}(M_t \times N_u)$$
$$F_{XY}(t, \infty) = P_{XY}(M_t \times R_2)$$
and
$$F_{XY}(\infty, u) = P_{XY}(R_1 \times N_u)$$

The region $M_t \times N_u$ is the region which lies on or to the left of a vertical line through the point (t, u) and, at the same time, on or below the horizontal line through that point. Such a region is indicated in Fig. 3-6-1a. $F_{XY}(t, u)$ is the mass distributed in this region in the plane. The region $M_t \times R_2$ is the region on and to the left of the vertical line through the

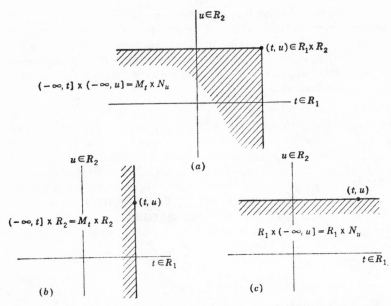

Fig. 3-6-1   The joint and marginal distribution functions and regions in the plane. (a) $F_{XY}(t,\ u)$ is the probability mass in the shaded region $M_t \times N_u$; (b) $F_{XY}(t,\ \infty) = F_X(t)$ is the probability mass in the shaded half plane $M_t \times R_2$; (c) $F_{XY}(\infty,\ u) = F_Y(u)$ is the probability mass in the shaded half plane $R_1 \times N_u$.

point (t, u). Thus $F_{XY}(t,\ \infty) = F_X(t)$ is the mass located in that region of the plane. Similarly, $F_{XY}(\infty,\ u) = F_Y(u)$ is the mass in the region of the plane lying on or below the horizontal line through the point (t, u). These regions are shown in Fig. 3-6-1*b* and *c*, respectively.

As in the case of a real-valued random variable, the distribution function $F_{XY}(\cdot,\ \cdot)$ determines the joint probability measure $P_{XY}(\cdot)$ induced by the random vector $[X, Y](\cdot)$. The distribution function determines probability mass assigned to each set in the class $\mathcal{G}$ of all two-dimensional semi-infinite intervals $M_t \times N_u = \{(x, y): x \leq t,\ y \leq u\}$. It is known that determination of $P_{XY}(\cdot)$ on this class determines it uniquely on the class $\mathcal{B}$ of Borel sets. These considerations extend immediately to any finite number of dimensions.

From this mass picture, we can visualize a number of properties analogous to those for the distribution function for a single random variable. For any fixed u, $F_{XY}(t, u)$ is monotone-increasing in t, and for any fixed t, $F_{XY}(t, u)$ is monotone-increasing in u. Also, $F_{XY}(-\infty,\ -\infty) = 0$ and $F_{XY}(\infty,\ \infty) = 1$. In the plane, we may have point mass concentrations or mass concentrations along lines and curves. In the case of vector-valued random variables, we may extend the concept of absolute continuity of the induced probability measure in a manner analogous to that for a single real-valued random variable. In the two-dimensional case, for example, Lebesgue measure is a generalization of the area of sets in the plane. We say that the probability measure $P_{XY}(\cdot)$ is absolutely continuous (with respect to Lebesgue measure on the plane) if zero probability is assigned to sets of zero area (zero Lebesgue measure). Again, we may appeal to the basic Radon-Nikodym theorem of measure theory to ensure the existence of the function defined as follows:

Definition 3-6b

If the joint probability measure $P_{XY}(\cdot)$ induced by $X(\cdot)$ and $Y(\cdot)$ is absolutely continuous, a function $f_{XY}(\cdot, \cdot)$ exists such that

$$\iint\limits_{Q} f_{XY}(t, u)\, dt\, du = P_{XY}(Q)$$

for each Borel set Q on the plane. The function $f_{XY}(\cdot, \cdot)$ is called a *joint probability density function for* $X(\cdot)$ *and* $Y(\cdot)$.

It is apparent that properties similar to those for the density function for a single variable must hold:

(f1) $f_{XY}(\cdot, \cdot) \geq 0$ (almost everywhere)

(f2) $\displaystyle\iint\limits_{-\infty}^{\infty} f_{XY}(x, y)\, dx\, dy = 1$

(f3) $\displaystyle\int_{-\infty}^{t} \int_{-\infty}^{u} f_{XY}(x, y)\, dy\, dx = F_{XY}(t, u)$

(f4) $f_{XY}(t, u) = \dfrac{\partial^2 F}{\partial u\, \partial t} = \dfrac{\partial^2 F}{\partial t\, \partial u}$ (under suitable regularity conditions)

The density function may be visualized graphically as producing a surface over the plane. The probability that the pair of values for $X(\cdot)$ and $Y(\cdot)$ lies in any region Q in the plane is the volume in the cylinder whose base is Q and which is bounded on the top by the f_{XY} surface. The situation is represented in Fig. 3-6-2.

Some very simple examples should serve to illustrate these ideas.

Example 3-6-1   *Discrete Random Variables*

There is a set of number pairs $\{(t_i, u_j): i \in I, j \in J\}$, with $P(X = t_i, Y = u_j) = p(i, j) \geq 0$ and $\displaystyle\sum_{i,j} p(i, j) = 1$.

DISCUSSION   In this case there is no density function. The distribution function is a step function over the plane, which could be represented by a steplike surface over the plane. It may be expressed analytically by the expression

$$F_{XY}(t, u) = \sum_{i,j} p(i, j) u_+(t - t_i) u_+(u - u_j)$$

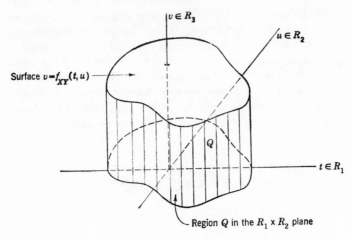

Probability mass in region Q is the volume in the
cylindrical region over Q and under the f_{XY} surface

Fig. 3-6-2   The joint density function and probability mass distributions.

For any fixed t, u, the product $u_+(t - t_i)u_+(u - u_j)$ has the value unity for each pair
t_i, u_j such that $t_i \leq t$ and $u_j \leq u$, so that the sum consists of all the probability mass
lying at points in the appropriate region. ∎

Rather than plot the functional surface, it is simpler, because of the step
character of the function, to indicate its value in various rectangular
regions of the plane.   This is illustrated in Fig. 3-6-3 by a simple case.
Values in circles are values of the joint distribution function $F_{XY}(\cdot, \cdot)$.
Values of the point mass are indicated at appropriate points on the
plane.   A typical point (t, u) is shown.   Dotted lines show the boundary
of the region in which mass contributing to the function value is located.
In the case shown, this mass is $p(1, 1) + p(2, 1) = 0.1 + 0.3 = 0.4$.
Throughout the rectangular region containing this point the value of the
function does not change.   If the point (t, u) shown on the figure is moved
vertically into the next region above it, two additional mass points are
included, to give the value $p(1, 1) + p(2, 1) + p(1, 2) + p(2, 2) = 0.8$,
as shown.   Values in other regions are determined in a similar manner.

Example 3-6-2

Uniform distribution over a rectangle (Fig. 3-6-4).

DISCUSSION   In this case, the density function $f_{XY}(\cdot, \cdot)$ is defined, as well as the
distribution function.   Since the rectangle has unit total mass, the uniform mass
density must be $1/ab$ (i.e., the reciprocal of the area of the rectangle).   The density

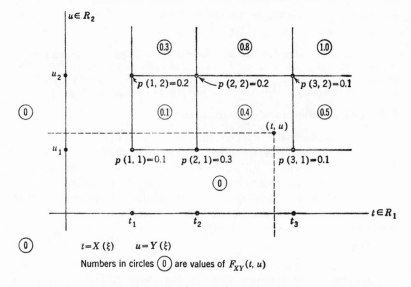

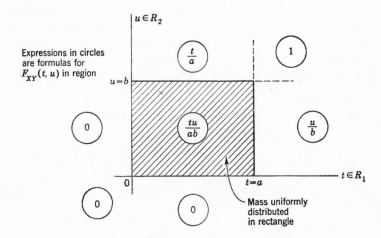

Fig. 3-6-3   Values of the joint distribution function for a pair of discrete random variables.

Fig. 3-6-4   Joint distribution function for a pair of random variables which are uniformly distributed over a rectangle.

is zero elsewhere. This may be expressed analytically:

$$f_{XY}(t, u) = \frac{1}{ab}[u(t) - u(t - a)][u(u) - u(u - b)]$$

Values of the distribution function are indicated on the diagram by formulas in terms of t and u for the value at the point (t, u). The individual variables $X(\cdot)$ and $Y(\cdot)$ are uniformly distributed over the indicated intervals. The marginal distribution and density functions may be determined by reference to Example 3-4-1. ∎

3-7 Independent random variables

The notion of stochastic independence, introduced in Sec. 2-6 for events, may be extended to random variables. A real-valued random variable $X(\cdot)$ defines an event $X^{-1}(M)$ corresponding to any Borel set M on the real line. A second real-valued random variable $Y(\cdot)$ also defines an event $Y^{-1}(N)$ for any Borel set N on the real line. Independence of the random variables is defined in terms of the independence of such events in the following manner:

Definition 3-7a

The random variables $X(\cdot)$ and $Y(\cdot)$ are said to be (*stochastically*) *independent* iffi, for each choice of Borel sets M and N, the events $X^{-1}(M)$ and $Y^{-1}(N)$ are independent events.

This definition is equivalent to the statement that the *product rule*

$$P(X \in M, Y \in N) = P(X \in M)P(Y \in N)$$

holds for each choice of the Borel sets M and N. The independence condition may also be stated in terms of the sigma fields generated by the random variables. $X(\cdot)$, $Y(\cdot)$ form an independent pair of random variables iffi $\{\mathcal{E}(X), \mathcal{E}(Y)\}$ is an independent family. This means that if any $A \in \mathcal{E}(X)$ and $B \in \mathcal{E}(Y)$ are selected, the pair $\{A, B\}$ is independent.

The following two special cases are of interest.

Example 3-7-1

The indicator functions $I_A(\cdot)$ and $I_B(\cdot)$ for events A and B are independent iffi the events A and B are independent.

DISCUSSION $I_A{}^{-1}(M)$ is one of the sets in the class $\mathcal{C} = \{A, A^c, S, \emptyset\}$, as shown in Example 3-1-3. Similarly, $I_B{}^{-1}(N)$ is one of the sets in the class $\mathcal{B} = \{B, B^c, S, \emptyset\}$. Thus $I_A(\cdot)$ and $I_B(\cdot)$ are independent random variables iffi any member of class $\mathcal{C}$ is independent of any member of class $\mathcal{B}$. By Theorems 2-6A and 2-6B this condition holds iffi A and B are independent. ∎

Example 3-7-2

Consider two simple random variables $X(\cdot)$ and $Y(\cdot)$, expressed in canonical form as follows:

$$X(\cdot) = \sum_{i=1}^{n} t_i I_{A_i}(\cdot) \qquad Y(\cdot) = \sum_{j=1}^{m} u_j I_{B_j}(\cdot)$$

The classes $\mathfrak{C} = \{A_i: 1 \leq i \leq n\}$ and $\mathfrak{B} = \{B_j: 1 \leq j \leq m\}$ are partitions. The simple random variables $X(\cdot)$ and $Y(\cdot)$ are independent iffi $P(A_iB_j) = P(A_i)P(B_j)$ for each i and j.

DISCUSSION   For any Borel set M, the inverse image $X^{-1}(M)$ is the set of all those ξ mapped into any t_i which is a member of M. Thus, if $J_{XM} = \{i: t_i \in M\}$, we must have $X^{-1}(M) = \biguplus\limits_{i \in J_{XM}} A_i$. Similarly, $Y^{-1}(N) = \biguplus\limits_{j \in J_{YN}} B_i$, where $J_{YN} = \{j: u_j \in N\}$. If $P(A_iB_j) = P(A_i)P(B_j)$, Theorem 2-6G shows that $X^{-1}(M)$ and $Y^{-1}(N)$ must be independent. The argument holds for any choice of M and N, so that $X(\cdot)$ and $Y(\cdot)$ must be independent random variables. On the other hand, if the random variables are independent, the choice of the special Borel sets $M = \{t_i\}$ and $N = \{u_j\}$ for each i and j yields the product rule $P(A_iB_j) = P(A_i)P(B_j)$. ∎

The concept of independence may be applied to any finite or infinite class of random variables as follows:

Definition 3-7b

A class $\{X_i(\cdot): i \in J\}$ of random variables is said to be an independent class iffi, for each class $\{M_i: i \in J\}$ of Borel sets, arbitrarily chosen, the class of events $\{X_i^{-1}(M_i): i \in J\}$ is an independent class.

It should be emphasized that the independence of the class of events is to be understood in terms of the *product rule*.

Mass distributions

The independence condition for two random variables can be studied in terms of the mass distributions induced by the mappings. In Fig. 3-7-1, the joint mass distribution and the marginal mass distributions are indicated in the manner discussed in Sec. 3-5. We consider Borel sets M, N, and $M \times N$ on R_1, R_2, and $R_1 \times R_2$, respectively. According to the

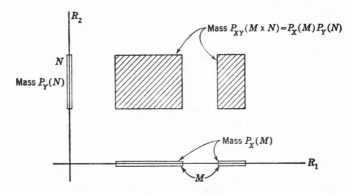

Fig. 3-7-1   Mass distributions for two independent random variables.

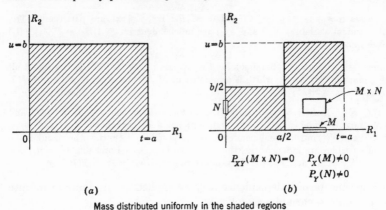

Mass distributed uniformly in the shaded regions

Fig. 3-7-2   Mass distributions as a test for independence.   Marginal mass distributions are uniform in both cases.   (*a*) Independent; (*b*) not independent.

relations developed in Sec. 3-5, we have independence iffi

$$P_{XY}(M \times N) = P[X^{-1}(M)Y^{-1}(N)] = P[X^{-1}(M)]P[Y^{-1}(N)]$$

Since

$$P_X(M) = P[X^{-1}(M)] \qquad \text{and} \qquad P_Y(N) = P[Y^{-1}(N)]$$

we have proved

Theorem 3-7A

Random variables $X(\cdot)$ and $Y(\cdot)$ are independent iffi

$$P_{XY}(M \times N) = P_X(M)P_Y(N)$$

for each Borel set M in R_1 and N in R_2.

Geometrically, this means that the mass in any rectangle set $M \times N$ must be the product of the marginal masses assigned to the coordinate sets M and N.   This geometrical fact can often be used as a test of independence.

Example 3-7-3

Consider the two joint mass distributions in Fig. 3-7-2.   The mass is distributed uniformly over the shaded areas in each case.   According to the geometrical interpretation of the marginal mass as the total joint mass projected onto the line, the marginal distributions for both $X(\cdot)$ and $Y(\cdot)$ are uniform distributions in each case. The marginal mass in a given interval where the mass is located is proportional to the length of the interval.   The distribution in Fig. 3-7-2a satisfies the product condition, so that the random variables producing the joint distribution must be independent.   On the other hand, the product condition is not met in the distribution of Fig. 3-7-2*b*. For example, consider the rectangle set $M \times N$ in the lower right-hand quarter of the rectangular area, as shown. This has mass $P_{XY}(M \times N) = 0$, since there is no mass in that region of the plane. On the other hand, each of the marginal masses must differ from zero, since the joint distribution is such that positive mass is projected onto each of the coordinate lines. This means that $P_X(M)P_Y(N) \neq 0$. Therefore the random variables producing this joint distribution cannot be independent, even though they produce the same marginal distributions as in the case described in Fig. 3-7-2*a*. ∎

For two independent simple random variables $X(\cdot)$ and $Y(\cdot)$, in canonical form (see Example 3-7-2)

$$X(\cdot) = \sum_{i=1}^{n} t_i I_{A_i}(\cdot) \qquad \text{and} \qquad Y(\cdot) = \sum_{j=1}^{m} u_j I_{B_j}(\cdot)$$

we have $X(\cdot)$ and $Y(\cdot)$ are independent iffi $p(i, j) = p(i, *)p(*, j)$, where $p(i, j) = P(X = t_i, Y = u_j)$, $p(i, *) = P(X = t_i)$, and

$$p(*, j) = P(Y = u_j)$$

That is, the joint mass at any point (t_i, u_j) is the product of the marginal masses located at $t = t_i$ and $u = u_j$.

It is not necessary that the mass distributions be purely point masses or purely continuously distributed masses.

Example 3-7-4

Suppose the pair of random variables $X(\cdot)$ and $Y(\cdot)$ induce the joint mass distribution shown in Fig. 3-7-3. One-half of the mass is distributed uniformly over the rectangular area shown. At each of the points indicated by the heavy dots, there is a concentrated mass of $\frac{1}{8}$. Then $P(X \leq 1, Y > 0)$ is the total mass at or to the left of

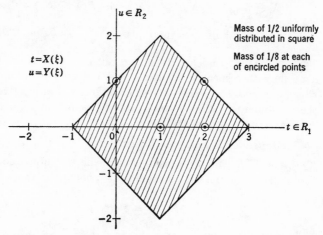

Fig. 3-7-3 A joint mass distribution which is neither continuous nor discrete.

the line $t = 1$ and above the line $u = 0$. This is a mass of $\frac{1}{8}$ uniformly distributed in one-quarter of the square plus the point mass at $t = 0$, $u = 1$. Thus the probability is $\frac{1}{4}$. $P(X \leq 1)$ is the total mass on or to the left of the line $t = 1$. This totals to $\frac{1}{2}$. $P(Y > 0)$ is the total mass in the region above $u = 0$. This also has value $\frac{1}{2}$. Thus $P(X \leq 1,\ Y > 0) = P(X \leq 1)P(Y > 0)$. We cannot conclude that $X(\cdot)$ and $Y(\cdot)$ are independent, however. Consider $P(X < 1,\ Y \geq 0)$. This is the mass to the left of $t = 1$ and on or above the line $u = 0$, which totals to $\frac{1}{4}$. $P(X < 1)$ is the mass to the left of $t = 1$, which amounts to $\frac{3}{8}$. $P(Y \geq 0)$ is the mass $\frac{3}{4}$. Thus $P(X < 1,\ Y \geq 0) \neq P(X < 1)P(Y \geq 0)$, which shows that the two random variables are *not* independent. ∎

An equivalent condition for independence

We have seen that $P_X(\cdot)$, $P_Y(\cdot)$, and $P_{XY}(\cdot)$ are determined uniquely by specifying values for sets M_t, N_u, and $M_t \times N_u$, respectively. From the product rule it follows immediately that if $X(\cdot)$, $Y(\cdot)$ form an independent pair, then for each real t, u we have $P_{XY}(M_t \times N_u) = P_X(M_t)P_Y(N_u)$. It is also known that if this condition ensures the product rule $P_{XY}(M \times N) = P_X(M)P_Y(N)$ holds for all $M \times N$ where M and N are Borel sets. Thus we have

Theorem 3-7B

Any two real-valued random variables $X(\cdot)$ and $Y(\cdot)$ are independent iffi, for all semi-infinite, half-open intervals M_t and N_u, defined above,

$$P_{XY}(M_t \times N_u) = P_X(M_t)P_Y(N_u)$$

By definition, the latter condition is equivalent to the condition

$$P(X \in M_t,\ Y \in N_u) = P(X \in M_t)P(Y \in N_u)$$

for all such half-open intervals.

To examine independence, then, it is not necessary to examine the product rule with respect to all Borel sets, but only with those special sets which are the semi-infinite, half-open intervals. In fact, we could replace the semi-infinite, half-open intervals with semi-infinite, open intervals or even with finite intervals of any kind.

Independence and joint distributions

Theorem 3-7B can be translated immediately into a condition on the distribution functions for independent random variables. Since

$$F_X(t) = P_X(M_t) \qquad F_Y(u) = P_Y(N_u)$$

and

$$F_{XY}(t,\ u) = P_{XY}(M_t \times N_u)$$

the independence condition becomes the following product rule for the distribution functions: $F_{XY}(t, u) = F_X(t)F_Y(u)$ for each pair of real numbers t, u. If the joint and marginal density functions exist, the rules of differentiation for multiple integrals show that this is equivalent to the product rule $f_{XY}(t, u) = f_X(t)f_Y(u)$ for the density functions. Thus we have the important

Theorem 3-7C

Two random variables $X(\cdot)$ and $Y(\cdot)$ are independent iffi their distribution functions satisfy the *product rule* $F_{XY}(t, u) = F_X(t)F_Y(u)$.

If the density functions exist, then independence of the random variables is equivalent to the product rule $f_{XY}(t, u) = f_X(t)f_Y(u)$ for the density functions.

In the discrete case, as we have already established, the independence condition is $p(i, j) = p(i, *)p(*, j)$. This could be considered to be a limiting situation, in which the probability masses are concentrated in smaller and smaller regions, which shrink to a point. The analytical expressions above, in terms of distribution or density functions, simply provide an analytical means for expressing the mass distribution property for independence.

Independent approximating simple functions

The problem of approximating random variables by simple random variables is discussed in Sec. 3-3. If $X(\cdot)$ and $Y(\cdot)$ are two independent random variables, and if $X_i(\cdot)$ and $Y_j(\cdot)$ are approximating simple random variables, formed in the manner discussed in connection with Theorem 3-3A, we must have independence of $X_i(\cdot)$ and $Y_j(\cdot)$. This follows easily from the results obtained in Example 3-7-2. The inverse image $X_i^{-1}(\{t_r\})$ for any point t_r in the range of $X_i(\cdot)$ is the inverse image $X^{-1}(M)$ for an appropriate interval; the inverse image $Y_j^{-1}(\{u_s\})$ for any point u_s in the range of $Y_j(\cdot)$ is the inverse image $Y^{-1}(N)$ for an appropriate interval. These must be independent events.

As a consequence of these facts, we may state the following theorem:

Theorem 3-7D

Suppose $X(\cdot)$ and $Y(\cdot)$ are independent random variables, each of which is nonnegative. Then there exist nondecreasing sequences of nonnegative simple random variables $\{X_n(\cdot): 1 \leq n < \infty\}$ and $\{Y_m(\cdot): 1 \leq m < \infty\}$ such that

$$\lim_{n \to \infty} X_n(\xi) = X(\xi) \qquad \lim_{m \to \infty} Y_m(\xi) = Y(\xi) \qquad \text{for all } \xi$$

and

$\{X_n(\cdot),\ Y_m(\cdot)\}$ is an independent pair for any choice of m, n

This theorem is used in Sec. 4-4 to develop an important property for integrals of independent random variables.

Independence of vector-valued random variables

Many results on the independence of real-valued random variables may be extended to the case of vector-valued random variables. The essential ideas of the proofs are usually quite similar to those used in the real-valued case, but are complicated by notational requirements for stating the relationships in higher dimensional space. We state two results for a pair of vector-valued random variables.

Let $Z(\cdot)$ and $W(\cdot)$ be vector-valued random variables with coordinate variables $X_i(\cdot)$, $1 \leq i \leq n$, and $Y_j(\cdot)$, $1 \leq j \leq m$, respectively. Let $Z_*(\cdot)$ and $W_*(\cdot)$ be random vectors whose coordinates comprise subclasses of the coordinate variables for $Z(\cdot)$ and $W(\cdot)$, respectively.

Theorem 3-7E

If $Z(\cdot)$, $W(\cdot)$ form an independent pair of random vectors, then $Z_*(\cdot)$, $W_*(\cdot)$ is an independent pair.

This follows from the fact that $Z_*(\cdot)$ is measurable $\mathcal{E}(Z)$ and $W_*(\cdot)$ is measurable $\mathcal{E}(W)$. This result holds in the special case that $Z_*(\cdot)$ and $W_*(\cdot)$ may each consist of only one coordinate, and thus be real-valued.

Theorem 3-7F

If the coordinate random variables for $Z(\cdot)$ and $W(\cdot)$ together form an independent class of random variables, then $Z(\cdot)$, $W(\cdot)$ is an independent pair.

The proof of this theorem involves ideas similar to those used for establishing Theorem 3-7B in the real-valued case.

It should be apparent that these two theorems may be extended to the case of more than two random vectors.

3-8   Functions of random variables

It frequently occurs that it is desirable to consider not the random variable observed directly, but some variable derived therefrom. For example:

1. $X(\cdot)$ is the observed value of a physical quantity. When a value t is observed, the desired quantity is t^2, the square of the directly observed quantity.

2. Suppose $X(\cdot)$ and $Y(\cdot)$ are random variables which are the diameters and lengths of cylindrical shafts manufactured on an assembly line.

When ξ is such that $X(\xi) = t$ and $Y(\xi) = u$, the number $(\rho\pi/4)t^2u$ is the weight of the shaft.

3. Suppose $X_k(\cdot)$, $k = 1, 2, \ldots, 24$, is the hourly recorded temperature in a room, throughout a single day. If values $t_1, t_2, \ldots, t_{24}$ are observed, the number

$$t_0 = \frac{1}{24} \sum_{k=1}^{24} t_k$$

is the 24-hour mean temperature.

It is natural to introduce the

Definition 3-8a

If $g(\cdot)$ is a real-valued function of a single real variable t, the function $Z(\cdot) = g[X(\cdot)]$ is defined to be the function on the basic space S which has the value $v = g(t)$ when $X(\xi) = t$.

Similarly, for two variables, if $h(\cdot, \cdot)$ is a real-valued function of two real variables t, u, the function $Z(\cdot) = h[X(\cdot), Y(\cdot)]$ is the function on the basic space which has the value $v = h(t, u)$ when the pair $X(\cdot)$, $Y(\cdot)$ have the values t, u, respectively.

The extension to more than two variables is immediate. The function of interest in example (1) is $Z(\cdot) = X^2(\cdot)$; that in example (2) is

$$Z(\cdot) = (\rho\pi/4)X^2(\cdot)Y(\cdot); \text{ and that in example (3) is } Z(\cdot) = \frac{1}{24} \sum_{k=1}^{24} X_k(\cdot).$$

Before referring to a function of a random variable as a random variable, it is necessary to consider the measurability condition; i.e., it is necessary to show that $Z^{-1}(M)$ is an event for each Borel set M. In order to see what is involved, we consider the mapping situation set up by a function of a random variable, as diagramed in Fig. 3-8-1. Figure 3-8-1a shows the direct mappings produced in the case of a single variable. The random variable $X(\cdot)$ maps ξ into t on the real line R_1. The function $g(\cdot)$ maps t into v on the real line R_2. The resultant is a single mapping from ξ into v, which is the mapping characterizing the function $Z(\cdot) = g[X(\cdot)]$. Figure 3-8-1b represents the inverse mappings. If M is any set in R_2, its inverse image $N = g^{-1}(M)$ is the set of points in R_1 which are mapped into M by the function $g(\cdot)$. The set $E = X^{-1}(N)$ is the set of those ξ which are mapped into N by the random variable $X(\cdot)$. But these are precisely the ξ which are mapped into M by the composite mapping. Thus $Z^{-1}(M) = X^{-1}(N) = X^{-1}[g^{-1}(M)]$. It is customary to indicate the last expression more simply by $X^{-1}g^{-1}(M)$.

The function $Z(\cdot) = g[X(\cdot)]$ is a random variable if $g(\cdot)$ has the property that $g^{-1}(M)$ is a Borel set in R_1 for every Borel set M in R_2.

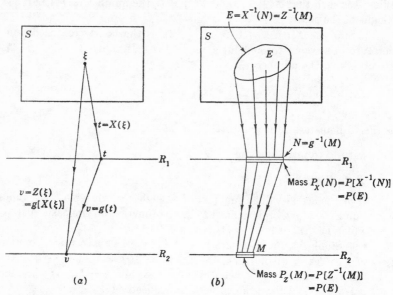

Fig. 3-8-1   (a) Direct mappings and (b) inverse mappings associated with a function of a single random variable.

Definition 3-8b

Let $g(\cdot)$ be a real-valued function, mapping points in the real line R_1 into points in the real line R_2. The function $g(\cdot)$ is called a *Borel function* iffi, for every Borel set M in R_2, the inverse image $N = g^{-1}(M)$ is a Borel set in R_1. An exactly similar definition holds for a *Borel function* $h(\cdot, \cdot)$, mapping points in the plane $R_1 \times R_2$ into points on the real line R_3.

The situation for functions of more than one variable is quite similar. The joint mapping from the basic space, in the case of two variables, is to points on the plane $R_1 \times R_2$. The second mapping $v = h(t, u)$ carries a point t, u in the plane into the point v on the real line, as illustrated in Fig. 3-8-2. If $h(\cdot, \cdot)$ is Borel, so that the inverse image Q of any Borel set M on R_3 is a Borel set on the plane, then the inverse image E of the Borel set Q is an event in the basic space S. Thus, the situation with respect to measurability is not essentially different in the case of two variables. For functions of more than two variables, the only difference is that the first mapping must be to higher-dimensional euclidean spaces.

These results may be summarized in the following

Theorem 3-8A

If $W(\cdot)$ is a random vector and $g(\cdot)$ is a Borel function of the appropriate number of variables, then $Z(\cdot) = g[W(\cdot)]$ is a random variable measurable $\mathcal{E}(W)$.

It is known from more advanced measure theory that if $Z(\cdot)$ is an $\mathcal{E}(W)$-measurable random variable, then there is a Borel function $g(\cdot)$ such that $Z(\cdot) = g[W(\cdot)]$.

The class of Borel functions is sufficiently general to include most functions encountered in practice. For this reason, it is possible in most applications to assume that a function of one or more random variables is itself a random variable, without examining the measurability question. Suppose $g(\cdot)$ is continuous and there is a countable number of intervals on each of which $g(\cdot)$ is monotone; then $g(\cdot)$ is a Borel function, since the inverse image of any interval is a countable union of intervals and hence a Borel set.

The mappings for a function of a single random variable produce probability mass distributions on the lines R_1 and R_2, as indicated in Fig. 3-8-1b. The assignments are according to the scheme

Distribution on R_1 by $X(\cdot)$:

$$P_X(N) = P[X^{-1}(N)] = P(E)$$

Distribution on R_2 by $Z(\cdot) = g[X(\cdot)]$:

$$P_Z(M) = P[Z^{-1}(M)] = P[X^{-1}g^{-1}(M)] = P[X^{-1}(N)] = P(E)$$

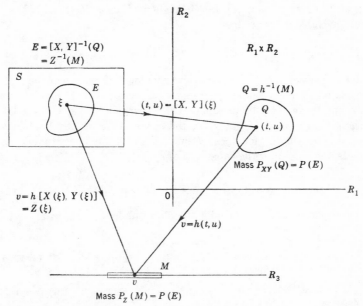

Fig. 3-8-2 Mappings and probability mass distributions for a function of two random variables.

Thus the probability mass distributions on the two real lines are related by the expression

$$P_Z(M) = P_X[g^{-1}(M)]$$

The situation for two random variables is shown on Fig. 3-8-2.   The joint mapping from the basic space S is indicated by $[X, Y](\cdot)$.   Sets M, Q, and E in R_3, $R_1 \times R_2$, and S, respectively, stand in the relation

$$Q = h^{-1}(M) \qquad E = [X, Y]^{-1}(Q) = Z^{-1}(M)$$

Each of these sets is assigned the same probability mass

$$P_Z(M) = P_{XY}(Q) = P(E)$$

It is apparent in later developments that it is sometimes convenient to work with the mass distribution on the plane $R_1 \times R_2$ and sometimes with the distribution on the line R_3.   Either course is at our disposal.

Simple random variables

For simple random variables, the formulas for functions of the random variables are of interest.   Let $X(\cdot)$ and $Y(\cdot)$ be simple random variables which are expressed as follows:

$$X(\cdot) = \sum_{i=1}^{n} t_i I_{A_i}(\cdot) \text{ where } \alpha = \{A_i : 1 \leq i \leq n\} \text{ is a partition}$$

$$Y(\cdot) = \sum_{j=1}^{m} u_j I_{B_j}(\cdot) \text{ where } \mathcal{B} = \{B_j : 1 \leq j \leq m\} \text{ is a partition}$$

If $g(\cdot)$ and $h(\cdot, \cdot)$ are any Borel functions, then $g[X(\cdot)]$ and $h[X(\cdot), Y(\cdot)]$ are given by the formulas

$$g[X(\cdot)] = \sum_{i=1}^{n} g(t_i) I_{A_i}(\cdot) \quad \text{and} \quad h[X(\cdot), Y(\cdot)] = \sum_{i,j} h(t_i, u_j) I_{A_i B_j}(\cdot)$$

The expansion for $g[X(\cdot)]$ is in canonical form iffi no two of the t_i have the same image point under the mapping $v = g(t)$.   This requires that the expansion for $X(\cdot)$ be in canonical form and no two distinct t_i have the same image point.   If they do, the canonical form may be achieved by combining the A_i for those t_i having the same image point.   Similar statements hold for the expansion of $h[X(\cdot), Y(\cdot)]$.

If the A_i do not form a partition, in general $g[X(\cdot)] \neq \sum_{i=1}^{n} g(t_i) I_{A_i}(\cdot)$.

To illustrate this, we consider the following

Example 3-8-1

Suppose $g(t) = t^2$, and consider $X(\cdot) = 2I_A(\cdot) - I_B(\cdot) + 3I_C(\cdot)$ with $D = AB \neq \emptyset$, but $ABC = \emptyset$.   Compare $X^2(\cdot)$ with $Y(\cdot) = 2^2 I_A(\cdot) + (-1)^2 I_B(\cdot) + 3^2 I_C(\cdot)$.

SOLUTION   For $\xi \in D$, $X(\xi) = 2 - 1 = 1$. $X^2(\xi) = 1$. $Y(\xi) = 4 + 1 = 5 \neq X^2(\xi)$. ∎

Equality can hold only for very special functions $g(\cdot)$.   For one thing, we should require that $g(t_i + t_j) = g(t_i) + g(t_j)$ for any t_i and t_j in the range of $X(\cdot)$.

Independence of functions of random variables

We have defined independence of random variables $X(\cdot)$ and $Y(\cdot)$ in terms of independence of the sigma fields $\mathcal{E}(X)$ and $\mathcal{E}(Y)$ determined by the random variables, with immediate extensions to arbitrary classes of random variables (real-valued or vector-valued). In view of the results on measurability of Borel functions of random variables, the following theorem, although trivial to prove, has far reaching consequences.

Theorem 3-8-B

If $\{X_i(\cdot): i \in J\}$ is an independent class of random variables and if, for each $i \in J$, $W_i(\cdot)$ is $\mathcal{E}(X_i)$ measurable, then $\{W_i(\cdot): i \in J\}$ is an independent class of random variables.

We give some examples which are important in themselves and which illustrate the usefulness of the previous development.

1. Suppose $X(\cdot)$ and $Y(\cdot)$ are independent random variables and $A \in \mathcal{E}(X)$ and $B \in \mathcal{E}(Y)$. Then $I_A(\cdot)X(\cdot)$ and $I_B(\cdot)Y(\cdot)$ are independent random variables.

2. As an application of (1), let $A = \{\xi: X(\xi) \leq a\}$ and $B = \{\xi: Y(\xi) \leq b\}$ Then A and $A^c \in \mathcal{E}(X)$ and B and $B^c \in \mathcal{E}(Y)$. We let $X_1(\cdot) = I_A(\cdot)X(\cdot)$ and $X_2(\cdot) = I_{A^c}(\cdot)X(\cdot)$, with similar definitions for $Y_1(\cdot)$ and $Y_2(\cdot)$. Then $X(\cdot) = X_1(\cdot) + X_2(\cdot)$ with $X_1(\cdot)X_2(\cdot) = 0$ and $Y(\cdot) = Y_1(\cdot) + Y_2(\cdot)$ with $Y_1(\cdot)Y_2(\cdot) = 0$. If $X(\cdot)$ and $Y(\cdot)$ are independent, so also are the pairs $\{X_1(\cdot), Y_1(\cdot)\}$, $\{X_1(\cdot), Y_2(\cdot)\}$, $\{X_2(\cdot), Y_1(\cdot)\}$, and $\{X_2(\cdot), Y_2(\cdot)\}$. This decomposition and the resulting independence is important in carrying out a classical argument known as the *method of truncation*.

3. If $X(\cdot)$ and $Y(\cdot)$ are independent and $X_n(\cdot)$ and $Y_m(\cdot)$ are two approximating simple functions of the kind described in Sec. 3-3, then $X_n(\cdot)$ must be measurable $\mathcal{E}(X)$ and $Y_m(\cdot)$ must be measurable $\mathcal{E}(Y)$, so that $X_n(\cdot)$ and $Y_m(\cdot)$ are independent. This is a restatement of the proof of Theorem 3-7D.

Example 3-8-2

Suppose $\{X(\cdot), Y(\cdot), Z(\cdot)\}$ is an independent class of random variables. Then $U(\cdot) = X^2(\cdot) + 3Y(\cdot)$ and $V(\cdot) = |Z(\cdot)|$ are independent random variables. If $W(\cdot) = 2X(\cdot)Z(\cdot)$, then $U(\cdot)$ and $W(\cdot)$ are *not*, in general, independent random variables, since they involve a common function $X(\cdot)$. ∎

3-9 Distributions for functions of random variables

We consider first the problem of determining the probability that $Z(\cdot) = g[X(\cdot)]$ takes on values in a given set M when the probability distribution for $X(\cdot)$ is known. In particular, we consider the problem of determining $F_Z(\cdot)$, and $f_Z(\cdot)$ if it exists, when $F_X(\cdot)$ is known. We then extend the discussion to random variables which are functions of two or more random variables.

We shall develop a basic strategy with the aid of the mechanical picture of the probability mass distribution induced by the random variable $X(\cdot)$, or the joint distribution in the case of several random variables. When the problem is understood in terms of this mechanical picture, appropriate techniques for special problems may be discovered. Often the problem may not be amenable to straightforward analytical operations, but may be handled by approximate methods or by special methods which exploit some peculiarity of the distribution.

Functions of a single variable

Suppose $X(\cdot)$ is a real-valued random variable with distribution function $F_X(\cdot)$, and $g(\cdot)$ is a Borel function of a single real variable. We suppose that the domain of $g(\cdot)$ contains the range of $X(\cdot)$; that is, $g(\cdot)$ is defined for every value which $X(\cdot)$ can assume.

We begin by recalling the fundamental relationship

$$\{\xi: Z(\xi) \in M\} = \{\xi: X(\xi) \in g^{-1}(M)\}$$

This may be seen by referring back to Fig. 3-8-1. We thus have the fundamental probability relationship

$$P(Z \in M) = P[X \in g^{-1}(M)] = P_X[g^{-1}(M)]$$

To determine the probability that $Z(\cdot)$ takes a value in M, we determine the probability mass assigned to the t set $g^{-1}(M)$; this is the set of those t which are mapped into M by the mapping $v = g(t)$.

For the determination of the distribution function $F_Z(\cdot)$, we consider the particular sets

$$M_v = (-\infty, v] \qquad \text{and} \qquad Q_v = g^{-1}(M_v)$$

Now

$$g(t) \leq v \qquad \text{iffi } t \in Q_v$$

so that

$$Z(\xi) = g[X(\xi)] \leq v \qquad \text{iffi } X(\xi) \in Q_v$$

Hence we have as fundamental relationships

$$F_Z(v) = P(Z \leq v) = P(X \in Q_v) = P_X(Q_v)$$

The value of the distribution function $F_Z(\cdot)$ for any particular v can be determined if Q_v can be determined and the probability mass $P_X(Q_v)$ assigned to it can be evaluated. This determination may be made in any manner. We shall illustrate the basic strategy and the manner in which special methods arise by considering several simple examples.

Example 3-9-1

Suppose $g(t) = t^2$, so that $Z(\cdot) = X^2(\cdot)$. Determine $F_Z(\cdot)$ and $f_Z(\cdot)$.

SOLUTION We note that $Z(\cdot)$ cannot take on negative values, so that $F_Z(v) = 0$ for $v < 0$. For nonnegative values of v, $F_Z(v) = P(Z \in (-\infty, v]) = P(Z \in [0, v])$.

inverse image is an event. This freedom is possible because the class of Borel sets on the real line is a very general class, and the class of events on the basic space S is usually a very general class.

The major problem in dealing with random variables in practice is to determine the probabilities that random variables take on given values or sets of values. We make a major step in the direction of analytical description in the next section.

3-2 Random variables and mass distributions

Before we can make much progress in dealing with random variables, we must find ways to represent the distribution of values of such functions in a manner that facilitates the making of probability statements. In this section, we wish to extend the auxiliary model provided by the mechanical picture of probability as mass assigned to sets of points in the basic space S. We may do this in the following manner, which provides an important mental image to aid thinking about random variables. As a function, the random variable $X(\cdot)$ relates each point ξ in S to a point t on the real line R. *We may consider this mapping as producing a transfer of the probability mass from the basic space S to the real line R.* The resulting distribution of probability mass on the real line may be visualized most readily in the case of a finite set of discrete image points on the real line. To each point t in the range of $X(\cdot)$ is transferred all the mass of that set of ξ which is mapped into the given point t. More generally, if M is any t set and E is the set of points of S which are mapped into it [i.e., if $E = X^{-1}(M)$], the mass associated with E in S should also be associated with M in R. This relation may be visualized with the aid of a diagram such as that provided in Fig. 3-2-1.

This situation may be formulated precisely in analytical terms. Suppose M is any Borel set. We consider $P(\{\xi: X(\xi) \in M\})$, for which we adopt the shorter notation $P(X \in M)$. Then

$$P(X \in M) = P(\{\xi: X(\xi) \in M\}) = P[X^{-1}(M)]$$

For each Borel set M we put $P_X(M) = P[X^{-1}(M)]$. Since $X^{-1}(R) = S$ and $X^{-1}(\bigcup_i M_i) = \bigcup_i X^{-1}(M_i)$, we have

(P1) $\quad P_X(R) = 1$

(P2) $\quad P_X(M) \geq 0$

(P3) $\quad P_X\left(\bigcup_{i=1}^{\infty} M_i\right) = \sum_{i=1}^{\infty} P_X(M_i)$

so that $P_X(\cdot)$ is a probability measure on the class $\mathfrak{B}_R$ of Borel sets.

Measurability of a random variable $X(\cdot)$ is the condition that the class $X^{-1}(\mathcal{B}_R)$ of inverse images of Borel sets is a subclass of the class $\mathcal{E}$ of events. Arguments based on properties of inverse images in Theorem 3-1A show the class $X^{-1}(\mathcal{B}_R)$ to be a sigma field. Since this class is determined by the random variable $X(\cdot)$, it is convenient to adopt the following

Definition 3-1b.

If $X(\cdot)$ is a random variable and $\mathcal{B}_R$ is the class of Borel sets on the real line, the sigma field $\mathcal{E}(X) = X^{-1}(\mathcal{B}_R)$ is called the *sigma field determined by* $X(\cdot)$.

From Theorem 3-1B and Definition 3-1a, it follows that $X(\cdot)$ is a random variable iff $\mathcal{E}(X) \subset \mathcal{E}$. The class $\mathcal{E}(X)$ may not contain all possible events. The discussion of Example 3-1-3 shows that if $X(\cdot)$ is the indicator function for set E, then $\mathcal{E}(X) = \{\emptyset, E, E^c, S\}$. This is a sigma field, every set of which is an event if E is an event; it certainly does not exhaust the class of events in most probability systems. For the random variable representing the number of successes in a sequence of n repeated trials, described in Example 3-1-4, $\mathcal{E}(X)$ is the class of all finite unions of the events A_m, including the impossible event $\emptyset$ as the degenerate union of no members. This, too, is a sigma field. In the general case of Example 3-1-5, we can give no better description of $\mathcal{E}(X)$ than to say it is the class of inverse images of the Borel sets.

In some mathematical developments, it is necessary to know whether the inverse images of the Borel sets are contained in some subclass of the events. In such discussions, it is convenient to utilize the following terminology.

Definition 3-1c.

If $\mathcal{E}_0$ is a sigma field of events (i.e., a sub sigma field of the class $\mathcal{E}$ of events), we say $X(\cdot)$ is *measurable-$\mathcal{E}_0$* iff $\mathcal{E}(X) \subset \mathcal{E}_0$. We say random variable $X(\cdot)$ is *measurable with respect to random variable* $Y(\cdot)$ iff $\mathcal{E}(X) \subset \mathcal{E}(Y)$.

It is certainly true that $X(\cdot)$ is measurable with respect to itself, and if $X(\cdot)$ is a random variable, then it is measurable-$\mathcal{E}$.

As we noted earlier, one does not ordinarily worry about questions of measurability in practical applications. A numerical-valued phenomenon is represented by an appropriate function $X(\cdot)$ on the basic space. Any t set M arising in practice is assumed to be such that its

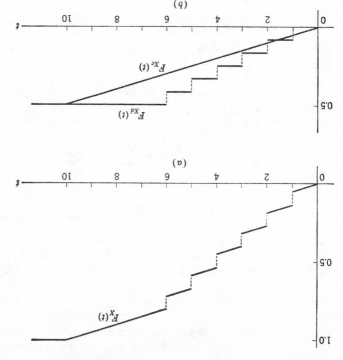

(a)

(b)

Fig. 3-4-5 Distribution function $F_X(\cdot)$ and components $F_{Xc}(\cdot)$ and $F_{Xd}(\cdot)$ for the random variable of Example 3-4-4.

Basic properties

A number of properties, which are illustrated in the simple examples above, may be established as general properties of the distribution function $F_X(\cdot)$. These may be visualized in terms of the mass picture, which is quite precise. On the other hand, they may be established analytically without reference to the mass representation. We shall establish the results analytically and interpret them in terms of the mass picture, in order that the double approach give the firmest possible grasp of the properties and their foundation.

(F1) $F_X(t)$ is monotonically increasing with increasing t.

PROOF Let $b > a$; then $F_X(b) = P(X \leq b) = P(X \leq a) + P(a < X \leq b)$. The events $\{\xi : X \leq a\}$ and $\{\xi : a < X \leq b\}$ are disjoint, and their union is $\{\xi : X \leq b\}$, so that additivity may be employed. Since the probability of an event is nonnegative, we can assert

$$F_X(b) = P(X \leq b) \geq P(X \leq a) = F_X(a) \quad \blacksquare$$

In terms of the mass picture, this says that the total mass at or to the left of b cannot be less than the total mass at or to the left of a. The proof amounts to arguing that the mass to the right of a but at or to the right of the left of b cannot be negative. The nondecreasing character of the distribution function may be observed in any of the examples.

In dealing with random variables, we shall always assume that the probabilities $P(X = \infty)$ and $P(X = -\infty)$ are zero. More precisely, we suppose $\lim_{t \to \infty} P(X > t) = 0$ and $\lim_{t \to -\infty} P(X \leq t) = 0$. Under these conditions, we must have

$$(\mathrm{F2}) \quad F_X(-\infty) = \lim_{t \to -\infty} F_X(t) = 0 \quad \text{and} \quad F_X(\infty) = \lim_{t \to \infty} F_X(t) = 1$$

If all the probability mass is located within a given finite interval, the distribution function must be 0 to the left of this interval and have the value 1 to the right of this interval. It is quite possible to have the mass distributed over the entire line, in which case the bounding values 0 and 1 are taken on as limits. These facts are illustrated in the various examples of this section.

$$(\mathrm{F3}) \quad P(a < X \leq b) = P(X \in (a, b]) = F_X(b) - F_X(a)$$

PROOF   Rearrange the expression in the argument for (F1). ∎

This says that the probability that $X(\cdot)$ is greater than a but no greater than b is equal to the probability mass lying to the right of point $t = a$ but at or to the left of the point $t = b$. We may consider a limiting case to obtain the following property:

$(\mathrm{F4})$ $F_X(\cdot)$ has a jump discontinuity of magnitude $\delta > 0$ at $t = a$ iff $P(X = a) = \delta$. $F_X(\cdot)$ is continuous at $t = a$ iff $P(X = a) = 0$.

PROOF   Using the type of argument employed in Example 3-1-5, we have $\{\xi : X = a\} = \bigcup_{n=1}^{\infty} \{\xi : a - 1/n < X \leq a\}$. The sets $A_n = \{\xi : a - 1/n < X \leq a\}$ form a decreasing sequence. We have by (P9) and (F3) $\lim_{n \to \infty} P(A_n) = \lim_{n \to \infty} [F_X(a) - F_X(a - 1/n)] = P(X = a)$.

The argument amounts to saying that the limiting value of the mass to the right of every point $t = a - 1/n$ but at or to the left of point $t = a$ is just the mass at $t = a$. A jump discontinuity in the distribution function occurs iff there is a point mass located there. Otherwise, the distribution function is continuous. An extension of this type of argument shows that the value taken on at a jump discontinuity by the

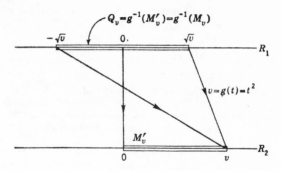

Fig. 3-9-1 Mapping which gives the probability distribution for $Z(\cdot) = X^2(\cdot)$.

Now $g^{-1}([0, v]) = [-\sqrt{v}, \sqrt{v}]$. Hence $F_Z(v) = P(X \in [-\sqrt{v}, \sqrt{v}]) = F_X(\sqrt{v}) - F_X(-\sqrt{v}) + P(X = -\sqrt{v})$, for $v \geq 0$. A single formula can be used if the last expression is multiplied by $u_+(v)$, to make it zero for negative v.

If the distribution for $X(\cdot)$ is absolutely continuous,

$$P(X = -\sqrt{v}) = 0$$

and

$$f_Z(v) = \frac{d}{dv} F_Z(v) = \frac{u(v)}{2\sqrt{v}} [f_X(\sqrt{v}) + f_X(-\sqrt{v})] \blacksquare$$

The essential facts of the argument are displayed geometrically in Fig. 3-9-1. The set $M_v = (-\infty, v]$ has the same inverse as does the set $M_v' = [0, v]$, since the inverse image of $(-\infty, 0)$ is $\emptyset$; that is, $g(\cdot)$ does not take on negative values. The inverse of M_v' is $Q_v = [-\sqrt{v}, \sqrt{v}]$ for $v \geq 0$ and $Q_v = \emptyset$ for $v < 0$. The probability mass $P_X(Q_v)$ assigned to this interval for nonnegative v is

$$F_X(\sqrt{v}) - F_X(-\sqrt{v}) + P(X = -\sqrt{v})$$

In the continuous case, the last term is zero, since there can be no concentration of probability mass on the real line.

Example 3-9-2

The current through a resistor with resistance R ohms is known to vary in such a manner that if the value of current is sampled at an arbitrary time, the probability distribution is gaussian. The power dissipated in the resistor is given by $w = i^2R$, where i is the current in amperes, R is the resistance in ohms, and w is the power in watts. Suppose R is 1 ohm, $I(\cdot)$ is the random variable whose observed value is the current, and $W(\cdot)$ is the random variable whose value is the power dissipated in the resistor. We suppose

$$f_I(t) = \frac{1}{\sqrt{2\pi a}} e^{-t^2/2a} \qquad a > 0$$

This is the density function for a gaussian random variable with the parameters $\mu = 0$ and $\sigma^2 = a$ (Example 3-4-3). Now $W(\cdot) = I^2(\cdot)$, since $R = 1$. According to the result in Example 3-9-1, we must have

$$f_W(v) = \frac{u(v)}{2\sqrt{v}}\frac{1}{\sqrt{2\pi a}}(e^{-v/2a} + e^{-v/2a}) = \frac{u(v)}{\sqrt{2\pi av}}e^{-v/2a}$$

This density function is actually undefined at $v = 0$. The rate of growth is sufficiently slow, however, so that the mass in an interval, $[0, v]$ goes to zero as v goes to zero. The unit step function $u(v)$ ensures zero density for negative v, as is required physically by the fact that the power is never negative in the simple system under study. ■

In the case of a discrete random variable $X(\cdot)$, the resulting random variable $Z(\cdot) = g[X(\cdot)]$ is also discrete. In this case it may be simpler to work directly with the probability mass distributions. The following simple example will illustrate the situation.

Example 3-9-3

A discrete positioning device may take the correct position or may be 1, 2, . . . , n units off the correct position in either direction. Let p_0 be the probability of taking the correct position. Let p_i be the probability of an error of i units to the right; also, let p_i be the probability of an error of i units to the left. In the design of positioning devices, position errors are often weighted according to the square of the magnitude. A negative error is as bad as a positive error; large errors are more serious than small errors. Let $E(\cdot)$ be the random variable whose value is the error on any reading. The range of $E(\cdot)$ is the set of integers running from $-n$ through n. The probability $P(E = i) = p_i$, with $p_{-i} = p_i$. We wish to find the distribution for $E^2(\cdot)$, the square of the position error. The result may be obtained with the aid of Fig. 3-9-2. $E^2(\cdot)$ has range $T = \{v_i: 0 \leq i \leq n\}$, with $v_i = i^2$. We have $P(E^2 = 0) = p_0$, and $P(E^2 = i^2) = 2p_i$ for $1 \leq i \leq n$. The distribution function $F_{E^2}(\cdot)$ may easily be written if desired. The density function does not exist. ■

The function considered in the next example is frequently encountered. It provides a change in origin and a change in scale for a random variable.

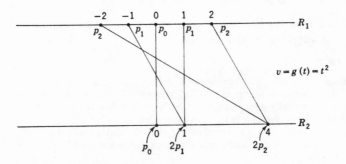

Fig. 3-9-2 Discrete probability distribution produced by the mapping $g(t) = t^2$ for the random variable in Example 3-9-3.

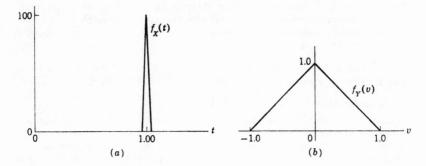

Fig. 3-9-3   Density functions for the random variables $X(\cdot)$ and $Y(\cdot) = 100X(\cdot) -$ 1.00 in Example 3-9-5.

Example 3-9-4

Suppose $g(t) = at + b$, so that $Z(\cdot) = aX(\cdot) + b$.

DISCUSSION   We need to consider two cases: (1) $a > 0$ and (2) $a < 0$ (the case $a = 0$ is trivial).

1. For $a > 0$,

$$F_Z(v) = P(aX + b \le v) = P\left(X \le \frac{v - b}{a}\right) = F_X\left(\frac{v - b}{a}\right)$$

2. For $a < 0$, so that $a = -|a|$,

$$-|a|X + b \le v \qquad \text{iffi} \;\; -X \le \frac{v - b}{|a|} \qquad \text{iffi} \;\; X \ge \frac{v - b}{a}$$

so that

$$F_Z(v) = 1 - F_X\left(\frac{v - b}{a}\right) + P\left(X = \frac{v - b}{a}\right)$$

In the absolutely continuous case, differentiation shows that for either sign of a, we have

$$f_Z(v) = \frac{1}{|a|} f_X\left(\frac{v - b}{a}\right) \;\blacksquare$$

As a simple application, consider the following

Example 3-9-5

A random variable $X(\cdot)$ is found to have a triangular distribution, as shown in Fig. 3-9-3a.   The triangle is symmetrical about the value $t = 1.00$.   The base extends from $t = 0.99$ to 1.01.   This means that the values of the random variable are clustered about the point $t = 1.00$.   By subtracting off this value and expanding the scale by a factor of 100, we obtain the random variable $Y(\cdot) = 100[X(\cdot) - 1.00]$. The new random variable has a density function

$$f_Y(v) = \frac{1}{100} f_X\left(\frac{v + 100}{100}\right) = 0.01 f_X(0.01v + 1)$$

The new density function $f_Y(\cdot)$ is thus obtained from $f_X(\cdot)$ by three operations: (1) scaling down the ordinates by a factor 0.01, (2) moving the graph to the left by 1.00 unit, and (3) expanding the scale by a factor of 100. The resulting graph is found in Fig. 3-9-3b. ∎

The function in the following example is interesting from a theoretical point of view and is sometimes useful in practice.

Example 3-9-6

Suppose $X(\cdot)$ is uniformly distributed over the interval [0, 1]. Let $F(\cdot)$ be any probability distribution function which is continuous and strictly increasing except possibly where it has the value zero or one. In this case, the inverse function $F^{-1}(\cdot)$ is defined as a point function at least for the open interval (0, 1). Consider the random variable

$$Y(\cdot) = F^{-1}[X(\cdot)]$$

We wish to show that the distribution function $F_Y(\cdot)$ for the new random variable is just the function $F(\cdot)$ used to define $Y(\cdot)$.

SOLUTION Because of the nature of an inverse function

$$Y(\xi) = F^{-1}[X(\xi)] \leq a \qquad \text{iffi } X(\xi) \leq F(a)$$

Thus $P(Y \leq a) = P[X \leq F(a)]$. Because of the uniform distribution of $X(\cdot)$ over [0, 1], $P[X \leq F(a)] = F(a)$. Hence we have $F_Y(a) = P(Y \leq a) = F(a)$, which is the desired result. ∎

It is often desirable to be able to produce experimentally a sampling of numbers which vary according to some desired distribution. The following example shows how this may be done with the results of Example 3-9-6 and a table of random numbers.

Example 3-9-7

Suppose $\{X_i(\cdot): 1 \leq i \leq n\}$ is an independent class of random variables, each distributed uniformly over the integers 0, 1, . . . , 9. This class forms a model for the choice of n *random digits* (decimal). Consider the function

$$Y_n(\cdot) = \sum_{k=1}^{n} X_k(\cdot)10^{-k}$$

which is a random variable since it is a linear combination of random variables. For each choice of a set of values of the $X_k(\cdot)$ we determine a unique value of $Y_n(\cdot)$ on the set of numbers $\{0, 10^{-n}, 2 \cdot 10^{-n}, \ldots, 1-10^{-n}\}$. The probability of any combination of values of the $X_k(\cdot)$ and hence of any value of $Y_n(\cdot)$ is 10^{-n}, because of the independence of the $X_k(\cdot)$. This means that the graph of $F_{Y_n}(\cdot)$ takes a step of magnitude 10^{-n} at points separated by 10^{-n}, beginning at zero. Thus, $F_{Y_n}(t) \approx F_X(t)$, where $X(\cdot)$ is uniformly distributed [0, 1]. If $Z_n(\cdot) = F^{-1}[Y_n(\cdot)]$, then $F_{Z_n}(t) \approx F(t)$ for all real t. ∎

Functions of two random variables

For functions of two random variables $X(\cdot)$ and $Y(\cdot)$, we suppose the joint distribution function $F_{XY}(\cdot, \cdot)$ of the joint probability measure

$P_{XY}(\cdot)$ induced on the plane is known.   If $h(\cdot, \cdot)$ is a Borel function of two real variables, we wish to determine the distribution function $F_Z(\cdot)$ for the random variable $Z(\cdot) = h[X(\cdot), Y(\cdot)]$.   The basic attack is the same as in the single-variable case.   If

$$M_v = (-\infty, v] \qquad \text{and} \qquad Q_v = h^{-1}(M_v) = \{(t, u): h(t, u) \leq v\}$$

then

$$F_Z(v) = P(h[X, Y] \leq v) = P([X, Y] \in Q_v) = P_{XY}(Q_v)$$

The problem amounts to determining the set Q_v of points (t, u) in the plane $R_1 \times R_2$, for which $h(t, u) \leq v$, and then determining the probability mass $P_{XY}(Q_v)$ assigned to that set of points.   Once the problem is

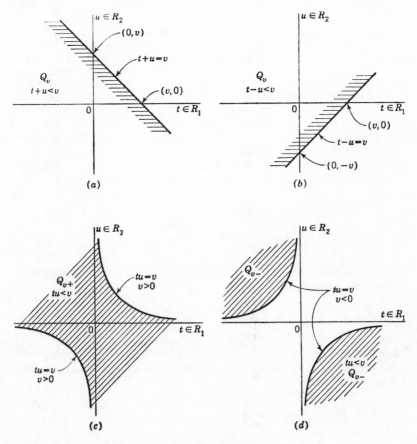

Fig. 3-9-4   Regions Q_v in which $h(t, u) \leq v$ for several functions $h(\cdot, \cdot)$.   (a) $h(t, u) = t + u$; (b) $h(t, u) = t - u$; (c) $h(t, u) = tu, v > 0$; (d) $h(t, u) = tu, v < 0$.

thus understood, the particular techniques of solution may be determined for a given problem.

We shall use a number of simple examples to illustrate some possibilities. As a first step, we consider the regions Q_v for various values of v and several common functions $h(\cdot, \cdot)$; these are shown on Fig. 3-9-4. Corresponding regions may be determined for other functions $h(\cdot, \cdot)$ encountered in practice by the use of analytical geometry.

As a first example, we consider a somewhat artificial problem designed to demonstrate the basic approach.

Example 3-9-8

Suppose $h(t, u) = t + u$, so that $Z(\cdot) = X(\cdot) + Y(\cdot)$. Determine $F_Z(\cdot)$ when the joint mass distribution is that shown in Fig. 3-9-5a.

SOLUTION   By simple graphical operations, the distribution function $F_Z(\cdot)$ shown in Fig. 3-9-5b may be determined. At $v = -2$, the point mass of $\frac{1}{8}$ is picked up. The continuously distributed mass in the region Q_v increases with the square of the increase in v until the two point masses of $\frac{1}{8}$ each are picked up simultaneously to give a jump of $\frac{1}{4}$ at $v = 0$. Then $F_Z(v)$ must vary as a constant minus the square of the distance from v to the value 2. At $v = 2$, the final point mass is picked up, to give a jump of $\frac{1}{8}$. Since all the mass is included in Q_2, further increase in v does not increase $F_Z(v)$. ∎

Example 3-9-9

Suppose $X(\cdot)$ and $Y(\cdot)$ have an absolutely continuous joint distribution. Determine the density function $f_Z(\cdot)$ for the random variable $Z(\cdot) = X(\cdot) + Y(\cdot)$.

SOLUTION

$$F_Z(v) = P_{XY}(Q_v) = \iint_{Q_v} f_{XY}(t, u) \, dt \, du = \int_{-\infty}^{\infty} \left[\int_{-\infty}^{v-u} f_{XY}(t, u) \, dt \right] du$$

Differentiating with respect to the variable v, which appears only in the upper limit for one of the integrals, we get

$$f_Z(v) = \int_{-\infty}^{\infty} f_{XY}(v - u, u) \, du$$

We have used the formula $\dfrac{d}{dv} \displaystyle\int_{a}^{\varphi(v)} f(t) \, dt = f[\varphi(v)]\varphi'(v)$. If we make the change of variable $t = v - u$, for any fixed v, the usual change-of-variable techniques show that we may also write

$$f_Z(v) = \int_{-\infty}^{\infty} f_{XY}(t, v - t) \, dt \quad ∎$$

Again we use a simple illustration to demonstrate how the previous result may be employed analytically.

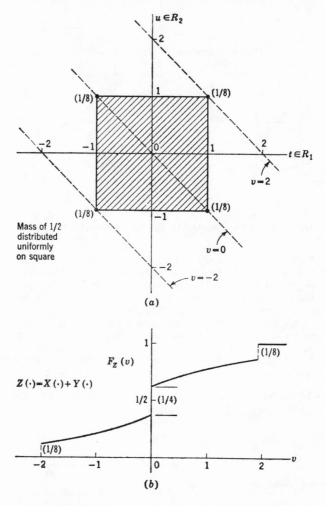

Fig. 3-9-5   A joint mass distribution and the probability distribution function for the sum of two random variables.   (*a*) Joint mass distribution for *X*, *Y*; (*b*) mass distribution function $F_Z(\cdot)$.

Example 3-9-10

Suppose, for the problem posed generally in the preceding example, the joint density function is that shown in Fig. 3-9-6.   We wish to evaluate

$$f_Z(v) = \int_{-\infty}^{\infty} f_{XY}(t, v - t)\, dt$$

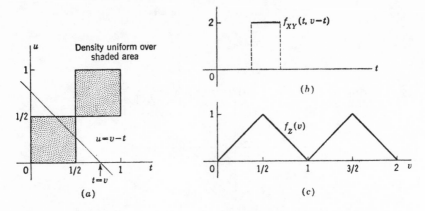

Fig. 3-9-6   Various probability density functions for Example 3-9-10.

We are aided graphically by noting that the points $(t, v - t)$ lie on a slant line of the as a function of t for fixed v, is a step function.   The length of the step is $1/\sqrt{2}$ times the length of that portion of the slant line in the region of positive density.   The integral of this step function is twice the length of the positive part of the step function.   This step function is twice the length of the positive part of the step function.   This length obviously increases linearly with v for $0 \leq v \leq \frac{1}{2}$; then it decreases linearly with v for $\frac{1}{2} \leq v \leq 1$; another cycle is completed for $1 \leq v \leq 2$.   The density function must be zero for $v < 0$ and $v > 2$.   The resulting function is graphed in Fig. 3-9-6c.   The same result could have been obtained by determining the distribution function, as in Example 3-9-8, and then differentiating. ∎

The integration procedure in the preceding two examples can be given a simple graphical interpretation.   If the joint density function $f_{XY}(\cdot, \cdot)$ is

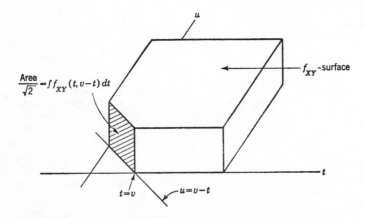

Fig. 3-9-7   Graphical interpretation of the integral $\int f_{XY}(t, v - t)\, dt$.

visualized graphically as producing a surface over the plane, in the manner discussed in Sec. 3-6 (Fig. 3-6-2), the value of the integral is $1/\sqrt{2}$ times the area under that surface and over the line $u = v - t$. This is illustrated in Fig. 3-9-7 for the probability distribution, which is uniform over a rectangle. The region under the f_{XY} surface may be viewed as a solid block. For any given v, the block is sectioned by a vertical plane through the line $u = v - t$. The area of the section (shown shaded in Fig. 3-9-7) is $\sqrt{2}$ times the value of $f_Z(v)$. The simple distribution was chosen for ease in making the pictorial representation. The process is quite general and may be applied to any distribution for which a satisfactory representation can be made.

Similar techniques may be applied to the difference of two random variables. The following may be verified easily:

Example 3-9-11

Suppose $X(\cdot)$ and $Y(\cdot)$ have an absolutely continuous joint distribution. The density function $f_W(\cdot)$ for the random variable $W(\cdot) = X(\cdot) - Y(\cdot)$ is given by

$$f_W(v) = \int_{-\infty}^{\infty} f_{XY}(v + u, u)\, du = \int_{-\infty}^{\infty} f_{XY}(t, t - v)\, dt \quad \blacksquare$$

If the two random variables $X(\cdot)$ and $Y(\cdot)$ are independent, the product rule on the density functions may be utilized to give alternative forms, which may be easier to work with.

Example 3-9-12

Suppose $X(\cdot)$ and $Y(\cdot)$ are independent random variables, each of which has an absolutely continuous distribution. Let $Z(\cdot) = X(\cdot) + Y(\cdot)$ and $W(\cdot) = X(\cdot) - Y(\cdot)$. Because of the independence, we have

$$f_{XY}(t, u) = f_X(t)f_Y(u)$$

The results of Examples 3-9-9 and 3-9-11 may be written (with suitable change of the dummy variable of integration) as follows:

$$f_Z(v) = \int_{-\infty}^{\infty} f_X(v - u)f_Y(u)\, du = \int_{-\infty}^{\infty} f_Y(v - u)f_X(u)\, du$$

$$f_W(v) = \int_{-\infty}^{\infty} f_X(v + u)f_Y(u)\, du = \int_{-\infty}^{\infty} f_Y(u - v)f_X(u)\, du$$

We may integrate these expressions with respect to v from $-\infty$ to t to obtain

$$F_Z(t) = \int_{-\infty}^{\infty} F_X(t - u)f_Y(u)\, du = \int_{-\infty}^{\infty} F_Y(t - u)f_X(u)\, du$$

$$F_W(t) = \int_{-\infty}^{\infty} F_X(t + u)f_Y(u)\, du = \int_{-\infty}^{\infty} [1 - F_Y(u - t)]f_X(u)\, du \quad \blacksquare$$

The integrals for $f_Z(v)$ are known as the *convolution* of the two densities $f_X(\cdot)$ and $f_Y(\cdot)$. This operation is well known in the theory of Laplace and Fourier transforms. Techniques employing these transforms are often useful in obtaining the convolution. Since a knowledge of these transform methods lies outside the scope of this study, we shall not illustrate them. The following example from reliability theory provides an interesting application.

Example 3-9-13

A system is provided with *standby redundancy* in the following sense. There are two subsystems, only one of which operates at any time. At the beginning of operation, system 1 is turned on. If system 1 fails before a given time t, system 2 is turned on. Let $X(\xi)$ be the length of time system 1 operates and $Y(\xi)$ be the length of time system 2 operates. We suppose these are independent random variables. The system operates successfully if $X(\xi) + Y(\xi) \geq t$, and fails otherwise. If F is the event of system failure, we have

$$P(F) = F_{X+Y}(t) = F_Z(t) = \int_{-\infty}^{\infty} F_X(t - u) f_Y(u) \, du$$

Experience has shown that for a large class of systems the probability distribution for "time to failure" is exponential in character. Specifically, we assume

$$F_X(t) = u(t)[1 - e^{-\alpha t}] \qquad \text{and} \qquad F_Y(u) = u(u)[1 - e^{-\beta u}]$$

where the unit step functions ensure zero values for negative values of the arguments. This means that $f_Y(u) = u(u)\beta e^{-\beta u}$. The limits of integration may be adjusted to account for the fact that the integrand is zero for $u < 0$ or $u > t$ (note that t is fixed for any integration). We thus have

$$P(F) = F_Z(t) = \int_0^t [1 - e^{-\alpha(t-u)}]\beta e^{-\beta u} \, du \qquad t > 0$$

Combining the exponentials and evaluating the integrals, we obtain the result

$$\begin{aligned} F_Z(t) &= u(t)\left[1 - e^{-\beta t} - \frac{\beta}{\beta - \alpha} (e^{-\alpha t} - e^{-\beta t}) \right] \qquad \text{for } \alpha \neq \beta \\ &= u(t)[1 - e^{-\beta t} - \beta t e^{-\beta t}] \qquad \text{for } \alpha = \beta \end{aligned}$$

The corresponding density function is given by

$$\begin{aligned} f_Z(t) &= u(t) \frac{\alpha\beta}{\beta - \alpha} (e^{-\alpha t} - e^{-\beta t}) \qquad \text{for } \alpha \neq \beta \\ &= u(t)\beta^2 t e^{-\beta t} \qquad \text{for } \alpha = \beta \end{aligned}$$

It is interesting to compare the reliability for the standby-redundancy case and the parallel case in which both subsystems operate simultaneously. For the former, we have $R = 1 - P(F) = P(X + Y \geq t)$. Now for the first subsystem we have $R_1 = P(X \geq t)$, and for the second subsystem we have $R_2 = P(Y \geq t)$. The reliability for parallel operation is $R_p = P(X \geq t \text{ or } Y \geq t)$.

The event $\{X \geq t\} \cup \{Y \geq t\}$ implies the event $\{X + Y \geq t\}$. Thus $R_p \leq R$, by property (P6). We cannot say, however, that the second event implies the first. We may have $X(\xi) = 2t/3$ and $Y(\xi) = 2t/3$, for example. Figure 3-9-8 shows plots of the density functions for the case $\alpha = \beta$. The density functions $f_X(\cdot) = f_Y(\cdot)$ for the subsystems begin at value β for $t = 0$ and drop to $\beta/e = 0.37\beta$ at $t = 1/\beta$. The

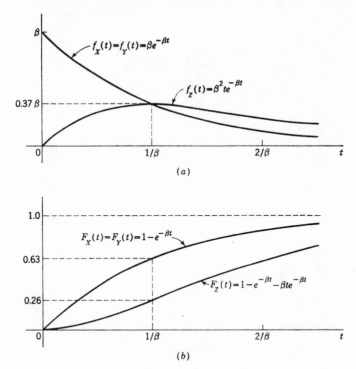

Fig. 3-9-8   Density and distribution functions for the standby-redundancy system of Example 3-9-13.

density function for the sum increases to a maximum value of $\beta/e = 0.37\beta$ at $t = 1/\beta$. The distribution-function curves, which at any time t give the probability of failure on or before that time, are shown in Fig. 3-9-8b. At $t = 1/\beta$, the probability of either subsystem having failed is $1 - 1/e = 0.63$. The probability that the standby system has failed is $1 - 2/e = 0.26$. The probability that the parallel system has failed is the product of the probabilities that either of the two subsystems has failed; this is $(1 - 1/e)^2 = 1 - 2/e + 1/e^2 = 0.26 + 0.14 = 0.40$. ∎

Example 3-9-14

If $X(\cdot)$ and $Y(\cdot)$ are independent and both are uniformly distributed over the interval $[a, b]$, the joint distribution is that shown in Fig. 3-9-9a. Use of the methods already discussed in this section shows that the sum $Z(\cdot) = X(\cdot) + Y(\cdot)$ is distributed according to the curves shown in Fig. 3-9-9a and b. The difference $W(\cdot) = X(\cdot) - Y(\cdot)$ has distribution function and density function whose graphs are identical in shape but which are symmetrical about $v = 0$, with the probability mass in the interval $[a - b, b - a]$. Note that $a < b$. ∎

As an application of the result of Example 3-9-14, consider the following situation:

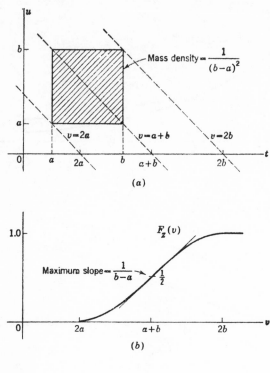

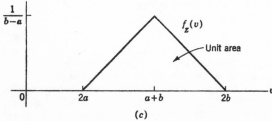

Fig. 3-9-9 Distribution and density for the sum of two uniformly distributed random variables. (*a*) Joint distribution; (*b*) distribution function for sum; (*c*) density function for sum.

Example 3-9-15

In the manufacture of an electric circuit, it is necessary to have a pair of resistors matched to within 0.05 ohm. The resistors are selected from a lot in which the values are uniformly distributed between $R_0 - 0.05$ ohms and $R_0 + 0.05$ ohms. Two resistors are chosen. What is the probability of a satisfactory match?

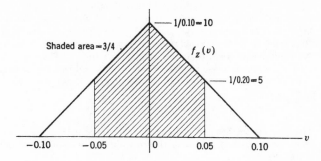

Fig. 3-9-10 Density function for the difference in resistor values in Example 3-9-15.

SOLUTION Let $X(\xi)$ be the value of the first resistor chosen and $Y(\xi)$ be the value of the second resistor chosen. Let $W(\cdot) = X(\cdot) - Y(\cdot)$. The event of a satisfactory match is $\{\xi: -0.05 \leq W(\xi) \leq 0.05\}$. By Example 3-9-14, the density function for $W(\cdot)$ is that given in Fig. 3-9-10. The desired probability is equal to the shaded areas shown on that figure, which is equal to 1 minus the unshaded area in the triangle. Simple geometry shows this to be $1 - 0.05/0.20 = 0.75$. ∎

The following example of the breaking strength of a chain serves as an important model for certain types of systems in reliability theory. Such a system is a type of series system, which fails if any subsystem fails. We discuss the system in terms of a chain, but analogs may be visualized readily. For example, the links in the chain might be "identical" electronic-circuit units in a register of a digital computer. The system fails if any one of the units fails. Each unit is subjected to the same overvoltage, due to a variation in power-supply voltage. Being "identical" units, each unit has the same probability distribution for failure as a function of voltage.

Example 3-9-16 *Chain Model*

Consider a chain with n links manufactured "the same." The same stress is applied to all links. What is the probability of failure? We let $X_i(\cdot)$ be the breaking strength of the ith link and let $Y(\cdot)$ be the applied stress. We suppose these are all random variables whose values are nonnegative (the chain does not have compressive strength, only tensile strength). We assume $\{X_i(\cdot): 1 \leq i \leq n\}$ is an independent class, all members of which have the same distribution. We let $W(\cdot)$ be the breaking strength of the n-link chain. Then

$$W(\cdot) = \min \left[X_1(\cdot), X_2(\cdot), \ldots, X_n(\cdot)\right]$$

Now

$$P(W > v) = P(X_1 > v, X_2 > v, \ldots, X_n > v)$$

$$= P(\bigcap_{i=1}^{n} \{X_i > v\}) = \prod_{i=1}^{n} P(X_i > v)$$

$$= [1 - F_X(v)]^n$$

where $F_X(\cdot)$ is the common distribution function for the $X_i(\cdot)$. From this it follows that $F_W(v) = P(W \leq v) = 1 - [1 - F_X(v)]^n$. Note that $F_W(v) = 0$ for $v < 0$. The problem is to determine the probability that the breaking strength $W(\cdot)$ is greater than the value of the applied stress $Y(\cdot)$; that is, it is desired to determine $P(W > Y)$. Now this is equivalent to determining $P(Z > 0) = 1 - F_Z(0)$, where $Z(\cdot) = W(\cdot) - Y(\cdot)$. According to the result of Example 3-9-12 if we suppose the breaking strength $W(\cdot)$ and the applied stress $Y(\cdot)$ to be independent random variables, we have

$$F_Z(v) = \int_{-\infty}^{\infty} F_W(v + u) f_Y(u) \, du = \int_0^{\infty} F_W(v + u) f_Y(u) \, du$$

The limits in the last integral are based on the fact that $f_Y(u)$ is zero for negative u. Since the integral of $f_Y(\cdot)$ over the positive real line must have the value 1, we may write

$$1 - F_Z(v) = \int_0^{\infty} [1 - F_W(v + u)] f_Y(u) \, du$$
$$= \int_0^{\infty} [1 - F_X(v + u)]^n f_Y(u) \, du$$

On putting $v = 0$, we have

$$P(W > Y) = 1 - F_Z(0) = \int_0^{\infty} [1 - F_X(u)]^n f_Y(u) \, du$$

The problem is determined when the common distribution for the $X_i(\cdot)$ and the distribution for $Y(\cdot)$ are known. Let us suppose once more (Example 3-9-13) that the strength at failure is distributed exponentially; that is, we suppose $F_X(u) = u(u)[1 - e^{-\alpha u}]$. Then $1 - F_X(u) = e^{-\alpha u}$ for $u > 0$. We thus have

$$P(W > Y) = \int_0^{\infty} e^{-n\alpha u} f_Y(u) \, du$$

If $Y(\cdot)$ is distributed uniformly over the interval $[0, f_0]$, it is easy to show that

$$P(W > Y) = \frac{1}{f_0 \alpha n} [1 - e^{-f_0 \alpha n}] \quad \blacksquare$$

It may be noted that the integral expression for $P(W > Y)$ is the Laplace transform of the density function $f_Y(\cdot)$, evaluated for the parameter $s = \alpha n$. Tables of the Laplace transform may be utilized to determine the probability, once $f_Y(\cdot)$ is known.

We consider one more function of two random variables.

Example 3-9-17

Suppose $X(\cdot)$ and $Y(\cdot)$ are absolutely continuous. Determine the density function for the random variable $Z(\cdot) = X(\cdot)Y(\cdot)$. We have $h(t, u) = tu$, and the region Q_v is that shown in Fig. 3-9-4c and d. The most difficult part of the problem is to determine the limits of integration in the expression $F_Z(v) = \iint\limits_{Q_v} f_{XY}(t, u) \, dt \, du$. It is necessary to divide the problem into two parts, one for $v > 0$ and one for $v < 0$. In the first case, examination of the region Q_v in Fig. 3-9-4c shows that the proper limits are given in the following expression:

$$F_Z(v) = \int_0^{\infty} \left[\int_{-\infty}^{v/u} f_{XY}(t, u) \, dt \right] du + \int_{-\infty}^0 \left[\int_{v/u}^{\infty} f_{XY}(t, u) \, dt \right] du$$

Differentiating with the aid of rules for differentiation with respect to limits of integration gives

$$f_Z(v) = \int_0^\infty \frac{1}{u} f_{XY}\left(\frac{v}{u}, u\right) du - \int_{-\infty}^0 \frac{1}{u} f_{XY}\left(\frac{v}{u}, u\right) du$$

Making use of the fact that in the second integral $u < 0$, we may combine this into a single-integral expression.

$$f_Z(v) = \int_{-\infty}^\infty \frac{1}{|u|} f_{XY}\left(\frac{v}{u}, u\right) du$$

For the case $v < 0$, the regions are different, but an examination shows that the limits have the same formulas. Thus, the same formula is derived for $f_Z(\cdot)$ in either case. ∎

Many other such formulas may be developed. Some of these are included in the problems at the end of this chapter. The strategy is simple. The difficulty comes in handling the details. Although there is no way to avoid some of these difficulties, a clear grasp of the task to be performed often makes it possible to discover how special features of the problem at hand may be exploited to simplify analysis and computations. Extensions to functions of higher numbers of random variables are equally simple, in principle, but generally difficult to carry out.

3-10 Almost-sure relationships

We have had occasion to note that $P(E) = 0$ does not imply that E is impossible. But for practical purposes, and in so far as probability calculations are concerned, such an event E is "almost impossible." Similarly, $P(A) = 1$ does not imply that A is the sure event S. But A is "almost sure," and for many purposes we need not distinguish between such an almost-sure event and the sure event. If two events have the same elements except possibly for a set whose total probability mass is zero, these could be considered essentially the same for many purposes in probability theory. And two random variables which have the same values for all elementary outcomes except for possibly a set of ξ whose total probability mass is zero could be considered essentially the same. In this section, we wish to examine and formalize such relationships.

Events and classes of events equal with probability 1

We begin by considering two events.

Definition 3-10a

Two *events* A and B are said to be *equal with probability 1*, designated in symbols by

$$A = B \ [P]$$

iffi

$$P(A) = P(B) = P(AB)$$

We also say A and B are *almost surely equal.*

This condition could have been stated in any of several equivalent ways, as the following theorem shows.

Theorem 3-10A

$A = B$ [P] iffi any one of the following conditions holds:

1. $P(AB^c \cup A^cB) = P(AB^c) + P(A^cB) = 0$
2. $P(A \cup B) = P(AB)$
3. $A^c = B^c$ [P]

PROOF   Conditions 1 and 2 follow from the fact that

$$P(A) = P(AB) + P(AB^c) \quad P(B) = P(AB) + P(A^cB)$$
$$P(A \cup B) = P(AB) + P(A^cB) + P(AB^c)$$

Condition 3 follows from condition 2 and the fact that

$$P(A^cB^c) = 1 - P(A \cup B) = 1 - P(AB) = P(A^c \cup B^c)$$

The condition of equality with probability 1 is illustrated in the Venn diagrams of Fig. 3-10-1. Figure 3-10-1a shows two sets A and B with the total probability mass in set $A \cup B$ actually located entirely in the common part AB. Figure 3-10-1b shows a case of probability mass concentrated at discrete points. Any two sets A and B which contain the same mass points must be equal with probability 1.

This concept may be extended to classes of events as follows:

Definition 3-10b

Two classes of events $\mathcal{A}$ and $\mathcal{B}$ are said to be *equal with probability 1* or *almost surely equal*, designated $\mathcal{A} = \mathcal{B}$ [P], iffi their members may be put into a one-to-one correspondence such that $A_i = B_i$ [P] for each corresponding pair.

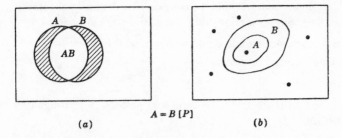

$$A = B \text{ [}P\text{]}$$

(a)   (b)

Fig. 3-10-1   Events equal with probability 1. (*a*) Probability mass in the shaded region is zero; (*b*) probability mass concentrated at discrete points.

We note immediately the following theorem, whose proof is given in Appendix D-1.

Theorem 3-10B

Let $\mathfrak{a}$ and $\mathfrak{B}$ be countable classes such that $\mathfrak{a} = \mathfrak{B} \ [P]$. Then the following three conditions must hold:

1. $\displaystyle\bigcup_{i \in J_1} A_i = \bigcup_{i \in J_1} B_i \ [P]$ for every $J_1 \subset J$

2. $\displaystyle\bigcap_{i \in J_2} A_i = \bigcap_{i \in J_2} B_1 \ [P]$ for every $J_2 \subset J$

3. $A_i{}^c = B_i{}^c \ [P]$ for each $i \in J$

This theorem shows that calculations of the probabilities of events involving the usual combinations of member events of the class $\mathfrak{a}$ are not changed if any of the A_i are replaced by corresponding B_i. It is easy to show, also, that if $\mathfrak{a}$ is an independent class, so also is $\mathfrak{B}$, since the product rule for the latter is a ready consequence of the product rule for the class $\mathfrak{a}$.

We have seen that partitions play an important role in probability theory. A finite or countably infinite partition has the properties

$$(1) \ \sum_{i \in J} P(A_i) = 1 \quad\quad (2) \ P(A_i A_j) = 0 \quad\quad \text{for } i \neq j$$

Now it may be that the class $\mathfrak{a}$ is not a partition, even though it has the two properties listed above. From the point of view of probability calculations, however, it has the essential character of a partition. In fact, we can make the following assertion, which is proved in Appendix D-1.

Theorem 3-10C

If a countable class $\mathfrak{a}$ of events has the properties 1 and 2 noted above, then there exists a partition $\mathfrak{B}$ such that $\mathfrak{a} = \mathfrak{B} \ [P]$.

Random variables

The concept of almost-sure equality extends readily to random variables as follows:

Definition 3-10c

Two random variables $X(\cdot)$ and $Y(\cdot)$ are said to be equal with probability 1, designated $X(\cdot) = Y(\cdot) \ [P]$, iffi the elements ξ for which they differ all belong to a set D having zero probability. We also say in this case that $X(\cdot)$ and $Y(\cdot)$ are *almost surely equal*.

This means that if we rule out the set of ξ on which $X(\cdot)$ and $Y(\cdot)$ differ, we rule out a set which has zero probability mass. Over any other ξ set the two random variables have the same values; hence any point ξ in such a set must be mapped into the same t by both $X(\cdot)$ and $Y(\cdot)$. The mass picture would make it appear that two random variables that are almost surely equal must induce the same mass distribution on the real line.

The case of simple random variables is visualized easily. If two simple random variables are equal with probability 1, they have essentially the same range T and assign the same probability mass to each $t_i \in T$. If there are values of either variable which do not lie in the common range T, these points must be assigned 0 probability. Thus, for practical purposes, these do not represent values of the function to be encountered. These statements may be sharpened to give the following theorem, whose proof is found in Appendix D-2.

Theorem 3-10D

Consider two simple random variables $X(\cdot)$ and $Y(\cdot)$ with ranges T_1 and T_2, respectively. Let $T = T_1 \cup T_2$, with $t_i \in T$. Put $A_i = \{\xi : X(\xi) = t_i\}$ and $B_i = \{\xi : Y(\xi) = t_i\}$. Then $X(\cdot) = Y(\cdot)$ $[P]$ iffi $A_i = B_i$ $[P]$ for each i.

For the general case, we have the following theorem, which is also proved in Appendix D-2.

Theorem 3-10E

Two random variables $X(\cdot)$ and $Y(\cdot)$ are equal with probability 1 iffi $X^{-1}(M) = Y^{-1}(M)$ $[P]$ for each Borel set M on the real line.

This theorem shows that equality with probability 1 requires $P_X(M) = P_Y(M)$ for any Borel set M, so that two almost surely equal random variables $X(\cdot)$ and $Y(\cdot)$ induce the same probability mass distribution on the real line. The converse does not follow, however. We have, in fact, illustrated in Example 3-2-3 that two quite different random variables may induce the same mass distribution.

In dealing with random variables, it is convenient to extend the ideas discussed above to various types of properties and relationships. We therefore make the following

Definition 3-10d

A *property* of a random variable or a *relationship* between two or more random variables is said to *hold with probability 1* (indicated by the symbol $[P]$ after the appropriate expression) iffi the elements ξ for which the property or relationship fails to hold belong to a set D having 0 probability.   In this case we may also say that the property or the relationship *holds almost surely.*

A property or relationship is said to *hold with probability 1 on (event) E* (indicated by *"[P] on E"*) iffi the points of E for which the property or relationship fails to hold belong to a set having 0 probability.   We also use the expression *"almost surely on E."*

Thus we may say $X(\cdot) = 0$ $[P]$, $X(\cdot) \leq Y(\cdot)$ $[P]$ on E, etc.   Other examples of this usage appear in later discussions, particularly in Chaps. 4 and 6.

Problems

3-1. Suppose $X(\cdot)$ is a simple random variable given by

$$X(\cdot) = I_A(\cdot) - 4I_B(\cdot) + 2I_C(\cdot) + 5I_D(\cdot)$$

where $\{A, B, C, D\}$ is a partition of the whole space S.
   (a) What is the range of $X(\cdot)$?
   (b) Express in terms of members of the partition the set $X^{-1}(M)$, where
   (1) $M = (0, 1)$ i.e., the interval $0 < t < 1$
   (2) $M = \{-1, 3, 5\}$
   (3) $M = (-\infty, 4]$ i.e., the interval $-\infty < t \leq 4$
   (4) $M = (2, \infty)$ i.e., the interval $2 < t < \infty$

3-2. Consider the function $X(\cdot) = I_A(\cdot) + 3I_B(\cdot) - 4I_C(\cdot)$.   The class $\{A, B, C\}$ is a partition of the whole space S.
   (a) What is the range of $X(\cdot)$?
   (b) What is the inverse image $X^{-1}(M)$ when
   (1) $M = (-\infty, 3]$
   (2) $M = (1, 4]$ ANSWER: B
   (3) $M = (2, 5)^c$

3-3. Suppose $X(\cdot)$ is a random variable.   For each real t let $E_t = \{\xi: X(\xi) \leq t\}$. Express the following sets in terms of sets of the form E_t for appropriate values of t.

   (1) $\{\xi: X(\xi) < a\}$ ANSWER: $\displaystyle\bigcup_{n=1}^{\infty} E_{a-\frac{1}{n}}$

   (2) $\{\xi: X(\xi) \geq a\}$
   (3) $\{\xi: X(\xi) \in [a, b)\}$

   (4) $\{\xi: X(\xi) \in (a, b)\}$ ANSWER: $E_a{}^c\left[\displaystyle\bigcup_{n=1}^{\infty} E_{b-\frac{1}{n}}\right]$

3-4. Consider the random variable

$$X(\cdot) = -4I_A(\cdot) + I_C(\cdot) + 3I_D(\cdot) + 5I_E(\cdot)$$

where $\{A, C, D, E\}$ is a disjoint class whose union is B^c.   Suppose $P(A) = 0.1$, $P(B) = 0.2$, $P(C) = 0.2$, and $P(D) = 0.2$.   Show the probability mass distribution produced on the real line by the mapping $t = X(\xi)$.
   ANSWER: Probability masses 0.1, 0.2, 0.2, 0.2, 0.3 at $t = -4, 0, 1, 3, 5$, respectively

3-5. Suppose $X(\cdot)$ is a simple random variable given by

$$X(\cdot) = 3I_A(\cdot) - 2I_B(\cdot) - I_C(\cdot)$$

where $\{A, B, C\}$ is a class which generates a partition, none of whose minterms are empty.

(a) Determine the range of $X(\cdot)$.
(b) Express the function in canonical form.
(c) Express the function in reduced canonical form.

3-6. A man stands in a certain position (which we may call the origin). He tosses a coin. If a head appears, he moves one unit to the left. If a tail appears, he moves one unit to the right.

(a) After 10 tosses of the coin, what are his possible positions and what are the probabilities?

(b) Show that the distance at the end of 10 trials is given by the random variable

$$X(\cdot) = \sum_{i=1}^{10} I_{A_i{}^c}(\cdot) - \sum_{i=1}^{10} I_{A_i}(\cdot)$$

where the distance to the left is considered negative. A_i is the event that a head appears on the ith trial. Make the usual assumption concerning coin-flipping experiments.

ANSWER: $t = 2r - 10$, where $0 \leq r \leq 10$ is the number of tails; $P(X = t) = C_r{}^{10}2^{-10}$

3-7. The random variable $X(\cdot)$ has a distribution function $F_X(\cdot)$, which is a step function with jumps of $\frac{1}{4}$ at $t = 0$, $\frac{3}{8}$ at $t = 1$, $\frac{1}{4}$ at $t = 2$, and $\frac{1}{8}$ at $t = 3$.

(a) Sketch the mass distribution produced by the variable $X(\cdot)$.
(b) Determine $P(1 \leq X \leq 2)$, $P(X > 1.5)$.   ANSWER: $\frac{3}{8} + \frac{1}{4}$, $\frac{1}{4} + \frac{1}{8}$

3-8. Suppose a random variable $X(\cdot)$ has distribution function

$$F_X(t) = \sum_{k=0}^{12} p_k u_+(t - k) \qquad \text{with} \quad \sum_{k=0}^{12} p_k = 1$$

In terms of $p_0, p_1, \ldots, p_{12}$, express the probabilities:

(a) $P(X = 12)$ (d) $P(X < 12)$ (g) $P(X < 2)$
(b) $P(X \leq 12)$ (e) $P(X > 12)$ (h) $P(X \leq 15)$
(c) $P(X \geq 12)$ (f) $P(X < 0)$ (i) $P(X > 2)$

ANSWERS: (a) p_{12} (b) 1   (c) p_{12} (d) $1 - p_{12}$

3-9. An experiment consists of a sequence of tosses of an honest coin (i.e., to each elementary event corresponds an infinite sequence of heads and tails). Let A_k be the event that a head appears for the first time at the kth toss in a sequence, and let H_k be the event of a head at the kth toss. Suppose the H_k form an independent class with $P(H_k) = \frac{1}{2}$ for each k. For a given sequence corresponding to the elementary outcome ξ, let $X(\xi)$ be the number of the toss in the sequence for which the first head appears.

(a) Express $X(\cdot)$ in terms of indicator functions for the A_k.
(b) Determine the distribution function $F_X(\cdot)$.

ANSWER: $F_X(\cdot)$ has jump $1/2^k$ at $t = k$, $k = 1, 2, \ldots$.

3-10. A game is played consisting of n successive trials by a single player. The outcome of each trial in the sequence is denoted a success or a failure. The outcome at each trial is independent of all others, and there is a probability p of success. A success, or a win, adds an amount a to the player's account, and a failure, or loss, subtracts an amount b from the player's account.

(a) Let A_k be the event of a success, or win, on the kth trial. Let $X_n(\xi)$ be the net winnings after n trials. Write a suitable expression for $X_n(\cdot)$ in terms of the indicator functions for the events A_k and $A_k{}^c$.

(b) Suppose $n = 4$, $p = \frac{1}{4}$, $a = 3$, and $b = 1$. Plot the distribution function $F_n(\cdot)$ for $X_n(\cdot)$.

3-11. For each of the six functions $F_X(\cdot)$ listed below

(a) Verify that $F_X(\cdot)$ is a probability distribution function. Sketch the graph of the function.

(b) If the distribution is discrete, determine the probability mass distribution; if the distribution is absolutely continuous, determine the density function $f_X(\cdot)$ and sketch its graph.

Note: Where formulas are given over a finite range, assume $F_X(t) = 0$ for t to the left of this range and $F_X(t) = 1$ to the right of this range.

(1) $F_X(t) = t^2/4$ for $0 \leq t \leq 2$

(2) $F_X(t) = \begin{cases} \frac{1}{2}e^t & \text{for } t < 0 \\ \frac{1}{2} & \text{for } 0 \leq t \leq 1 \\ 1 - \frac{1}{2}e^{-(t-1)} & \text{for } t > 1 \end{cases}$

(3) $F_X(t) = 1 - a^{n+1}$ for $n \leq t < n + 1, n = 0, 1, 2, \ldots, 0 < a < 1$

(4) $F_X(t) = \frac{1}{2}(1 - \cos at)$ for $0 \leq t \leq \dfrac{\pi}{a}, a > 0$

(5) $F_X(t) = 1 - \dfrac{1}{(1 + t/2)^2}$ for $0 \leq t$

(6) $F_X(t) = 1 - e^{-at}$ for $0 \leq t, a > 0$

3-12. A random variable $X(\cdot)$ has a density function $f_X(\cdot)$ described as follows: it is zero for $t < 1$; it rises linearly between $t = 1$ and $t = 2$ to the value $\frac{1}{3}$; it remains constant for $2 < t < 4$; it drops linearly to zero between $t = 4$ and $t = 5$.

(a) Plot the distribution function $F_X(\cdot)$.

(b) Determine the probability $P(1.5 \leq X < 3)$.   ANSWER: $P = \frac{11}{24}$

3-13. A random variable $X(\cdot)$ has a density function $f_X(\cdot)$ described as follows: it is zero for $t < 1$; it has the value $\frac{1}{4}$ for $1 < t < 4$; it drops linearly to zero between $t = 4$ and $t = 6$.

(a) Plot the distribution function $F_X(\cdot)$.

(b) Determine $P(2 < X \leq 4.5)$.   ANSWER: $P = \frac{39}{64}$

3-14. A random variable $X(\cdot)$ has density function $f_X(\cdot)$ given by

$$f_X(t) = \begin{cases} \alpha t & 0 \leq t < 0.5 \\ \alpha(1 - t) & 0.5 \leq t < 1 \\ 0 & \text{otherwise} \end{cases}$$

Let A be the event $X < 0.5$, B be the event $X > 0.5$, and C be the event $0.25 < X < 0.75$.

(a) Find the value of α to make $f_X(\cdot)$ a probability density function.

(b) Find $P(A)$, $P(B)$, $P(C)$, and $P(A|B)$.

(c) Are A and C independent events? Why?

      ANSWER: A and C are independent

3-15. The density function of a continuous random variable $X(\cdot)$ is proportional to $t(1 - t)$ for $0 < t < 1$ and is zero elsewhere.

(a) Determine $f_X(t)$.

(b) Find the distribution function $F_X(t)$.

(c) Determine $P(X < \frac{1}{2})$.

3-16. Let $X(\cdot)$ be a random variable with uniform distribution between 10 and 20. A random sample of size 5 is chosen. From this, a single value is chosen at random. What is the probability that the final choice results in a value between 10 and 12? *Interpretative note:* Let E_j be the event that exactly j of the five values in the sample lie between 10 and 12. Let C be the event that the final value chosen has the appropriate magnitude. The selection of a random sample means that if $X_k(\cdot)$ is the kth value in the sample, the class $\{X_k(\cdot): 1 \le k \le 5\}$ is a class of independent random variables, each with the same distribution as $X(\cdot)$. The selection of one value from the sample of five at random means that $P(C|E_j) = j/5$.   ANSWER: $P(C) = 0.20$

3-17. The distribution functions listed below are for mixed probability distributions. For each function

(a) Sketch the graph of the function.

(b) Determine the point mass distribution for the discrete part.

(c) Determine the density function for the absolutely continuous part.

$$(1) \quad F_X(t) = \begin{cases} 0.4(t + 1) & \text{for } -1 \le t < 0 \\ 0.6 + 0.4t & \text{for } 0 \le t < 1 \end{cases}$$

$$(2) \quad F_X(t) = \begin{cases} \frac{1}{4}e^{2t} & \text{for } -\infty < t < 0 \\ 1 - \frac{1}{4}e^{-2t} & \text{for } 0 \le t < \infty \end{cases}$$

$$(3) \quad F_X(t) = \begin{cases} \frac{1}{2}u_+(t + 1) & \text{for } t < 0 \\ \dfrac{1}{\sqrt{2\pi}} \displaystyle\int_{-\infty}^{t} e^{-u^2/2}\, du & \text{for } t \ge 0 \end{cases}$$

3-18. A recording pen is recording a signal which has the following characteristics. If the signal is observed at a time chosen at random, the observed value is a random variable $X(\cdot)$ which has a gaussian distribution with $\mu = 0$ and $\sigma = 4$. The recorder will follow the signal faithfully if the value lies between -10 and 10. If the signal is more negative than -10, the pen stops at -10; if the signal is more positive than 10, the pen stops at 10. Let $Y(\cdot)$ be the random variable whose value is the position of the recorder pen at the arbitrary time of observation. What is the probability distribution function for $Y(\cdot)$? Sketch a graph of the function. Determine the point mass distribution and the density for the absolutely continuous part.

3-19. A truck makes a run of 450 miles at an essentially constant speed of 50 miles per hour, except for two stops of 30 minutes each. The first stop is at 200 miles, and the second is at 350 miles. A radio phone call is made to the driver at a time chosen at random in the period 0 to 10 hours. Let $X(\cdot)$ be the distance the truck has traveled at the time of the call. What is the distribution function $F_X(\cdot)$? ANSWER: Point masses 0.05 at $t = 200, 350$. $F_{Xc}(\cdot)$ has constant slope $0 \le t \le 450$

3-20. Two random variables $X(\cdot)$ and $Y(\cdot)$ produce a joint mass distribution under the mapping $(t, u) = [X, Y](\xi)$ which is uniform over the rectangle $1 \le t \le 2$, $0 \le u \le 2$.

(a) Describe the marginal mass distributions for $X(\cdot)$ and $Y(\cdot)$.

(b) Determine $P(X \le 1.5)$, $P(1 < Y \le 1.6)$, $P(1.1 \le X \le 1.2, 0 \le Y < 1)$.

ANSWER: 0.5, 0.3, 0.05

3-21. Two random variables $X(\cdot)$ and $Y(\cdot)$ produce a joint mass distribution under the mapping $(t, u) = [X, Y](\xi)$ which may be described as follows: (1) $\frac{3}{4}$ of the probability mass is distributed uniformly over the triangle with vertices $(0, 0)$, $(2, 0)$, and $(0, 2)$, and (2) a mass of $\frac{1}{4}$ is concentrated at the point $(1, 1)$.

(a) Describe the marginal mass distributions.

(b) Determine $P(X > 1)$, $P(-3 < Y \leq 1)$, $P(X = 1, Y \geq 1)$, $P(X = 1, Y < 1)$.

ANSWER: $\frac{3}{16}$, $1\frac{3}{16}$, $\frac{1}{4}$, 0

3-22. For two discrete random variables $X(\cdot)$ and $Y(\cdot)$, let

$$p(i, j) = P(X = t_i, Y = u_j)$$

have the following values:

$$p(1, 1) = 0.1 \qquad p(2, 1) = 0.2 \qquad p(3, 1) = 0.2$$
$$p(1, 2) = 0.3 \qquad p(2, 2) = 0.1 \qquad p(3, 2) = 0.1$$

If $t_1 = -1$, $t_2 = 1$, $t_3 = 2$, $u_1 = -2$, $u_2 = 2$, plot the joint distribution function $F_{XY}(t, u)$ by giving the values in each appropriate region of the t, u plane.

3-23. Let $X(\cdot)$ and $Y(\cdot)$ be two discrete random variables. $X(\cdot)$ has range $t_i = i - 3$, $i = 1, 2, 3, 4, 5$, and $Y(\cdot)$ has range $u_j = j - 1$ for $j = 1, 2, 3$. Values of the joint probabilities $p(i, j)$ are given as follows:

	$i = 1$	$i = 2$	$i = 3$	$i = 4$	$i = 5$
$j = 1$	0.1	0.1	0.05	0.15	0.1
$j = 2$	0.05	0.05	0.05	0.15	0
$j = 3$	0.05	0.05	0	0.1	0

(a) Determine the marginal probabilities, and show the joint and marginal mass distributions on the plane and on the coordinate lines.

(b) Show values of the joint distribution function $F_{XY}(t, u)$ by indicating values on the appropriate regions of the plane.

3-24. Two random variables $X(\cdot)$ and $Y(\cdot)$ are said to have a joint gaussian (or normal) distribution iffi the joint density function is of the form

$$f_{XY}(t, u) = \frac{1}{2\pi\sigma_X\sigma_Y \sqrt{1 - \rho^2}} e^{-Q(t, u)}$$

where

$$Q(t, u) = \frac{1}{2(1 - \rho^2)} \left[\left(\frac{t - \mu_X}{\sigma_X} \right)^2 - 2\rho \left(\frac{t - \mu_X}{\sigma_X} \right) \left(\frac{u - \mu_Y}{\sigma_Y} \right) + \left(\frac{u - \mu_Y}{\sigma_Y} \right)^2 \right]$$

and $\sigma_X > 0$, $\sigma_Y > 0$, $|\rho| < 1$, and μ_X, μ_Y are constants which appear as parameters.

Show that $X(\cdot)$ is normal with parameters μ_X and σ_X. Because of the symmetry of the expression, we may then conclude that $Y(\cdot)$ is normal with parameters μ_Y and σ_Y. [*Suggestion:* Let $\varphi(\cdot)$ be defined by $\varphi(x) = (1/\sqrt{2\pi})e^{-x^2/2}$. It is known that

$$\int_{-\infty}^{\infty} \frac{1}{b} \varphi\left(\frac{x - a}{b} \right) dx = 1 \qquad \text{for any } a \text{ and any positive } b$$

Show that

$$f_{XY}(t, u) = \frac{1}{\sigma_X} \varphi \left(\frac{t - \mu_X}{\sigma_X} \right) \cdot \frac{1}{\sigma_Y \sqrt{1 - \rho^2}} \varphi \left(\frac{u - r}{\sigma_Y \sqrt{1 - \rho^2}} \right)$$

where r depends upon ρ, μ_X, μ_Y, σ_X, σ_Y, and t.   Integrate to obtain $f_X(t)$.]

3-25. Let $X(\cdot)$ and $Y(\cdot)$ be two discrete random variables which are stochastically independent.   These variables have the following distributions of possible values:

$$P(X = 1) = \tfrac{1}{3} \qquad P(X = 2) = \tfrac{2}{3} \qquad P(Y = 0) = \tfrac{1}{2} \qquad P(Y = 1) = \tfrac{1}{2}$$

Determine the mass distribution on the plane produced by the mapping

$$(t, u) = [X, Y](\xi)$$

Show the locations and amounts of masses.

3-26. Consider two discrete random variables $X(\cdot)$ and $Y(\cdot)$ which produce a joint mass distribution on the plane.   Let $p(i, j) = P(X = t_i, Y = u_j)$, and suppose the values are

$$\begin{array}{ll} p(1, 1) = 0.2 & p(1, 2) = 0.1 \\ p(2, 1) = 0.1 & p(2, 2) = 0.2 \\ p(3, 1) = 0.1 & p(3, 2) = 0.3 \end{array}$$

(a) Calculate $P(X = t_i)$, $i = 1, 2, 3$, and $P(Y = u_j)$, $j = 1, 2$.

(b) Show whether or not the random variables are independent.

ANSWER: Not independent

3-27. For the random variables in Prob. 3-26, determine the conditional probabilities $P(X = t_i | Y = u_j)$.

3-28. A discrete random variable $X(\cdot)$ has range $(1, 2, 5) = (t_1, t_2, t_3)$, and a discrete random variable $Y(\cdot)$ has range $(0, 2) = (u_1, u_2)$.
Suppose $p(i, j) = P(X = t_i, Y = u_j) = \alpha(i + j)$.

(a) Determine the $p(i, j)$, and show the mass distribution on the plane.

(b) Are the random variables $X(\cdot)$ and $Y(\cdot)$ independent?   Justify your answer.

ANSWER: Not independent

3-29. Two independent random variables $X(\cdot)$ and $Y(\cdot)$ are uniformly distributed between 0 and 10.   What is the probability that simultaneously $1 \le X \le 2$ and $5 \le Y \le 10$?

3-30. Addition modulo 2 is defined by the following addition table:

$$\begin{array}{ll} 0 \oplus 0 = 0 & 0 \oplus 1 = 1 \\ 1 \oplus 0 = 1 & 1 \oplus 1 = 0 \end{array}$$

The disjunctive union $A \oplus B$ of two sets is defined by $A \oplus B = AB^c \uplus A^cB$.

(a) Show that $I_A(\cdot) \oplus I_B(\cdot) = I_{A \oplus B}(\cdot)$, where the $\oplus$ in the left-hand member indicates addition modulo 2 and in the right-hand member indicates disjunctive union.

(b) Express the function $I_A(\cdot) \oplus 1$ in terms of $I_A(\cdot)$ or $I_{A^c}(\cdot)$.

(c) Suppose A and B are independent events with $P(A) = \tfrac{1}{2}$.   Show that $I_B(\cdot)$ and $I_{A \oplus B}(\cdot)$ are independent random variables.   (Note that it is sufficient to show that B and $A \oplus B$ are independent events.)

3-31. An experiment consists in observing the values of n points distributed uniformly and independently in the interval [0, 1]. The n points may be considered to be observed values of n independent random variables, each of which is uniformly distributed in the interval [0, 1]. Let a be a number lying between 0 and 1. What is the probability that among the n points, the point farthest to the right lies to the right of point a?   ANSWER: $1 - a^n$

3-32. The location of 10 points may be considered to be independent random variables, each having the same triangular distribution function. This function rises linearly from $t = 1$ to $t = 2$, then decreases linearly to zero at $t = 3$. The resulting graph is a triangle, symmetric about the value $t = 2$.

(a) What is the probability that all 10 points lie within a distance $\frac{1}{2}$ of the position $t = 2$?   ANSWER: $P = (\frac{3}{4})^{10}$

(b) What is the probability that exactly 3 of the 10 points lie within a distance $\frac{1}{2}$ of the position $t = 2$?

3-33. Random variables $X(\cdot)$ and $Y(\cdot)$ have joint probability density function

$$f_{XY}(t, u) = \begin{cases} \dfrac{\pi^2}{8} \sin \dfrac{\pi}{2}(t + u) & 0 \leq t \leq 1, 0 \leq u \leq 1 \\ 0 & \text{elsewhere} \end{cases}$$

(a) Find the marginal density functions

$$f_X(t) = f_{XY}(t, *) \quad \text{and} \quad f_Y(u) = f_{XY}(*, u)$$

and show whether or not $X(\cdot)$ and $Y(\cdot)$ are independent.
ANSWER: $X(\cdot)$ and $Y(\cdot)$ are not independent.

(b) Calculate $P(X > \frac{3}{4})$.
ANSWER: $P(X > \frac{3}{4}) = \frac{1}{2}[1 - \sin(3\pi/8) + \cos(3\pi/8)] = 0.229$

3-34. If $X(\cdot)$ and $Y(\cdot)$ have a joint normal distribution (Prob. 3-24), show that they are independent iffi the parameter $\rho = 0$.

3-35. Two random variables $X(\cdot)$ and $Y(\cdot)$ produce a joint probability mass distribution which is uniform over the circle of unit radius, center at the origin. Show whether or not the random variables are independent. Justify your answer.
ANSWER: Not independent

3-36. Two random variables $X(\cdot)$ and $Y(\cdot)$ produce a joint probability mass distribution as follows: one-half of the mass is spread uniformly over the rectangle having vertices $(0, 0)$, $(1, 0)$, $(1, 1)$, and $(0, 1)$. A mass of $\frac{1}{4}$ is placed at each of the points $(0.75, 0.25)$ and $(0.25, 0.75)$. Show whether or not the random variables are independent. Justify your answer.

3-37. Suppose $X(\cdot)$ and $Y(\cdot)$ have a joint density function $f_{XY}(\cdot, \cdot)$.

(a) Show that $X(\cdot)$ and $Y(\cdot)$ are independent iffi it is possible to express the joint density function as

$$f_{XY}(t, u) = kg(t)h(u) \quad \text{where } k \text{ is a nonzero constant}$$

(b) Show that if $X(\cdot)$ and $Y(\cdot)$ are independent, the region of nonzero density must be the rectangle $M \times N$, where M is the set of t for which $f_X(t) > 0$ and N is the set of u for which $f_Y(u) > 0$.

3-38. Consider the simple random variable $X(\cdot) = -I_A(\cdot) + I_B(\cdot) + 2I_C(\cdot)$. Let m_i be the ith minterm in the partition generated by $\{A, B, C\}$, and let $p_i = P(m_i)$.

Values of these probabilities are

$$p_0 = 0 \qquad p_1 = 0.1 \qquad p_2 = 0.1 \qquad p_3 = 0.2$$
$$p_4 = 0 \qquad p_5 = 0.2 \qquad p_6 = 0.3 \qquad p_7 = 0.1$$

(a) Determine the probability mass distribution produced by the mapping $t = X(\xi)$. Show graphically the locations and amounts of masses.

ANSWER: Masses 0.3, 0.3, 0.2, 0.2 at $t = 0, 1, 2, 3$, respectively

(b) Determine the probability mass distribution produced by the mapping $u = X^2(\xi) + 2$.

3-39. The random variable $X(\cdot)$ is uniformly distributed between 0 and 1. Let $Z(\cdot) = X^2(\cdot)$.

(a) Sketch the distribution function $F_Z(\cdot)$.

(b) Sketch the density function $f_Z(\cdot)$.

3-40. What is the distribution function $F_Y(\cdot)$ in terms of $F_X(\cdot)$ if $Y(\cdot) = -X(\cdot)$? In the continuous case, express the density function $f_Y(\cdot)$ in terms of $f_X(\cdot)$.

3-41. Suppose $X(\cdot)$ is any random variable with distribution function $F_X(\cdot)$. Define a quasi-inverse function $F_X^{-1}(\cdot)$ by letting $F_X^{-1}(u)$ be the smallest t such that $F_X(t) \geq u$. Show that if $X(\cdot)$ is an absolutely continuous random variable, the new random variable $Y(\cdot) = F_X[X(\cdot)]$ is uniformly distributed on the interval $[0, 1]$. (Compare these results with Example 3-9-6.)

3-42. Consider the discrete random variables

$$X(\cdot) \text{ with range } (-2, 0, 1, 3) = (t_1, t_2, t_3, t_4)$$

and

$$Y(\cdot) \text{ with range } (-1, 0, 1) = (u_1, u_2, u_3)$$

Let $p(i, j) = P(X = t_i, Y = u_j)$ be given as follows:

	$j = 1$	$j = 2$	$j = 3$
$i = 1$	0.10	0.05	0.15
$i = 2$	0.02	0.08	0.05
$i = 3$	0	0.10	0.10
$i = 4$	0.23	0.12	0

(a) Sketch to scale graphs for $F_X(\cdot)$ and $F_Y(\cdot)$, and show values thereon.

(b) Let $Z(\cdot) = Y(\cdot) - X(\cdot)$. Sketch to scale the graph for $F_Z(\cdot)$, and show values thereon. ANSWER: $F_Z(\cdot)$ has jumps at $v = -4, -3, -1, 0, 1, 2, 3$ of magnitudes 0.23, 0.12, 0.12, 0.18, 0.15, 0.05, 0.15, respectively.

3-43. A pair of random variables $X(\cdot)$ and $Y(\cdot)$ produce the joint probability mass distribution under the mapping $(t, u) = [X, Y](\xi)$ as follows: mass of $\frac{1}{2}$ is uniformly distributed on the unit square $0 \leq t \leq 1$, $0 \leq u \leq 1$; mass of $\frac{1}{2}$ is uniformly distributed on the vertical line segment $t = \frac{1}{2}$, $0 \leq u \leq 1$. Define the new random variables $Z(\cdot) = Y^2(\cdot)$ and $W(\cdot) = 2X(\cdot)$. Determine the distribution functions $F_X(\cdot)$, $F_Y(\cdot)$, $F_Z(\cdot)$, and $F_W(\cdot)$. Sketch graphs of these functions.

3-44. The joint probability mass distribution induced by the mapping

$$(t, u) = [X, Y](\xi)$$

is described as follows: mass of $\frac{1}{2}$ is distributed uniformly over a square having vertices $(-1, 0)$, $(1, -2)$, $(3, 0)$, and $(1, 2)$; mass of $\frac{1}{8}$ is concentrated at each of the points $(1, 0)$, $(2, 0)$, $(0, 1)$, and $(2, 1)$.

(a) Let $A = \{\xi: X(\xi) \leq 1\}$ and $B = \{\xi: Y(\xi) > 0\}$. Show that A and B are independent events. However, consider the events $A_1 = \{\xi: X(\xi) < 1\}$ and $B_1 = \{\xi: Y(\xi) \geq 0\}$ to show that $X(\cdot)$ and $Y(\cdot)$ are *not* independent random variables.

(b) Let $Z(\cdot) = X(\cdot) - Y(\cdot)$. Determine the distribution function $F_Z(\cdot)$ for the random variable $Z(\cdot)$.

3-45. Random variables $X(\cdot)$ and $Y(\cdot)$ have the joint density functions listed below. For each of these

(a) Obtain the marginal density functions $f_X(\cdot)$ and $f_Y(\cdot)$.

(b) Obtain the density function for the random variable $Z(\cdot) = X(\cdot) + Y(\cdot)$.

(c) Obtain the density function for the random variable $W(\cdot) = X(\cdot) - Y(\cdot)$.

(1) $f_{XY}(t, u) = 4(1 - t)u$ for $0 \leq t \leq 1, 0 \leq u \leq 1$

(2) $f_{XY}(t, u) = 2t$ for $0 \leq t \leq 1, 0 \leq u \leq 1$

(3) $f_{XY}(t, u) = \dfrac{1}{\pi} e^{-(t^2+u^2)}$ for $-\infty < t < \infty, -\infty < u < \infty$

3-46. A pair of random variables $X(\cdot)$ and $Y(\cdot)$ produce a joint mass distribution on the plane which is uniformly distributed over the square whose vertices are at the points $(-1, 0)$, $(1, 0)$, $(0, -1)$, and $(0, 1)$. The mass density is constant over this square and is zero outside. Determine the distribution functions and the density functions for the random variables $X(\cdot)$, $Y(\cdot)$, $Z(\cdot) = X(\cdot) + Y(\cdot)$, and $W(\cdot) = X(\cdot) - Y(\cdot)$.   ANSWER: $f_Z(v) = f_W(v) = \frac{1}{2}$ for $|v| \leq 1$

3-47. On an assembly line, shafts are fitted with bearings. A bearing fits a shaft satisfactorily if the bearing diameter exceeds the shaft diameter by not less than 0.005 inch and not more than 0.035 inch. If $X(\cdot)$ is the shaft diameter and $Y(\cdot)$ is the bearing diameter, we suppose $X(\cdot)$ and $Y(\cdot)$ are independent random variables. Suppose $X(\cdot)$ is uniformly distributed over the interval $[0.74, 0.76]$ and $Y(\cdot)$ is uniformly distributed over $[0.76, 0.78]$. What is the probability that a bearing and a shaft chosen at random from these lots will fit satisfactorily?   ANSWER: $P = \frac{15}{16}$

3-48. Suppose $X(\cdot)$ and $Y(\cdot)$ are independent random variables, uniformly distributed in the interval $[0, 1]$. Determine the distribution function for
$Z(\cdot) = 3X(\cdot) + 2Y(\cdot)$.

3-49. Random variables $X(\cdot)$ and $Y(\cdot)$ are independent. The variable $X(\cdot)$ is uniformly distributed between $(-2, 0)$. The variable $Y(\cdot)$ is distributed uniformly between $(2, 4)$. Determine the density function for the variable

$Z(\cdot) = \frac{1}{2}[X(\cdot) + Y(\cdot)]$

3-50. Let $X(\cdot)$ and $Y(\cdot)$ be independent random variables. Suppose $X(\cdot)$ is uniformly distributed in the interval $[0, 2]$ and $Y(\cdot)$ is uniformly distributed in the interval $[1, 2]$. What is the probability that $Z(\cdot) = X(\cdot)Y(\cdot) \leq 1$?

ANSWER: $\frac{1}{2} \log_e 2$

3-51. Obtain the region Q_v for the function $h(t, u) = t/u$. Show that, under the appropriate conditions, the density function for the random variable $R(\cdot) = X(\cdot)/Y(\cdot)$

is given by

$$f_R(v) = \int_{-\infty}^{\infty} |u| f_{XY}(vu, u) \, du$$

3-52. Suppose A, B are independent events, and suppose $A = A_0$ $[P]$ and $B = B_0$ $[P]$. Show that A_0, B_0 is an independent pair.

3-53. Consider the simple random variable $X(\cdot)$ in canonical form as follows:

$$X(\cdot) = I_A(\cdot) + 2I_B(\cdot) + 3I_C(\cdot)$$

with $P(A) = P(B) = \frac{1}{4}$ and $P(C) = \frac{1}{2}$. Suppose $C = D \uplus E$ with

$$P(D) = P(E) = \frac{1}{4}$$

Construct at least three other simple random variables having the same probability distribution but which differ on a set of positive probability.

Selected references

BRUNK [1964], chaps. 3, 4.   Cited at the end of our Chap. 2.

FISZ [1963], chap. 2.   Cited at the end of our Chap. 2.

GNEDENKO [1962]: "The Theory of Probability," chap. 4 (transl. from the Russian). Although written primarily for the mathematician, the discussions are generally clear and readable.

GOLDBERG [1960], chap. 4, secs. 1, 2.   Cited at the end of our Chap. 1.

LLOYD AND LIPOW [1962]: "Reliability: Management, Methods, and Mathematics," chaps. 6, 9.   Discusses basic mathematical models in reliability engineering in a clear and interesting manner.

MCCORD AND MORONEY [1964], chap. 5.   Cited at the end of our Chap. 2.

PARZEN [1960], chap. 7.   Cited at the end of our Chap. 2.

WADSWORTH AND BRYAN [1960]: "Introduction to Probability and Random Variables," chaps. 3 through 6.   Gives a detailed discussion, with many examples, of a wide variety of useful probability distributions and techniques for handling them.

Handbook

National Bureau of Standards [1964]: "Handbook of Mathematical Functions," chap. 26.   A very useful, moderately priced work which provides an important collection of formulas, properties, relationships, and computing aids and techniques, as well as excellent numerical tables and an extensive bibliography. Much material in other chapters (e.g., combinatorial analysis) adds to the usefulness for the worker in probability.

chapter 4

Sums and integrals

The notion of a probability-weighted average plays an indispensable role in the development of probability theory and in the application of this theory to real-world problems. For a discrete random variable, the probability-weighted average of the values of the variable is expressed simply and naturally as a sum. For other distributions it is necessary to express the averaging operation in terms of integrals, which are generalized sums. The theory of mathematical expectation, which is the topic of the next chapter, deals extensively with such probability-weighted averages in a variety of forms. In this chapter we examine the concept of the integral, in order to extend the notion in a manner that allows a general formulation of the averaging operation. To do this, we must introduce the concept of an integral on an abstract space, upon which a suitable measure (probability measure, in our case) is defined.

Integrals are viewed essentially as sums, since they are defined in terms of limits of approximating sums. Alternative means of forming the approximating sums—the methods of Riemann and of Lebesgue—are examined. The method due to Lebesgue is seen to have the advantage that it can be extended to abstract domains for the integrand; this is done in a manner which makes possible the desired formulation of the probability-weighted average in the general case.

The integral of a simple function (simple random variable) is defined as a sum. Since a general random variable can be represented as the limit of a sequence of simple random variables, it is natural to investigate the possibility of defining the integral of a general random variable as the limit of the integrals of the approximating simple functions. A pair of limit theorems, borrowed from measure theory, ensure the success of such a program. Various properties of integrals are examined in the case of simple functions, then extended immediately to the general case by an appeal to these fundamental limit theorems.

Some attention is given to the Lebesgue-Stieltjes integral, a special case of the abstract integral which is important in the theory of mathematical expectation. These integrals are defined on euclidean spaces of various dimensions. Their relation to ordinary integrals and sums is discussed.

Integrals may be transformed from one space to another by a "change of variable." The manner in which these transformations affect the form of the integral supplies the connection between the various forms of mathematical expectation encountered in practice. Some general results on transformations of integrals, pertinent to this topic, are studied with the aid of the mass-mapping concept.

4-1   Integrals of Riemann and Lebesgue

In this section, we examine the notion of an integral as a generalization of a sum. Integrals are defined in terms of limiting operations applied to a class of approximating sums. The ordinary integral, with which most students of the calculus are familiar, involves a method of forming the sums that is generally credited to Riemann. We shall compare this with an alternative method due to Lebesgue and see that Lebesgue's method has the advantage that it allows an extension of considerable importance for the development of probability theory (as a part of more general measure theory). We shall be concerned principally with fundamental ideas and concepts rather than with details of mathematical justification.

Figure 4-1-1 shows the graph of an ordinary function $f(\cdot)$ and the subdivisions used in forming approximating sums for integrals. The function is supposed to produce a mapping from R_1, with elements t, to R_2, with elements u. For convenience, we call R_1 the *domain space* and R_2 the *range space*. To simplify exposition of the ideas, we suppose $f(\cdot)$ is bounded and that the integration is over a finite interval $a \leq t \leq b$, which may be considered the domain of $f(\cdot)$. On this interval, the values of $f(\cdot)$ range over an interval $\alpha \leq u \leq \beta$, which may be considered the

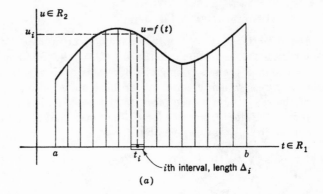

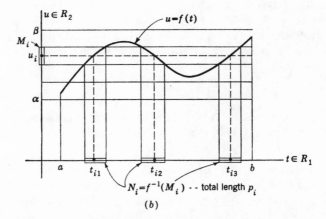

Fig. 4-1-1   Comparison of the methods of Riemann and Lebesgue in forming the partial sums for defining integrals.   (*a*) Riemann: partition of domain; (*b*) Lebesgue: partition of range produces partition of domain.

range of $f(\cdot)$. We consider, now, two ways of forming approximating sums for the integral of $f(\cdot)$ over the prescribed interval.

Riemann (refer to Fig. 4-1-1*a*)

Partition the *domain* [a, b] into nonoverlapping intervals of length Δ_i. The first interval has for its lower limit the value $t = a$; the last interval has as its upper limit the value $t = b$. Let t_i be any value of t in the ith interval, and put $u_i = f(t_i)$. The approximating sum S is given by

$$S = \sum_i u_i \Delta_i \approx \int_a^b f(t)\, dt \qquad \text{(Riemann)}$$

Lebesgue (refer to Fig. 4-1-1*b*)

Partition the *range* [α, β] into nonoverlapping intervals M_i.   This *implies a partition of the domain* into the sets $N_i = f^{-1}(M_i)$.   For Borel functions, for instance, the sets N_i must be Borel sets.   These sets can be assigned a "measure," which for intervals or finite unions of intervals is the length of the set.   Suppose the measure of the set N_i is p_i, and let u_i be any point in M_i [u_i is thus $f(t_i)$ for at least one t_i in N_i].   The approximating sum S, in this case, is given by

$$S = \sum_i u_i p_i \approx \int_a^b f(t)\ dt \qquad \text{(Lebesgue)}$$

For functions with the smooth properties exemplified by the graphs in Fig. 4-1-1, it seems highly plausible that the two sums should be quite close in value for sufficiently fine partitions, and that in the limit, as the partitions are refined, the sums should approach the same value.   This fact has been carefully established in the following theorem (cf. Munroe, [1953, p. 177]).

Theorem 4-1A

If a bounded function has an integral in the sense of Riemann over a finite interval, it also has an integral in the sense of Lebesgue; the two integrals have the same value.

The proof of this theorem is not easy.   We simply state it as a part of our interpretive discussion.   It shows that Lebesgue's form of the integral is at least as general as that of Riemann.   In fact, classic examples are cited in the mathematical literature to show that the Lebesgue integral is defined (in the sense of having a value) for functions for which the Riemann integral does not exist.   Thus the Lebesgue integral provides an extension of the Riemann integral to a broader class of functions.

The Lebesgue form of the integral points to other generalizations which are of considerable interest in developing the theory of probability. Such extensions are suggested by changing the representation of Fig. 4-1-1*b* to that of Fig. 4-1-2.   The domain space R_1 is represented by a copy of the real line; the range space R_2 is represented by a second copy of the real line.   The function $f(\cdot)$ provides a mapping from R_1 to R_2. We suppose the range of $f(\cdot)$ to be the interval $\alpha \leq u \leq \beta$, a subset of R_2. An interval M_i in the range space consists of points which are the images of the points in the domain set $N_i = f^{-1}(M_i)$.   The set N_i is assigned a weight p_i equal to its length (as the concept of length is generalized in terms of *Lebesgue measure*).   The approximating sum is that considered above.

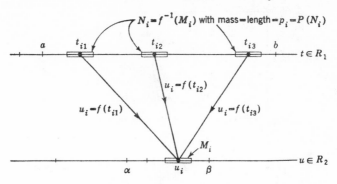

Fig. 4-1-2   Alternative representation of the situation in Fig. 4-1-1*b*.

As a first step in extending the Lebesgue integral, we replace the Lebesgue measure with more general measures defined on the real line. The essential property of the Lebesgue measure (generalized length) for forming the approximating sums is its *additivity* [analogous to property (P3) for probability measures]. This property is shared by all measure functions. In particular, we shall be interested in the probability measures induced on the real line by random variables. These have been interpreted in terms of mass distributions. The Lebesgue measure can also be visualized in terms of a mass distribution. It defines a mass which is uniformly distributed over the whole line (or over any part of it under consideration). It is an easy and natural step to relax the restriction on the type of measure, to allow measures of all sorts.

From this point, it is an easy second step to a full generalization of the Lebesgue integral. The domain space R_1 serves only as a "bearer" of the measure or mass. No geometrical or topological considerations are important for the integral just described. We may deal as well with a measure defined on a suitable class of subsets of an abstract space S. In order to exhibit the relevance of the topic to probability theory, we state the facts in terms of functions which are random variables and of measures which are probability measures. Extensions to more general measures should be apparent.

Suppose $X(\cdot)$ is a random variable defined on a basic space S and having values on the real line R. For simplicity of exposition, we suppose $X(\cdot)$ is bounded, so that the range is contained in some finite interval. We produce a partition of the range by partitioning the finite interval into nonoverlapping intervals M_i, in the manner indicated in Fig. 4-1-3. This, in turn, induces a partition of the basic space into sets $A_i = X^{-1}(M_i)$. We select a value t_i arbitrarily in each of the intervals M_i; we must have $t_i = X(\xi_i)$ for at least one ξ_i in A_i. Now consider

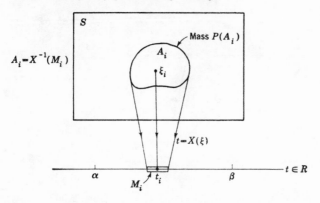

Fig. 4-1-3   Generalization of the situation in Fig. 4-1-2 to abstract spaces with probability (or other measures).

the sum

$$I = \sum_i t_i P(A_i)$$

For fine enough subdivisions of the range set into small intervals, we may expect, in many cases at least, that these sums will approximate a limiting value. If so, the value is called the integral of $X(\cdot)$ with respect to the probability measure $P(\cdot)$. It is designated by the symbol

$$\int X\, dP \qquad \text{or sometimes} \qquad \int X(\xi)\, dP(\xi)$$

Before turning to examine somewhat more systematically the development of the theory of such abstract integrals, it may be well to note several facts. In Riemann's theory of the integral, it is necessary to be able to divide the domain in some direct fashion that reduces the weights (lengths, in the case considered) of the members of the partition. This is possible in the case of the real line because of the geometrical properties of length. In Lebesgue's theory, only the range space (which is the real line) need have this property. Once the range space is partitioned systematically into small intervals, the domain is partitioned by the inverse mapping. The inverse mapping must have the character that a "weight," or value, of a measure may be assigned to the inverse images of the intervals. This measure need have only the additivity property. No geometrical properties of the domain are required. The condition that a value of the measure must be assignable to the inverse images is precisely the *measurability condition* required for random variables (Sec. 3-1 and Appendix C). For random variables, we require that the inverse image of any Borel set (and hence of any interval) must be an event.

An event is a member of that completely additive class $\mathcal{E}$ upon which the probability measure $P(\cdot)$ is defined.

4-2 Integral of a simple random variable

As in the previous section, we shall be concerned primarily with measures that are probability measures and with measurable functions that are random variables. For some purposes, however, we shall need a slightly more general measure defined below. Rather than approach the integral of a general random variable directly, we utilize an elegant approach which begins with a consideration of simple random variables (i.e., random variables characterized by finite sets of values). *Unless specifically noted otherwise, we shall suppose the function $X(\cdot)$ considered in this section to be of the form*

$$X(\cdot) = \sum_{i \in J} t_i I_{A_i}(\cdot)$$

where J is a finite index set, and the sets $A_i = \{\xi \colon X(\xi) = t_i\}$ form a partition. The expression above is thus in the canonical form for the simple random variable $X(\cdot)$ (see Sec. 3-3).

Definition 4-2a

The *integral* of the simple random variable $X(\cdot)$ *with respect to probability measure* $P(\cdot)$ is given by

$$\int X \, dP = \sum_{i \in J} t_i P(A_i)$$

This may be related, as shown in Fig. 4-2-1, to the approximating sums for the general case, described with the aid of Fig. 4-1-3. We suppose the index set J consists of the integers $1, 2, \ldots, n$ and that $t_1 < t_2 < \cdots < t_n$. We let $M_i = [t_i, t_{i+1})$ for $1 \le i \le n - 1$, and we put $M_n = \{t_n\}$, the set consisting of the single point t_n. The mapping $X(\xi) = t$ maps all points of the set A_i into the point t_i. Now $A_i = X^{-1}(M_i)$, in a somewhat trivial sense, since $X^{-1}(M_i) = X^{-1}(\{t_i\})$.

The integral can be given a simple but important interpretation. We rewrite the defining sum as follows:

$$\int X \, dP = \sum_{i \in J} t_i P(X = t_i)$$

Each of the possible values t_i is multiplied by the probability that the function $X(\cdot)$ takes on that value. The sum is thus the probability-weighted sum of the values of the function. Since the total probability mass is unity, this sum is *the probability-weighted average of the values*

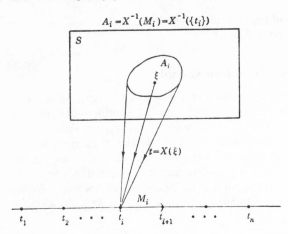

Fig. 4-2-1   Mappings and partitions for the integral of a simple function.

of the random variable.   It is this fact which makes the abstract integral important for the development of the theory of averages or expectations in probability theory.

The definition of the integral (over the whole space S) may be extended to the slightly more general concept of an integral over a given set or event.

Definition 4-2b

Let E be any event, and let $X(\cdot)$ be a simple random variable.   The *integral of $X(\cdot)$ with respect to $P(\cdot)$ over the event E* is

$$\int_E X \, dP = \int I_E X \, dP$$

where $I_E(\cdot)$ is the indicator function for the event E.

This definition actually introduces nothing new into the concept of the integral of a simple random variable.   The function $\varphi_E(\cdot) = I_E(\cdot)X(\cdot)$ is a simple random variable when $X(\cdot)$ is.   We wish to show that the integral of $X(\cdot)$ over E may be expressed by

$$\int_E X \, dP = \sum_{i \in J} t_i P(EA_i)$$

If $t_i \neq 0$, $\varphi_E^{-1}(\{t_i\}) = EA_i$.   For $t_i = 0$, the inverse image must contain all the points mapped into 0 by $\varphi_E(\cdot)$.   Every point in E^c is mapped into 0; so also is every point in $A_0 E$, where $A_0 = \{\xi \colon X(\xi) = 0\}$. We thus have $\varphi_E^{-1}(\{0\}) = E^c \uplus EA_0$. Hence

$$\int_E X \, dP = \sum_{\substack{i \in J \\ t_i \neq 0}} t_i P(EA_i) + 0 \cdot P(E^c \uplus EA_0)$$

Since the coefficient is 0, the quantity $P(E^c \uplus EA_0)$ may be replaced by $P(EA_0)$ without changing the value of the sum. Upon doing this, we have the desired expression.

For some purposes, we need a slightly more general measure and the concept of the integral with respect to it. For these purposes, it is sufficient to make the

Definition 4-2c

A measure $m(\cdot)$ defined on the class $\mathcal{E}$ of events is said to be a *totally finite measure* on $\mathcal{E}$ iffi

$$m(E) = a P_m(E) \qquad \text{for each event } E$$

where $a = m(S)$ is a positive constant, and $P_m(\cdot)$ is a probability measure.

The *integral of a simple random variable* $X(\cdot)$ *with respect to a totally finite measure* $m(\cdot)$ is given by

$$\int X \, dm = m(S) \int X \, dP_m$$

A totally finite measure describes a mass distribution on the basic space; the total mass is a finite constant $a = m(S)$. The associated probability measure $P_m(\cdot)$ is the normalized measure obtained by taking $a = m(S)$ as the unit of mass. It is apparent how arguments for integrals with respect to a probability measure carry over without essential change to the case of a totally finite measure.

We now develop some basic properties of integrals of simple random variables. As we shall see in Sec. 4-4, these properties are also valid in the general case. In the following treatment, we shall assume that the random variables under consideration are all simple. In order to simplify the statements of the properties, we shall leave this assumption unstated.

(I1) *Linearity with respect to the integrand.* If a and b are constants,

$$\int_E (aX + bY) \, dP = a \int_E X \, dP + b \int_E Y \, dP$$

PROOF It is sufficient to show this for $E = S$, since $I_E(\cdot)X(\cdot)$ and $I_E(\cdot)Y(\cdot)$ are simple random variables if $X(\cdot)$ and $Y(\cdot)$ are.

$$X(\cdot) = \sum_i t_i I_{A_i}(\cdot) = \sum_{i,j} t_i I_{A_i B_j}(\cdot)$$

$$Y(\cdot) = \sum_j u_j I_{B_j}(\cdot) = \sum_{i,j} u_j I_{A_i B_j}(\cdot)$$

$$aX(\cdot) + bY(\cdot) = \sum_{i,j} (at_i + bu_j) I_{A_i B_j}(\cdot)$$

$$\int (aX + bY)\, dP = \sum_{i,j} (at_i + bu_j) P(A_i B_j)$$

$$= a \sum_{i,j} t_i P(A_i B_j) + b \sum_{i,j} u_j P(A_i B_j)$$

$$= a \sum_i t_i P(A_i) + b \sum_j u_j P(B_j)$$

$$= a \int X\, dP + b \int Y\, dP \quad \blacksquare$$

(**12**) *Linearity with respect to measure.* If $m(\cdot)$, $m'(\cdot)$, and $m''(\cdot)$ are totally finite measures and a and b are positive constants such that $m(\cdot) = am'(\cdot) + bm''(\cdot)$, then

$$\int_E X\, dm = a \int_E X\, dm' + b \int_E X\, dm''$$

(**13**) *Additivity.* If $\{E_i \colon i \in J\}$ is a finite or countably infinite partition of E, then

$$\int_E X\, dP = \sum_{i \in J} \int_{E_i} X\, dP$$

PROOF   The finite case follows immediately from the fact that

$$I_E(\cdot) = \sum_{i \in J} I_{E_i}(\cdot)$$

For the infinite case, let

$$E = \overset{\infty}{\underset{i=1}{\biguplus}} E_i \qquad A_n = \overset{n}{\underset{i=1}{\biguplus}} E_i \qquad B_n = \overset{\infty}{\underset{i=n+1}{\biguplus}} E_i$$

Then

$$E = \lim_{n \to \infty} A_n \qquad \text{and} \qquad P(E) = \lim_{n \to \infty} P(A_n)$$

Since $P(E) = P(A_n) + P(B_n)$ for each n, it follows that $\lim_{n \to \infty} P(B_n) = 0$. To simplify writing, we abbreviate $\int_E X\, dP$ by $\int_E$, since only the domain of integration is of interest.   From the finite case we have

$$\int_E = \sum_{i=1}^n \int_{E_i} + \int_{B_n}$$

We wish to show

$$\lim \sum_{i=1}^{n} \int_{E_i} = \int_{E}$$

We have

$$\left| \int_{E} - \sum_{i=1}^{n} \int_{E_i} \right| = \left| \int_{B_n} \right|$$

Since there is a number b such that $|X| \leq b$, it is easy to see from the definition of the integral that $\left| \int_{B_n} \right| \leq bP(B_n)$. Thus we can make the difference above as small as desired by choosing n sufficiently large, so that the limit property is established. ■

(I4) $\left| \int_{E} X \, dP \right| \leq \int_{E} |X| \, dP$

PROOF   This follows immediately from a general inequality on sums, which asserts in this case that

$$\left| \sum_{i \in J} t_i P(EA_i) \right| \leq \sum_{i \in J} |t_i| P(EA_i) \quad ■$$

(I5)   If $P(E) = 0$, then $\int_{E} X \, dP = 0$.

PROOF   This follows from the fact that $\left| \int_{E} \right| \leq bP(E)$, as noted in the proof of property (I3). ■

(I6) $(A) \int_{E} X \, dP = 0$ for all E iffi (B) $X(\cdot) = 0$ $[P]$.

$(A') \int_{E} X \, dP = \int_{E} Y \, dP$ for all E iffi (B') $X(\cdot) = Y(\cdot)$ $[P.]$

PROOF   The equivalence of the two formulations of the theorem follows from the linearity property and the fact that $X(\cdot) = Y(\cdot)$ $[P]$ iffi $X(\cdot) - Y(\cdot) = 0$ $[P]$. We argue the first formulation.

1. To show that (B) implies (A).
Let $A = \{\xi: X \neq 0\}$; then $P(A) = 0$.
For $E \subset A^c$, we must have $I_E(\cdot)X(\cdot)$ identically zero, so that the integral is zero. For $E \subset A$ we must have $P(E) = 0$, so that the integral is zero.   For the general case we use the additivity property (I3) and the fact that $E = EA \uplus EA^c$.
2. To show that (A) implies (B).

Let $B = \{\xi: X > 0\}$ and $C = \{\xi: X < 0\}$.   We need to show that $P(B) = P(C) = 0$.

For each positive integer n, let $B_n = \{\xi: X \geq 1/n\}$ and $C_n = \{\xi: -X \geq 1/n\}$.   Then $B = \lim_{n \to \infty} B_n$ and $C = \lim_{n \to \infty} C_n$. Now we must have $\int_{B_n} X \, dP \geq \frac{1}{n} P(B_n)$, which implies $P(B_n) = 0$.   Since this holds for each n, it follows that the limit must be zero. A similar argument, with allowance for the sign, shows that $P(C) = 0$. ■

(17)   If $X(\cdot) \geq 0$ $[P]$, then $\int_E X \, dP \geq 0$ for all events E; equality holds iffi
$X(\cdot) = 0$ $[P]$.

More generally, if $X(\cdot) \geq |Y(\cdot)|$ $[P]$, then $\int_E X \, dP \geq \int_E Y \, dP$ for all events E.

PROOF   The first assertion follows immediately from the definition of the integral and property (16).   The second follows from the first by use of linearity and the fact that $X(\cdot) - Y(\cdot) \geq X(\cdot) - |Y(\cdot)| \geq 0$ $[P]$. ■

(18)   *Product rule for independent random variables.*   If $X(\cdot)$ and $Y(\cdot)$ are independent simple random variables, then

$$\int XY \, dP = \int X \, dP \int Y \, dP$$

PROOF $X(\cdot)Y(\cdot) = \sum_{i,j} t_i u_j I_{A_i B_j}(\cdot)$.   Because of the results of Example 3-7-2 and rules on summation, we may write

$$\int XY \, dP = \sum_{i,j} t_i u_j P(A_i B_j) = \sum_{i,j} t_i u_j P(A_i) P(B_j)$$
$$= \sum_i t_i P(A_i) \sum_j u_j P(B_j) = \int X \, dP \int Y \, dP \quad ■$$

The properties derived above for integrals of simple random variables are extended to integrals of general random variables in Sec. 4-4.

4-3   Some basic limit theorems

In this section we state, without proof, two limit theorems which make it possible to pass, in a simple and elegant manner, from the theory of integrals of simple random variables to the theory of integrals of general random variables.   For a proof of these theorems, reference may be made to a standard work on measure theory, such as Munroe [1953, sec. 23].

Theorem 4-3A

Consider a nondecreasing sequence of nonnegative simple random variables $\{X_n(\cdot) : 1 \leq n < \infty\}$.   Let $X_0(\cdot)$ be a simple random variable such that, for each $\xi \in E$,

$$\lim_{n \to \infty} X_n(\xi) = X_0(\xi)$$

Then

$$\lim_{n \to \infty} \int_E X_n \, dP = \int_E X_0 \, dP$$

Theorem 4-3B

If $\{X_n(\cdot) : 1 \leq n < \infty\}$ and $\{Y_n(\cdot) : 1 \leq n < \infty\}$ are two non-decreasing sequences of nonnegative simple random variables such that, for each $\xi \in E$,

$$\lim_{n \to \infty} X_n(\xi) = \lim_{n \to \infty} Y_n(\xi)$$

then

$$\lim_{n \to \infty} \int_E X_n \, dP = \lim_{n \to \infty} \int_E Y_n \, dP$$

The significance of these two theorems will appear in the following sections, where they play a very important role.

4-4   Integrable random variables

We now extend the definition of the integral to general integrable random variables. We recall that it is shown in Sec. 3-3 that a random variable can be expressed as the difference of two nonnegative random variables $X(\cdot) = X_+(\cdot) - X_-(\cdot)$.

Definition 4-4a

A nonnegative random variable $X(\cdot)$ is said to be *integrable on set E with respect to* $P(\cdot)$ iffi there is a nondecreasing sequence of simple random variables $\{X_n(\cdot) : 1 \leq n < \infty\}$, with

$$\lim_{n \to \infty} X_n(\xi) = X(\xi) \qquad \text{for each } \xi \in E$$

and such that

$$\lim_{n \to \infty} \int_E X_n \, dP < \infty$$

In this case we define the *integral of* $X(\cdot)$ *over E with respect to* $P(\cdot)$ to be

$$\int_E X \, dP = \lim_{n \to \infty} \int_E X_n \, dP$$

If $X(\cdot)$ is any random variable, it is said to be *integrable on E with respect to* $P(\cdot)$ iffi both $X_+(\cdot)$ and $X_-(\cdot)$ are integrable. In this case

$$\int_E X \, dP = \int_E X_+ \, dP - \int_E X_- \, dP$$

We note immediately that if $X(\cdot)$ is a simple random variable, Theorem 4-3A ensures the fact that the new definition of the integral is consistent with the former definition of the integral of a simple random variable. Also, Theorem 4-3B ensures the fact that the definition does not depend upon the particular sequence of simple random variables chosen. Theorem 3-3A shows the existence of a sequence of simple random variables approaching a nonnegative random variable. It is apparent that the question of integrability, when the measure $P(\cdot)$ is a probability measure, arises only when the random variable is unbounded. For more general measures, which may be infinite on the basic space, even bounded measurable functions might not be integrable.

It is noted in Sec. 4-2 that the integral of a simple random variable with respect to a probability measure provides a probability-weighted average of the values of the random variable. In the general case, each of the approximating integrals provides a probability-weighted average of the approximating simple random variable. The limiting case, which is the integral of the general random variable, can thus be conceived as a similar probability-weighted average for that random variable.

We now list a set of properties of the integral which, for the most part, are extensions of the corresponding rules developed for simple random variables. For the complex-valued case see Appendix E. We restrict the statements to the case of real-valued random variables. Most of the proofs involve simple employment of the rules for limits (e.g., the limit of a sum is the sum of the limits, etc.) and suitable combinations of $X_+(\cdot)$, $X_-(\cdot)$, $Y_+(\cdot)$, $Y_-(\cdot)$, etc. Where these are the only considerations, no proof is given.

(I1)   *Linearity with respect to the integrand.*   If $X(\cdot)$ and $Y(\cdot)$ are integrable random variables and E is any event,

$$\int_E (aX + bY)\, dP = a \int_E X\, dP + b \int_E Y\, dP$$

(I2)   *Linearity with respect to measure.*   If $m(\cdot)$, $m'(\cdot)$, and $m''(\cdot)$ are totally finite measures, if a and b are positive constants such that $m(\cdot) = am'(\cdot) + bm''(\cdot)$, and if $X(\cdot)$ is integrable with respect to each of these measures,

$$\int_E X\, dm = a \int_E X\, dm' + b \int_E X\, dm''$$

(I3)   *Additivity.*   If $\{E_i : i \in J\}$ is a finite or countably infinite partition of E, and if $X(\cdot)$ is an integrable random variable then

$$\int_E X\, dP = \sum_{i \in J} \int_{E_i} X\, dP$$

(I4)   A random variable $X(\cdot)$ is integrable iff $|X(\cdot)|$ is integrable, and

$$\left| \int_E X\, dP \right| \leq \int_E |X|\, dP \qquad \text{for all events } E$$

PROOF   The first assertion follows from the fact that $|X(\cdot)| = X_+(\cdot) + X_-(\cdot)$. The second assertion follows from the fact that

$$\left| \int_E X_+ \, dP - \int_E X_- \, dP \right| \leq \int_E X_+ \, dP + \int_E X_- \, dP = \int_E |X| \, dP \quad \blacksquare$$

The last inequality plays an important role in analysis. It may be extended to the case of complex-valued random variables (Appendix E).

(15)   If $P(E) = 0$, then $\int_E X \, dP = 0$.

(16) (A) $\int_E X \, dP = 0$ for all events E iffi (B) $X(\cdot) = 0$ [P].

(A') $\int_E X \, dP = \int_E Y \, dP$ for all events E iffi (B') $X(\cdot) = Y(\cdot)$ [P].

PROOF   Theorem 3-10F (in Appendix D-1) shows that $X(\cdot) = Y(\cdot)$ [P] iffi there are sequences of simple approximating random variables such that $X_n(\cdot) = Y_n(\cdot)$ [P] for each n. The present theorem follows in an obvious way from the corresponding theorem for simple random variables and the basic limit theorems of Sec. 4-3. $\blacksquare$

(17)   Let $X(\cdot)$ be an integrable random variable such that $X(\cdot) \geq 0$ [P]. Then $\int_E X \, dP \geq 0$ for all events E. If $P(E) > 0$, then $\int_E X \, dP = 0$ iffi $X(\cdot) = 0$ [P] on E.

More generally, if $X(\cdot)$ is an integrable random variable and $Y(\cdot)$ is a random variable such that $X(\cdot) \geq |Y(\cdot)|$ [P], then $Y(\cdot)$ is integrable and $\int_E X \, dP \geq \left| \int_E Y \, dP \right|$ for all E.

PROOF   From (17) for simple random variables, it follows that $X(\cdot) \geq 0$ [P] implies $\int_E X \, dP \geq 0$, since each of the approximating simple random variables must be nonnegative with probability 1. If $X(\cdot) = 0$ [P] on E, it follows that the integral is zero. To show the converse, we consider $A = \{\xi : X > 0\}$ and show that $P(AE) = 0$. Let $A_n = \{\xi : X > 1/n\}$; since $X(\xi) > 0$ iffi for some n, $X(\xi) > 1/n$, we must have

$$A = \bigcup_{n=1}^{\infty} A_n \quad \text{and} \quad AE = \bigcup_{n=1}^{\infty} A_n E$$

Noting that the A_n, and hence the $A_n E$, form an increasing sequence, we may assert that $AE = \lim_{n \to \infty} A_n E$. Since

$$\int_{A_n E} X \, dP \geq \frac{1}{n} P(A_n E)$$

we must conclude that $P(A_n E) = 0$ for any n, so that $P(AE) = \lim_{n \to \infty} P(A_n E) = 0$. The last assertion follows from the corresponding statement for simple random variables, since $X(\cdot) \geq 0$ and $|Y(\cdot)| = Y_+(\cdot) + Y_-(\cdot)$. $\blacksquare$

(18)   *Product rule for independent random variables.*   If $X(\cdot)$ and $Y(\cdot)$ are independent random variables which are integrable, the product $X(\cdot)Y(\cdot)$ is integrable and

$$\int XY \, dP = \int X \, dP \int Y \, dP$$

PROOF   This follows from the corresponding rule for simple random variables and the fact that the approximating simple random variables for $X(\cdot)$ and $Y(\cdot)$ are independent (Theorem 3-7D). $\blacksquare$

4-5   The Lebesgue-Stieltjes integral

Suppose $F(\cdot)$ is a finite, nondecreasing, real-valued function defined on the whole real line R, with the property that it is continuous from the right at every point. Distribution functions for random variables are functions of this type. The function $F(\cdot)$ may be used to define a set function $m_F(\cdot)$ on the class of all half-open intervals $M_{ab} = (a, b]$, by setting $m_F(M_{ab}) = F(b) - F(a)$. This set function obviously has the additivity property for unions of disjoint intervals. If the value $m_F(M_{ab})$ is visualized as mass assigned to the half-open intervals, then it seems plausible that a unique mass distribution on the line is determined. The methods of measure theory show that the set function thus defined on the intervals does, in fact, determine a measure function (and hence a mass distribution) for the class of all Borel sets on the real line.

Definition 4-5a

A measure $m_F(\cdot)$ defined on the Borel sets on the real line by the above procedure is called the *Lebesgue-Stieltjes measure induced by* $F(\cdot)$. The function $F(\cdot)$ is called a *distribution function* for $m_F(\cdot)$.

It should be noted that we do not say *the* distribution function, since we may add any real constant to $F(\cdot)$ to obtain a new distribution function which induces the same measure. In the case of the probability distribution function $F_X(\cdot)$ associated with a random variable $X(\cdot)$, the uniqueness follows from the fact that $F(-\infty) = 0$ or $F(+\infty) = 1$. The Lebesgue-Stieltjes measure induced by the probability distribution function $F_X(\cdot)$ is precisely the probability measure $P_X(\cdot)$ induced by the random variable $X(\cdot)$.

Definition 4-5b

Let $F(\cdot)$ be a distribution function, and let $m_F(\cdot)$ be the Lebesgue-Stieltjes measure induced by it. The *Lebesgue-Stieltjes integral* of a function $g(\cdot)$, defined on the real line, is given by

$$\int_a^b g(t)\, dF(t) = \int_{[a,b]} g\, dm_F$$

where $[a, b]$ is the closed interval $a \leq t \leq b$.

In the special case that $F(\cdot)$ is given by the relation $F(t) = t$, the measure induced is the ordinary *Lebesgue measure*, which assigns to each interval its length. The Lebesgue-Stieltjes integral reduces, in this case, to the ordinary Lebesgue integral.

For the remainder of the section, we consider a probability distribution function $F_X(\cdot)$ and the corresponding probability measure $P_X(\cdot)$

on the real line. Similar statements can be made for other distributions. First we consider three special cases which are combined to form the general case.

1. *Absolutely continuous mass distribution.* In this case, there are no point mass concentrations and every set of points on the real line having zero length (zero Lebesgue measure) is assigned zero probability (Sec. 3-4). The distribution function $F_X(\cdot)$ is continuous, and the density function $f_X(\cdot)$ exists. The Lebesgue-Stieltjes integral is expressible in the form

$$\int_a^b g(t)\, dF_X(t) = \int_a^b g(t) f_X(t)\, dt$$

The last integral is an integral in the sense of Lebesgue. In most cases of practical interest, however, this is equal to the corresponding Riemann integral, which may be treated by ordinary techniques of integration.

2. *Discrete mass distribution.* Mass is concentrated at discrete points. In this case, $F_X(\cdot)$ is a step function. Evaluation can be carried out either by an appeal to the Riemann-Stieltjes integral or by direct consideration of the integral with respect to $P_X(\cdot)$. The integrand $g(\cdot)$ is equal with probability 1 to any function which has value $g(t_i)$ at each of the discrete values $t = t_i$ where the mass is located. We may consider the simple function which has these values at the various t_i and is zero elsewhere. The integral is, by definition,

$$\int g(t)\, dP_X = \sum_i g(t_i) P_X(\{t_i\})$$

3. *Singular continuous mass distribution.* It is possible to construct mass distributions of such a character that the mass is located on a set of points of Lebesgue measure zero, but such that the distribution function $F_X(\cdot)$ is continuous (cf. Munroe [1953, sec. 27]). In this case, there is no density function and no discrete mass distribution. For a single real-valued random variable this situation is so unusual that it is completely ignored in most works directed to application. For vector-valued random variables, the situation is not so unusual, as is noted below.

4. *General case.* The mass distribution may be separated into a *continuous part* and a *discrete part*. This is done by writing the distribution function $F_X(\cdot)$ as the sum of a continuous function $F_{Xc}(\cdot)$ and a step function $F_{Xd}(\cdot)$. These component functions have the character of distribution functions and define Lebesgue-Stieltjes measures $m_{Xc}(\cdot)$ and $m_{Xd}(\cdot)$ on the Borel sets on the real line. It is apparent that $P_X(M) = m_{Xc}(M) + m_{Xd}(M)$ for any Borel set M. Elementary arguments show (cf. Loève [1963, sec. 11]) that $F_{Xd}(\cdot)$ has at most a countable infinity of jump points at which are located point mass concentrations.

A general decomposition theorem, due to Lebesgue, shows that there is a further decomposition of the measure $m_{Xc}(\cdot)$ for the continuous part into measures with the following properties. One measure $m_{Xac}(\cdot)$ is absolutely continuous (Def. 3-4b) in the sense that it assigns zero probability mass to each Borel set of Lebesgue measure (length) zero. We call this the *absolutely continuous part* of the mass distribution. The second component measure $m_{Xs}(\cdot)$ determines a mass distribution which is continuous and singular in the sense of case 3, described above. We refer to this as the *singular continuous part* of the mass distribution.

If $F_{Xac}(\cdot), F_{Xd}(\cdot)$, and $F_{Xs}(\cdot)$ are the distribution functions corresponding to the measures $m_{Xac}(\cdot)$, $m_{Xd}(\cdot)$, and $m_{Xs}(\cdot)$, respectively, we must have $F_X(\cdot) = F_{Xac}(\cdot) + F_{Xd}(\cdot) + F_{Xs}(\cdot)$ and $P_X(\cdot) = m_{Xac}(\cdot) + m_{Xd}(\cdot) + m_{Xs}(\cdot)$. The three parts of the mass distribution correspond to the three types of probability mass distributions described in cases 1 to 3, above. In almost all cases of interest in applications, the singular continuous part is absent; then the absolutely continuous part constitutes the entire continuous part of the distribution.

If t_i is one of the jump points of the step function $F_{Xd}(\cdot)$, there is a positive probability mass located at that point. We must therefore have $m_{Xd}(\{t_i\}) = P(X = t_i) = P_X(\{t_i\})$. Associated with the absolutely continuous measure $m_{Xac}(\cdot)$ for the absolutely continuous part, there is a density function $f_{Xac}(\cdot)$, with the property that its integral over any Borel set M gives the mass $m_{Xac}(M)$. In terms of these measures and their distribution functions, we have

$$\int_{-\infty}^{\infty} g(t)\, dF_X(t) = \int_{-\infty}^{\infty} g(t) f_{Xac}(t)\, dt + \sum_i g(t_i) P_X(\{t_i\})$$
$$+ \int_{-\infty}^{\infty} g(t)\, dF_{Xs}(t)$$

These formulas are of considerable importance in dealing with mathematical expectations, to be studied in the next chapter.

The ideas discussed above for a single real-valued random variable may be extended to the probability measures induced by vector-valued random variables. Distribution functions may be defined and Lebesgue-Stieltjes integrals may be introduced in a manner analogous to that for the one-dimensional case. Thus, for two dimensions,

$$\iint_Q g(t, u)\, dF_{XY}(t, u) = \int_Q g\, dP_{XY}$$

where Q is a Borel set in the plane.

Three cases may be identified, as in the discussion above. In the absolutely continuous case, a density function may be defined over the plane. It is known that the Lebesgue-Stieltjes integral can be evaluated in the absolutely continuous case by the double integral

$$\iint\limits_Q g(t, u)\, dF_{XY}(t, u) = \iint\limits_Q g(t, u) f_{XY}(t, u)\, dt\, du$$

In the case in which the probability mass is located at discrete points (e.g., the joint mapping produced by a pair of simple random variables), the integral has the value given by the sum

$$\int g\, dP_{XY} = \sum_{i,j} g(t_i, u_j) P_{XY}[\{(t_i, u_j)\}]$$

As in the one-dimensional case, distributions may be mixed, so that the probability distribution may be decomposed into an absolutely continuous part, a discrete part, and a singular continuous part. Since mass distributions on the plane, for example, may yield singularities by placing positive mass along a curve having zero Lebesgue measure (i.e., zero area), the singular part of the mass distribution may be important. In many situations, however, the mass distribution may be decomposed into an absolutely continuous part and a discrete part (point mass concentrations), with a consequent decomposition of the integrals into an integral involving a density function and a sum involving the point masses. These ideas are illustrated in the next chapter, where they are applied to the concept of mathematical expectation (probability-weighted average).

4-6 Transformation of integrals

In Chap. 3 it is shown that a function of a random variable gives rise to probability measures on the real lines R_1 and R_2, as shown in Fig. 4-6-1. We consider the mappings produced by

The random variable $X(\cdot)$: $S \to R_1$
The Borel function $g(\cdot)$: $R_1 \to R_2$
The random variable $Z(\cdot) = g[X(\cdot)]$: $S \to R_2$

If M is any Borel set on R_2, then $N = g^{-1}(M)$ is a Borel set on R_1, and $E = X^{-1}(N)$ is an event. On the Borel sets of R_1 and R_2 we define the probability measures $P_X(\cdot)$ and $P_Z(\cdot)$, respectively, by the requirement that

$$P(E) = P_X(N) = P_Z(M)$$

where E, N, and M are related as described above.

We wish to consider the relationship between certain integrals defined on the various spaces. Since $Z(\cdot) = g[X(\cdot)]$, by definition, we must have

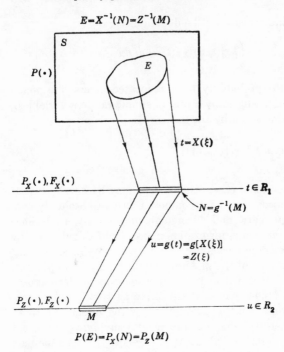

$E = X^{-1}(N) = Z^{-1}(M)$

Fig. 4-6-1   Mappings and probability measures induced by a Borel function of a random variable.

$\int Z \, dP = \int g[X] \, dP$

We wish to develop the basic transformation relationship

$$\int_S g[X] \, dP = \int_R g \, dP_X$$

The spaces of integration, S and R, are indicated for emphasis.   We shall argue the case for which $Z(\cdot) = g[X(\cdot)]$ and $X(\cdot)$ are nonnegative and bounded.   Removal of the restrictions on sign is quite simple; removal of the restriction of boundedness involves some technicalities but no new ideas.

Suppose $0 \leq Z(\xi) \leq a$.   We subdivide the interval $[0, a]$ in R_2 at points $u_0 = 0, u_1, u_2, \ldots, u_n = a$, with $u_0 < u_1, < \ldots < u_n$.

Let $M_i = [u_i, u_{i+1})$, for $i < n$, and $M_n = \{a\}$
  $N_i = g^{-1}(M_i)$
  $E_i = X^{-1}(N_i) = Z^{-1}(M_i)$

The probabilities assigned to each of the corresponding sets on their respective spaces is the same.   We define the simple functions

$$Z_0(\xi) = \sum_i u_i I_{E_i}(\xi) \quad \text{and} \quad g_0(t) = \sum_i u_i I_{N_i}(t)$$

Then $g_0(\cdot)$ and $Z_0(\cdot)$ are simple-function approximations to $g(\cdot)$ and $Z(\cdot)$, respectively. The approximations improve if the subdivision of $[0, a]$ is made finer. Now

$$\int Z_0 \, dP = \sum_i u_i P(E_i) \quad \text{and} \quad \int g_0 \, dP_X = \sum_i u_i P_X(N_i)$$

Since $P_X(N_i) = P(E_i)$ for each i, we have

$$\int Z_0 \, dP = \int g_0 \, dP_X$$

If the subdivision of $[0, a]$ is made increasingly finer,

$$\int Z_0 \, dP \to \int Z \, dP \quad \text{and} \quad \int g_0 dP_X \to \int g \, dP_X$$

Since the approximating integrals are the same for any subdivision of $[0, a]$, the limits are the same and the result is established.

When we use the alternate notation of the Lebesgue-Stieltjes integral, we have

$$\int Z \, dP = \int g[X] \, dP = \int_{-\infty}^{\infty} g(t) \, dF_X(t)$$

For the special case $g(t) = t$, we have

$$\int X \, dP = \int_{-\infty}^{\infty} t \, dF_X(t)$$

This formula may be applied to any random variable and its distribution function; in particular, it may be applied to $Z(\cdot)$ and $F_Z(\cdot)$ to give

$$\int Z \, dP = \int_{-\infty}^{\infty} u \, dF_Z(u)$$

We thus have the important relation

$$\int_{-\infty}^{\infty} u \, dF_Z(u) = \int_{-\infty}^{\infty} g(t) \, dF_X(t).$$

For the special cases of discrete or absolutely continuous distributions, we have the corresponding special formulas involving sums or density functions.

Similar results hold for functions of several random variables. For two random variables $X(\cdot)$ and $Y(\cdot)$, let $Z(\cdot) = g[X(\cdot), Y(\cdot)]$. A development similar to the above shows that the following integrals must satisfy

$$\int g(X, Y) \, dP = \int g \, dP_{XY} = \iint_{-\infty}^{\infty} g(t, u) \, dF_{XY}(t, u)$$

and

$$\int_{-\infty}^{\infty} v \, dF_Z(v) = \iint_{-\infty}^{\infty} g(t, u) \, dF_{XY}(t, u)$$

For simple random variables the integrals are ordinary sums. In the absolutely continuous case, the appropriate formulas using density functions may be used.

Before turning to applications, we make an important but simple extension of the fundamental transformation relation. As before, we let N be any Borel set on the real line R_1 and $E = X^{-1}(N)$. If $I_N(\cdot)$ is the indicator function for N, defined on R_1, then $I_N(\cdot)$ is a Borel function. Further, if $I_E(\cdot)$ is the indicator function for the event E, we must have

$$I_N[X(\xi)] = I_E(\xi) \text{ for all } \xi$$

since both have the value 1 iff $X(\xi) \in N$ and have the value zero otherwise. Applying the transformation theorem, we have

$$\int_E g(X) \, dP = \int I_E g(X) \, dP = \int I_N(X) g(X) \, dP = \int I_N g \, dP_X$$

$$= \int_N g \, dP_X \quad \text{where} \quad E = X^{-1}(N)$$

Similar relationships hold for functions of two or more variables.

These developments provide the background needed to understand the important concept of mathematical expectation, which is introduced in the next chapter.

Selected references

MUNROE [1953]: Cited in our Chap. 2. Although written for the mathematician, this work is characterized by careful exposition and helpful discussions. Among the standard works on measure theory, it is probably the most useful reference with respect to the discussion of the present chapter.

LOÈVE [1963]: "Probability Theory," 3d ed. A thorough treatise written at the graduate level, this work (in its third edition) is rapidly becoming a classic. Section 11 gives a clear and careful development of the decomposition of probability distributions. Many other topics of our Chap. 4 are dealt with in detailed and general fashion.

chapter 5

Mathematical expectation

The concept of mathematical expectation seems to have been introduced first in connection with the development of probability theory in the early study of games of chance. It also arose in a more general form in the applications of probability theory to statistical studies, where the idea of a probability-weighted average appears quite naturally. The concept of a probability-weighted average finds general expression in terms of the abstract integrals studied in Chap. 4. Use of the theorems on the transformation of integrals, developed in Sec. 4-6, makes it possible to derive from a single basic expression a variety of special forms of mathematical expectation employed in practice; the topic is thus given unity and coherence. The abstract integral provides the underlying basic concept; the various forms derived therefrom provide the tools for actual formulation and analysis of various problems. Direct application of certain basic properties of abstract integrals yields corresponding properties of mathematical expectation. These properties are expressed as rules for a calculus of expectations. In many arguments and manipulations it is not necessary to deal directly with the abstract integral or with the detailed special forms.

The clear identification of the character of a mathematical expectation as a probability-weighted average makes it possible to give concrete interpretations, in terms of probability mass distributions, to some of the more commonly employed expectations.

Brief attention is given to the concept of random sampling, which is fundamental to statistical applications. Some of the basic ideas of information theory are presented in terms of random variables and mathematical expectations. Moment-generating functions and the closely related characteristic functions for random variables are examined briefly.

5-1 Definition and fundamental formulas

The concept of mathematical expectation arose early in the history of the theory of probability. It seems to have been introduced first in connection with problems in gambling, which served as a spur to much of the early effort. Suppose a game has a number of mutually exclusive ways of being realized. Each way has its probability of occurring, and each way results in a certain return (or loss, which may be considered a negative return) to the gambler. If one sums the various possible returns multiplied by their respective probabilities, one obtains a number which, by intuition and a certain kind of experience, may be called the *expected return*. In terms of our modern model, this situation could be represented by introducing a simple random variable $X(\cdot)$ whose value is the return realized by the gambler. Suppose $X(\cdot)$ is expressed in canonical form as follows:

$$X(\cdot) = \sum_{i=1}^{n} t_i I_{A_i}(\cdot)$$

The values t_i are the possible returns, and the corresponding A_i are the events which result in these values of return. The expected return for the game described by this random variable is the sum

$$E = \sum_{i=1}^{n} t_i P(A_i)$$

This is a *probability-weighted average* of the possible values of $X(\cdot)$.

The concept of a *weighted average* plays a significant role in sampling statistics. A sample of size n is taken. The members of the sample are characterized by numbers. Either these values are from a discrete set of possible values or they are classified in groups, by placing all values in a given interval into a single classification. Each of these classifications is represented by a single number in the interval. Suppose the value assigned to the ith classification is t_i and the number of members of the sample in this classification is n_i. The average value is given by

$$\frac{1}{n} \sum_i t_i n_i = \sum_i t_i f_i$$

where $f_i = n_i/n$ is the relative frequency of occurrence of the ith classification in the sample.

To apply probability theory to the sampling problem, we may suppose that we are dealing with a random variable $X(\cdot)$ whose range T is the set of t_i. To a first approximation, at least, we suppose $P(X = t_i) = f_i$. In that case, the sum becomes

$$\sum_i t_i f_i = \sum_i t_i P(X = t_i) = \int X \, dP$$

In a similar way, if we deal with a function $g(\cdot)$ of the values of the sample, the weighted average becomes

$$\sum_i g(t_i) f_i = \sum_i g(t_i) P(X = t_i) = \int g(X) \, dP$$

The notion of mathematical expectation arose out of such considerations as these. Because of the manner in which integrals are defined in Chap. 4, it seems natural to make the following definition in the general case.

Definition 5-1a

If $X(\cdot)$ is a real-valued random variable and $g(\cdot)$ a Borel function, the *mathematical expectation $E[g(X)] = E[Z]$ of the random variable $Z(\cdot) = g[X(\cdot)]$* is given by

$$E[Z] = E[g(X)] = \int Z \, dP = \int g(X) \, dP$$

Similarly, if $X(\cdot)$ and $Y(\cdot)$ are two real-valued random variables and $h(\cdot)$ is a Borel function of two real variables, the *mathematical expectation $E[h(X, Y)] = E[Z]$ of the random variable $Z(\cdot) = h[X(\cdot), Y(\cdot)]$* is given by

$$E[Z] = E[h(X, Y)] = \int Z \, dP = \int h(X, Y) \, dP$$

The terms *ensemble average* (ensemble being a synonym for set) and *probability average* are sometimes used. The term *expectation* is frequently suggestive and useful in attempting to visualize or anticipate results. But as in the case of names for other concepts, such as independence, the name may tempt one to bypass a mathematical argument in favor of some intuitively or experientially based idea of "what may be expected." It is probably most satisfactory to think in terms of probability-weighted averages, in the manner discussed after the definitions of the integral in Secs. 4-2 and 4-4.

By virtue of the theoretical development of Chap. 4, we may proceed to state immediately a number of equivalent expressions and relations for mathematical expectations. These are summarized below:

> Single real-valued random variable $X(\cdot)$:
> Mapping $t = X(\xi)$ from S to R_1
> Probability measure $P_X(\cdot)$ induced on Borel sets of R_1
> Probability distribution function $F_X(\cdot)$ on R_1
> Probability density function $f_X(\cdot)$ in the absolutely continuous case
> Mapping $v = g(t)$ from R_1 to R_2 produces mapping
$v = Z(\xi) = g[X(\xi)]$ from S to R_2
> Probability measure $P_Z(\cdot)$ induced on the Borel sets of R_2
> Probability distribution function $F_Z(\cdot)$ on R_2
> Probability density function $f_Z(\cdot)$ in the absolutely continuous case

In dealing with continuous distributions and mixed distributions, we shall *assume throughout this chapter that there is no singular continuous part* (Sec. 4-5). Thus the continuous part is the absolutely continuous part, for which a density function exists.

General case:

$$E[g(X)] = \int g(X)\, dP = \int g(t)\, dF_X(t) = \int v\, dF_Z(v) = \int Z\, dP = E[Z]$$

Discrete case:

$$E[g(X)] = \sum_i g(t_i)P(X = t_i) = \sum_j v_j P(Z = v_j) = E[Z]$$

Absolutely continuous case:

$$E[g(X)] = \int g(t)f_X(t)\, dt = \int v f_Z(v)\, dv = E[Z]$$

Mixed case: Consider the probability measure $P_X(\cdot)$ on R_1. We may express $P_X(\cdot) = m_{Xd}(\cdot) + m_{Xc}(\cdot)$. The measure $m_{Xd}(\cdot)$ describes the discrete mass distribution on R_1, and the measure $m_{Xc}(\cdot)$ describes the (absolutely) continuous mass distribution on R_1. Corresponding to the first, there is a distribution function $F_{Xd}(\cdot)$, and corresponding to the second, there is a distribution function $F_{Xc}(\cdot)$ and a density function $f_{Xc}(\cdot)$. We may then write

$$
\begin{aligned}
E[g(X)] &= \int g\, dP_X = \int g\, dm_{Xd} + \int g\, dm_{Xc} \\
&= \int g(t)\, dF_{Xd}(t) + \int g(t)\, dF_{Xc}(t) \\
&= \sum_i g(t_i)m_{Xd}(\{t_i\}) + \int g(t)f_{Xc}(t)\, dt
\end{aligned}
$$

We note that $m_{Xd}(\{t_i\}) = P_X(\{t_i\}) = P(X = t_i)$.

A similar development may be made for the mass distribution on the line R_2 corresponding to $P_Z(\cdot)$.

We have adopted the notational convention that integrals over the whole real line (or over the whole euclidean space in the case of several random variables) are written without the infinite limits, in order to simplify the writing of expressions. Integrals over subsets are to be designated as before.

Let us illustrate the ideas presented above with some simple examples.

Example 5-1-1

Determine $E[X]$ when $X(\cdot)$ has the *geometric distribution* (Example 3-4-6). The range is $\{k: 0 \le k < \infty\}$, with $p_k = P(X = k) = pq^k$, where $q = 1 - p$.

$$E[X] = p \sum_{k=1}^{\infty} kq^k = \frac{pq}{(1-q)^2} = \frac{q}{p} \quad \blacksquare$$

As an example of the absolutely continuous case, consider

Example 5-1-2

Let $g(t) = t^n$, and let $X(\cdot)$ have the *exponential distribution* with $f_X(t) = \alpha e^{-\alpha t} u(t)$, $\alpha > 0$.

$$E[X^n] = \int_0^\infty t^n \alpha e^{-\alpha t}\, dt = \frac{n!}{\alpha^n} \qquad n = 0, 1, 2, \ldots$$

The lower limit of the integral is set to zero because the integrand is zero for negative t. $\blacksquare$

In the mixed case, each part is handled as if it were the expectation for a discrete or a continuous random variable, with the exception that the probabilities, distribution functions, and density functions are modified according to the total probability mass which is discretely or continuously distributed. As the next example shows, it is usually easier to calculate the expectation as $E[g(X)]$ rather than as $E[Z]$, since it is then not necessary to derive the mass distribution for $Z(\cdot)$ from that given for $X(\cdot)$.

Example 5-1-3

Suppose $g(t) = t^2$ and suppose $X(\cdot)$ produces a mass distribution $P_X(\cdot)$ on R_1 which may be described as follows. A mass of $\frac{1}{2}$ is uniformly distributed in the interval $[-2, 2]$. A point mass of $\frac{1}{10}$ is located at each of the points $t = -2, -1, 0, 1, 2$.

DISCUSSION   The density function for the continuous part of the distribution for $X(\cdot)$ is given by

$$f_{X_c}(t) = \begin{cases} \dfrac{\frac{1}{2}}{4} & \text{for } -2 \le t \le 2 \\ 0 & \text{elsewhere} \end{cases} \qquad \textstyle\int f_{X_c}(t)\, dt = \frac{1}{2}$$

The probability masses for the discrete part of the distribution for $X(\cdot)$ are described by

$$P_X(\{t_i\}) = P(X = t_i) = \frac{1}{10} \qquad \text{for } t_i = -2, -1, 0, 1, 2, \quad \sum_i P_X(\{t_i\}) = \frac{1}{2}$$

Hence

$$E[g(X)] = E[X^2] = \sum_i t_i{}^2 P_X(\{t_i\}) + \int t^2 f_{X_c}(t)\, dt$$

$$= \frac{1}{10}\,(4 + 1 + 0 + 1 + 4) + \frac{1}{8}\int_{-2}^{2} t^2\, dt = 1 + \tfrac{2}{3}$$

We may make an alternative calculation by dealing with the mass distribution induced on R_2 by $Z(\cdot) = g[X(\cdot)]$, as follows. Using the result of Example 3-9-1 for the continuous part, we have

$$f_{Z_c}(v) = \frac{u_+(v)}{2\sqrt{v}}\,[f_{X_c}(\sqrt{v}) + f_{X_c}(-\sqrt{v})]$$

$$= \begin{cases} \dfrac{1}{8\sqrt{v}} & \text{for } 0 \le v \le 4 \\ 0 & \text{elsewhere} \end{cases}$$

For the discrete part, we have range $v_k = 0, 1, 4$, with probabilities $\tfrac{1}{10}$, $\tfrac{2}{10}$, and $\tfrac{2}{10}$, respectively.

$$E[Z] = \sum_k v_k P_Z(\{v_k\}) + \int v f_{Z_c}(v)\, dv$$

$$= \frac{2}{10}\,(1 + 4) + \frac{1}{8}\int_0^4 v^{1/2}\, dv = 1 + \frac{2}{3}\ \blacksquare$$

The situation outlined above for a single random variable may be extended to two or more random variables. We write out the case for two random variables. As in the case of real-valued random variables, we shall assume that there is no continuous singular part, so that continuous distributions or continuous parts of distributions are absolutely continuous and have appropriate density functions.

Two random variables $X(\cdot)$ and $Y(\cdot)$ considered jointly:
Mapping $(t, u) = [X, Y](\xi)$ from S to the plane $R_1 \times R_2$
Probability measure $P_{XY}(\cdot)$ induced on the Borel sets of $R_1 \times R_2$
Joint probability distribution function $F_{XY}(\cdot, \cdot)$ on $R_1 \times R_2$
Joint probability density function $f_{XY}(\cdot, \cdot)$ in the continuous case
Mapping $v = h(t, u)$ from $R_1 \times R_2$ to R_3 produces the mapping $v = Z(\xi) = h[X(\xi), Y(\xi)]$ from S to R_3
Probability measure $P_Z(\cdot)$ induced on the Borel sets of R_3
Probability distribution function $F_Z(\cdot)$ on R_3
Probability density function $f_Z(\cdot)$ in the continuous case

General case:

$$E[h(X, Y)] = \smallint h(X, Y)\, dP = \smallint\smallint h(t, u)\, dF_{XY}(t, u) = \smallint v\, dF_Z(v) = E[Z]$$

Discrete case:

$$E[h(X, Y)] = \sum_{i,j} h(t_i, u_j) P(X = t_i, Y = u_j) = \sum_k v_k P(Z = v_k) = E[Z]$$

Absolutely continuous case:

$$E[h(X, Y)] = \iint h(t, u)f_{XY}(t, u) \, dt \, du = \int vf_Z(v) \, dv = E[Z]$$

Mixed cases for point masses and continuous masses may be handled in a manner analogous to that for the single variable. Suitable joint distributions must be described by the appropriate functions. We illustrate with a simple mixed case.

Example 5-1-4

Let $h(t, u) = tu$ so that $Z(\cdot) = X(\cdot)Y(\cdot)$. The random variables are those in Example 3-9-8 (Fig. 3-9-5). A mass of $\frac{1}{2}$ is continuously and uniformly distributed over the square, which has corners at $(t, u) = (-1, -1)$, $(-1, 1)$, $(1, 1)$, $(1, -1)$. A point mass of $\frac{1}{8}$ is located at each of these corners.

DISCUSSION The discrete distribution is characterized by $P_{XY}[\{(t_i, u_j)\}] = \frac{1}{8}$ for each pair (t_i, u_j) corresponding to a corner of the square. The continuous distribution can be described by a joint density function

$$f_{XYc}(t, u) = \begin{cases} \frac{1}{8} & \text{for } -1 \leq t \leq 1, \ -1 \leq u \leq 1 \\ 0 & \text{otherwise} \end{cases}$$

Thus

$$E[XY] = \sum_{i,j} t_i u_j P_{XY}[\{(t_i, u_j)\}] + \iint tu f_{XYc}(t, u) \, dt \, du$$

$$= \frac{1}{8}(1 - 1 + 1 - 1) + \frac{1}{8} \int_{-1}^{1}\!\!\int tu \, dt \, du = 0$$

Calculations for $E[Z]$ are somewhat more complicated because of the more complicated expressions for the mass distribution on R_3. ∎

Before considering further examples, we develop some general properties of mathematical expectation which make possible a considerable calculus of mathematical expectation, without direct use of the integral and summation formulas. This is the topic of the next section.

5-2 Some properties of mathematical expectation

A wide variety of mathematical expectations are encountered in practice. Two very important ones, known as the *mean value* and the *standard deviation*, are studied in the next two sections. Before examining these, however, we discuss in this section some general properties of mathematical expectation which do not depend upon the nature of any given probability distribution. These properties are, for the most part, direct consequences of the properties of integrals developed in Chap. 4. These properties provide the basis for a calculus of expectations which makes it unnecessary in many arguments and manipulations to deal directly with the abstract integral or with the detailed special forms derived therefrom.

Unless special conditions—particularly the use of inequalities—indicate otherwise, the properties developed below hold for complex-valued random variables as well as for real-valued ones.   For a brief discussion of the complex-valued case, see Appendix E.   The statements of the properties given below tacitly assume the existence of the indicated expectations.   This is equivalent to assuming the integrability of the random variables.

We first obtain a list of basic properties and then give some very simple examples to illustrate their use in analytical arguments.   As a first property, we utilize the definition of the integral of a simple function to obtain

(E1) $E[aI_A] = aP(A)$ where a is a real or complex constant

A constant may be considered to be a random variable whose value is the same for all ξ.   If the random variable is almost surely equal to a constant, then for most purposes in probability theory it is indistinguishable from that constant.   As an expression of the linearity property of the integral, we have

(E2)   *Linearity.*   If a and b are real or complex constants,

$$E[aX + bY] = aE[X] + bE[Y]$$

By a simple process of mathematical induction, the linearity property may be extended to linear combinations of any finite number of random variables.   The mathematical expectation thus belongs to the class of *linear operators* and hence shares the properties common to this important class of operators.

(E3)   *Positivity.*   If $X(\cdot) \geq 0$ $[P]$,   then $E[X] \geq 0$.   If $X(\cdot) \geq 0$ $[P]$, then $E[X] = 0$ iffi $X(\cdot) = 0$ $[P]$.   If $X(\cdot) \geq Y(\cdot)$ $[P]$, then $E[X] \geq E[Y]$.

Members of the class of linear operators which have the positivity property share a number of important properties.   If $X(\cdot)$ and $Y(\cdot)$ are real-valued or complex-valued random variables, the mathematical expectation $E[X\bar{Y}]$ has the character of an *inner product*, which plays an important role in the theory of linear metric spaces.   Note that the bar over Y indicates the complex conjugate.

A restatement of property (I4) for integrals (as extended to the complex case in Appendix E) yields the important result

(E4) $E[X]$ exists iffi $E[|X|]$ does, and $|E[X]| \leq E[|X|]$.

The importance of this inequality in analysis is well known.

Another inequality of classical importance is the *Schwarz inequality,* characteristic of linear operators with the positivity property.

(E5)   *Schwarz inequality.* $|E[XY]|^2 \leq E[|X|^2]E[|Y|^2]$.   In the real case, equality holds iff there is a real constant λ such that $\lambda X(\cdot) + Y(\cdot) = 0$ [P].

PROOF   We first prove the theorem in the real case and then extend it to the complex case.   By the positivity property (E3), for any real λ, $E[(\lambda X + Y)^2] \geq 0$, with equality iff $\lambda X(\cdot) + Y(\cdot) = 0$ [P].   Expanding the squared term and using linearity, we get

$$E[\lambda^2 X^2 + 2\lambda XY + Y^2] = E[X^2]\lambda^2 + 2E[XY]\lambda + E[Y^2] \geq 0$$

This is of the form $A\lambda^2 + B\lambda + C \geq 0$.   The strict inequality holds iff there is no real zero for the quadratic in λ.   Equality holds iff there is a second-order zero for the quadratic, which makes it a perfect square.   Use of the quadratic formula for the zeros shows there is no real zero iff $B^2 - 4AC < 0$ and a pair of real zeros iff $B^2 = 4AC$.   Examination shows these statements are equivalent to the Schwarz inequality.   To extend to the complex case, we use property (E4) and the result for the real case as follows:

$$|E[XY]|^2 \leq E^2[|XY|] \leq E[|X|^2]E[|Y|^2] \quad\blacksquare$$

The product rule for integrals of independent random variables may be restated to give the

(E6)   *Product rule for independent random variables.*   If $X(\cdot)$ and $Y(\cdot)$ are independent, integrable random variables, then $E[XY] = E[X]E[Y]$.

By a process of mathematical induction, this may be extended to any finite, independent class of random variables.

The next property is not a direct consequence of the properties of integrals, but may be derived with the aid of property (E3), above. It is so frequently useful in analytical arguments that we list it among the basic properties of mathematical expectation.

(E7)   If $g(\cdot)$ is a nonnegative Borel function and if $A = \{\xi : g(X) \geq a\}$, then $E[g(X)] \geq aP(A)$.

PROOF   If we show $g[X(\xi)] \geq aI_A(\xi)$ for all ξ, the statement follows from property (E3).   For $\xi \in A$, $g[X(\xi)] \geq a$ while $aI_A(\xi) = a$; for $\xi \in A^c$, $g[X(\xi)] \geq 0$ while $aI_A(\xi) = 0$.   Thus the inequality holds for all ξ. $\blacksquare$

This elementary inequality is useful in deriving certain classical inequalities, such as the celebrated Chebyshev inequality in Theorem 5-4B.

A somewhat similar inequality is an extension of an inequality attributed to A. A. Markov.

(E8) If $g(\cdot)$ is a nonnegative, strictly increasing, Borel function of a single real variable and c is a nonnegative constant, then

$$P(|X| \geq c) \leq E[g(|X|)]/g(c)$$

PROOF $|X| \geq c$ iffi $g(|X|) \geq g(c)$. In property (E7), replace the constant a by the constant $g(c)$. ■

For the special case $g(t) = t^k$, we have the so-called *Markov inequality*.

The next inequality deals with convex functions, which are defined as follows:

Definition 5-2a

A Borel function $g(\cdot)$ defined on a finite or infinite open interval I on the real line is said to be *convex* (on I) iffi for every pair of real numbers a, b belonging to I we have

$$g\left(\frac{a + b}{2}\right) \leq \frac{1}{2} [g(a) + g(b)]$$

It is known that a convex function must be continuous on I in order to be a Borel function.   If $g(\cdot)$ is continuous, it is convex on I iffi to every t_0 in I there is a number $\lambda(t_0)$ such that for all t in I

$$g(t) \geq g(t_0) + \lambda(t_0)(t - t_0)$$

In terms of the graph of the function, this means that the graph of $g(\cdot)$ lies above a line passing through the point $(t_0, g(t_0))$.   Using this inequality, we may extend a celebrated inequality for sums and integrals to mathematical expectation.

(E9) *Jensen's inequality.*   If $g(\cdot)$ is a convex Borel function and $X(\cdot)$ is a random variable whose expectation exists, then

$g(E[X]) \leq E[g(X)]$

PROOF   In the inequality above, let $t_0 = E[X]$ and $t = X(\cdot)$; then take expectations of both sides, utilizing property (E3). ■

We now consider some simple examples designed to illustrate how the basic properties may be used in analytical arguments.   The significance of mathematical expectation for real-world problems will appear more explicitly in later sections.

Example 5-2-1   *Discrete Random Variables*

Let $X(\cdot) = \sum_{k=1}^{n} a_k I_{A_k}(\cdot)$.   By the linearity property (E2) and property (E1), we have

$E[X] = \sum_{k} a_k P(A_k)$, whether or not the expression is in canonical form.   In general,

$g[X(\cdot)] \neq \sum_k g(a_k)I_{A_k}(\cdot)$ if the A_k do not form a partition (Example 3-8-1); therefore,

in general, $E[g(X)] \neq \sum_k g(a_k)P(A_k)$.   By virtue of the formulas in Sec. 3-8 for the

canonical form, equality does hold if the A_k form a partition. ∎

Example 5-2-2   *Sequences of n Repeated Trials*

Let E_k be the event of a success on the kth trial.   Then

$$N(\cdot) = \sum_{k=1}^{n} I_{E_k}(\cdot)$$

is the number of successes in a sequence of n trials.   According to the result in Example 5-2-1,

$$E[N] = \sum_{k=1}^{n} E[I_{E_k}] = \sum_{k=1}^{n} P(E_k)$$

If $P(E_k) = p$ for all k, this reduces to $E[N] = np$.   No condition of independence is required.   If, however, the E_k form an independent class, we have the case of Bernoulli trials studied in Examples 2-8-3 and 3-1-4.   In that case, if we let A_{rn} be the event of exactly r successes in the sequence of n trials, we know that the A_{rn} form a partition and $P(A_{rn}) = C_r^n p^r (1 - p)^{n-r}$.   Hence

$$E[N] = \sum_{r=0}^{n} rP(A_{rn}) = \sum_{r=0}^{n} rC_r^n p^r (1 - p)^{n-r} = np$$

The series is summed in Example 2-8-10, in another connection, to give the same result.   For purposes of computing the mathematical expectation, the first form for $N(\cdot)$ given above is the more convenient. ∎

Example 5-2-3

Two real-valued random variables $X(\cdot)$ and $Y(\cdot)$ have a joint probability mass distribution which is continuous and uniform over the rectangle $a \leq t \leq b$ and $c \leq u \leq d$. Find the mathematical expectation $E[XY]$.

SOLUTION   The marginal mass distributions are uniform over the intervals $a \leq t \leq b$ and $c \leq u \leq d$, respectively.   The induced measures have the product property, which ensures independence of the two random variables.   Hence, by property (E6), we have

$$E[XY] = E[X]E[Y]$$

Now

$$E[X] = \int tf_X(t)\, dt = \frac{1}{b - a} \int_a^b t\, dt = \frac{b^2 - a^2}{2(b - a)} = \frac{a + b}{2}$$

Similarly, $E[Y] = (c + d)/2$, so that $E[XY] = \frac{1}{4}(a + b)(c + d)$. ∎

The next example is of interest in studying the variance and standard deviation, to be introduced in Sec. 5-4.

Example 5-2-4

Suppose $X(\cdot)$ and $X^2(\cdot)$ are integrable and $E[X] = \mu$. Then

$$E[(X - \mu)^2] = E[X^2] - \mu^2 = E[X^2] - E^2[X] \geq 0.$$

SOLUTION

$$E[(X - \mu)^2] = E[X^2 - 2\mu X + \mu^2] = E[X^2] - 2\mu E[X] + \mu^2 = E[X^2] - \mu^2 \blacksquare$$

Another result which is sometimes useful is the following:

Example 5-2-5

For any two random variables $X(\cdot)$ and $Y(\cdot)$ whose squares are integrable, $|E[XY]| \leq$ max $\{E[|X|^2], E[|Y|^2]\}$.

SOLUTION By the Schwarz inequality, $|E[XY]|^2 \leq E[|X|^2]E[|Y|^2]$. The inequality is strengthened by replacing the smaller of the factors on the right-hand side by the larger. Hence $|E[XY]|^2 \leq$ max $\{E^2[|X|^2], E^2[|Y|^2]\}$. The asserted inequality follows immediately from elementary properties of inequalities for real numbers. $\blacksquare$

We turn, in the next two sections, to a study of two special expectations which are useful in describing the probability distributions induced by random variables.

5-3 The mean value of a random variable

Discussions in earlier sections show that a real-valued random variable is essentially described for many purposes, if the mass distribution which it induces on the real line is properly described. Complete analytical descriptions are provided by the probability distribution function. In many cases, a complete description of the distribution is not possible and may not be needed. Certain parameters provide a partial description, which may be satisfactory for many investigations. In this and the succeeding section, we consider two of the most common and useful parameters of the probability distribution induced by a random variable.

The simplest of the parameters is described in the following

Definition 5-3a

If $X(\cdot)$ is a real-valued random variable, its *mean value*, denoted by one of the following symbols μ, μ_X, or $\mu[X]$, is defined by $\mu[X] = E[X]$.

We shall use $P_X(\cdot)$ to denote the probability measure induced on the real line R, as in the previous sections. The mean value can be given a very simple geometrical interpretation in terms of the mass distribution. According to the fundamental forms for expectations, we may write

$$\mu[X] = \int t \, dP_X(t)$$

If $\{t_i \colon i \in J\}$ is the set of values of a simple random variable which approximates $X(\cdot)$ and if $M_i = [t_i, t_{i+1})$, then the approximating integral is

$$\sum_{i \in J} t_i P_X(M_i)$$

As the subdivisions become finer and finer, this approaches the first moment of the mass distribution on the real line about the origin. Since the total mass is unity, this moment is also the coordinate of the center of mass. Thus *the mean value is the coordinate of the center of mass of the probability mass distribution.* This mechanical picture is an aid in visualizing the location of the possible values of the random variable. As an example of the use of this interpretation, consider first the uniform distribution over an interval.

Example 5-3-1

Suppose $X(\cdot)$ is *uniformly distributed* over the interval $[a, b]$. This means that probability mass is uniformly distributed along the real line between the points $t = a$ and $t = b$. The center of mass, and hence the mean value, should be given by $\mu_X = (a + b)/2$. This fact may be checked analytically as follows:

$$\mu_X = E[X] = \frac{1}{b-a} \int_a^b t\, dt = \frac{b^2 - a^2}{2(b-a)} = \frac{a+b}{2} \ \blacksquare$$

As a second example, we may consider the gaussian, or normal, distribution.

Example 5-3-2 *The Normal Distribution*

A real-valued random variable $X(\cdot)$ is said to have the normal, or gaussian, distribution if it is continuously distributed with probability density function

$$f_X(t) = \frac{1}{\sigma \sqrt{2\pi}} \exp\left[-\frac{1}{2} \left(\frac{t - \mu}{\sigma} \right)^2 \right]$$

where μ is a constant and σ is a positive constant. A graph of the density function is shown in Fig. 3-4-4. This is seen to be a curve symmetrical about the point $t = \mu$, as can be checked easily from the analytical expression. This means that the probability mass is distributed symmetrically about $t = \mu$ and hence has its center of mass at that point. The mean value of a random variable with the normal distribution is therefore the parameter μ. Again, we may check these results by straightforward analytical evaluation. We have

$$\mu_X = \int t f_X(t)\, dt = \frac{1}{\sigma \sqrt{2\pi}} \int t \exp\left[-\frac{1}{2} \left(\frac{t - \mu}{\sigma} \right)^2 \right] dt$$

$$= \frac{\sigma}{\sqrt{2\pi}} \int \frac{t - \mu}{\sigma} \exp\left[-\frac{1}{2} \left(\frac{t - \mu}{\sigma} \right)^2 \right] d\left(\frac{t - \mu}{\sigma} \right)$$

$$+ \frac{\mu}{\sigma \sqrt{2\pi}} \int \exp\left[-\frac{1}{2} \left(\frac{t - \mu}{\sigma} \right)^2 \right] dt$$

The integrand in the first integral in the last expression is an odd function of $t - \mu$; since the integral exists, its value must be zero. The second term in this expression is

μ times the area under the graph for the density function and therefore must have the value μ. ■

It is not always possible to use the center-of-mass interpretation so easily. Since $\mu[X]$ is a mathematical expectation, the various forms of expression for and properties of the mean value are obtained directly from the expressions and properties of the mathematical expectation. Some of the examples in Sec. 5-2 are examples of calculation of the mean value. Example 5-2-1 gives a general formula for the mean value of a simple random variable. Example 5-2-2 gives the mean value for the number of successes in a sequence of n trials. In the Bernoulli case, in which the random variable has the *binomial distribution* (Example 3-4-5), the mean value is np. The constant μ in Example 5-2-4 is the mean value for the random variable considered in that example.

Example 5-3-3   *Poisson's Distribution*

A random variable having the Poisson distribution is described in Example 3-4-7. This random variable has values 0, 1, 2, . . . , with probabilities

$$P(X = k) = p_k = e^{-\mu}\mu^k/k!$$

The mean value is thus

$$\mu[X] = e^{-\mu} \sum_{k=0}^{\infty} k \frac{\mu^k}{k!} = e^{-\mu}\mu \sum_{k=1}^{\infty} \frac{\mu^{k-1}}{(k-1)!} = \mu$$

since the last series expression converges to e^{μ}. Again, the choice of the symbol μ for the parameter in the formulas is based on the fact that this is a traditional symbol for the mean value. ■

Example 5-3-4

A man plays a game in which he wins an amount t_1 with probability p or loses an amount t_2 with probability $q = 1 - p$. The net winnings may be expressed as a random variable $W(\cdot) = t_1 I_A(\cdot) - t_2 I_{A^c}(\cdot)$, where A is the event that the man wins and A^c is the event that he loses. The game is considered a "fair game" if the mathematical expectation, or mean value, of the random variable is zero. Determine p in terms of the amounts t_1 and t_2 in order that the game be fair.

    SOLUTION   We set $\mu[W] = t_1 p - t_2 q = 0$. Thus $p/q = t_2/t_1 = r$, from which we obtain, by a little algebra, $p = r/(1 + r) = t_2/(t_1 + t_2)$. ■

Example 5-3-5

The sales demand in a given period for an item of merchandise may be represented by a random variable $X(\cdot)$ whose distribution is *exponential* (Example 3-4-8). The density function is given by

$$f_X(t) = \alpha e^{-\alpha t} u(t)$$

Suppose $1/\alpha = 1,500$ units of merchandise. What amount must be stocked at the beginning of the period so that the probability is 0.99 that the sales demand will not exceed the supply? What is the "expected number" of sales in the period?

SOLUTION   The first question amounts to asking what value of t makes $F_X(t) = u(t)[1 - e^{-\alpha t}] = 0.99$. This is equivalent to making $e^{-\alpha t} = 0.01$. Use of a table of exponentials shows that this requires a value $\alpha t = 4.6$. Thus $t = 4.6/\alpha = 6{,}900$ units. The expected number of sales is interpreted as the mathematical expectation, or mean value

$$\mu_X = E[X] = \alpha \int_0^\infty te^{-\alpha t}\, dt = \frac{1}{\alpha} = 1{,}500 \text{ units}$$

Note that this expectation may be obtained from the result of Example 5-1-2, with $n = 1$. The result above may be expressed in a general formula for the exponential distribution. If it is desired to make the probability of meeting sales demand equal to $1 - p$, then $e^{-\alpha t} = p$ or $\alpha t = -\log_e p$, so that

$$t = -\frac{1}{\alpha}\log_e p = -2.3\,\frac{1}{\alpha}\log_{10} p = -2.3\mu_X \log_{10} p \quad\blacksquare$$

The next example involves some assumptions that may not be realistic in many situations, but it points to the role of probability theory in decision making. Even such an oversimplified model may often give insight into the alternatives offered in a decision process.

Example 5-3-6

A government agency has m research and development problems it would like to have solved in a given period of time and n contracts to offer n equally qualified laboratories that will work essentially independently on the problems assigned to them. Each laboratory has a probability p of solving any one of the problems which may be assigned to it. The agency is faced with the decision whether to assign contracts with a uniform number n/m (assumed to be an integer) of laboratories working on each problem or to let the laboratories choose (presumably in a manner that is essentially random) the problem that it will attack. Which policy offers the greater prospect of success in the sense that it will result in the solution of the larger number of problems?

SOLUTION AND DISCUSSION   We shall approach this problem by defining the random variable $N(\cdot)$ whose value is the number of problems solved and determine $E[N]$ under each of the assignment schemes. If we let A_j be the event that the jth problem is solved, then

$$N(\cdot) = \sum_{i=1}^m I_{A_j}(\cdot) \qquad \text{and} \qquad E[N] = \sum_{j=1}^m P(A_j)$$

The problem reduces to finding the $P(A_j)$. Consider the events

B_{ij} = event ith laboratory works on jth problem
C_i = event ith laboratory is successful

We suppose $P(C_i) = p$ and $P(B_{ij})$ depends upon the method of assignment. We suppose further that $\{C_i, B_{ij}: 1 \le i \le n, 1 \le j \le m\}$ is an independent class of events. Now

$$A_j = \bigcup_{i=1}^n B_{ij}C_i \qquad \text{or} \qquad A_j{}^c = \bigcap_{i=1}^n [B_{ij}C_i]^c$$

from which it follows

$$P(A_j) = 1 - \prod_{i=1}^{n} [1 - P(B_{ij}C_i)]$$

Uniform assignment: $P(B_{ij}) = 1$ for n/m of the i for any j and 0 for the others; this means that $1 - P(B_{ij}C_i) = 1 - p$ for n/m factors and $= 1$ for the others in the expression for $P(A_j)$. Thus

$$P(A_j) = 1 - (1 - p)^{n/m} \quad \text{for each } j$$
$$E[N_U] = m[1 - (1 - p)^{n/m}]$$

Random assignment: $P(B_{ij}) = 1/m$ for each i and j. $P(B_{ij}C_i) = p/m$. Therefore $P(A_j) = 1 - (1 - p/m)^n$, and

$$E[N_R] = m \left[1 - \left(1 - \frac{p}{m} \right)^n \right]$$

The following numerical values may be of some interest.

n	m	p	$E[N_U]/m$	$E[N_R]/m$
8	4	0.500	0.750	0.656
8	4	0.100	0.190	0.183
10	2	0.100	0.410	0.401

It would appear that for small p there is little advantage in a uniform assignment. When one allows for the fact that the probability of success is likely to increase if the laboratory chooses its problem, one would suspect that for difficult problems with low probability of success it might be better to allow the laboratories to choose. On the other hand, one problem might be attractive to all laboratories and the others unattractive to them all, resulting in several solutions to one problem and no attempt to solve the others. A variety of assumptions as to probabilities and corresponding methods of assignments must be tested before a sound decision can be made. ∎

We turn next to a second parameter which is important in characterizing most probability distributions encountered in practice.

5-4 Variance and standard deviation

The mean value, introduced in the preceding section, determines the location of the center of mass of the probability distribution induced by a real-valued random variable. As such, it gives a measure of the *central tendency* of the values of the random variable. We next introduce a parameter which gives an indication of the spread, or dispersion, of values of the random variable about the mean value.

Definition 5-4a

Consider a real-valued random variable $X(\cdot)$ whose square is integrable. The *variance of $X(\cdot)$*, denoted $\sigma^2[X]$, is given by

$$\sigma^2[X] = E[(X - \mu)^2]$$

where $\mu = \mu[X]$ is the mean value of $X(\cdot)$.

The *standard deviation for $X(\cdot)$*, denoted $\sigma[X]$, is the positive square root of the variance.

We sometimes employ symbols σ or σ_X, to simplify writing.

A mechanical interpretation

As in the case of the mean value, we may give a simple mechanical interpretation of the variance in terms of the mass distribution induced by $X(\cdot)$. In integral form

$$E[(X - \mu)^2] = \int (t - \mu)^2 \, dP_X$$

To form an approximating integral, we divide the t axis into small intervals; the mass located in each interval is multiplied by the square of the distance from the center of mass μ to a point in the interval; the sum is taken over all the intervals. The integral is thus the *second moment about the center of mass;* this is the moment of inertia of the mass about its center of mass. Since the total mass is unity, this is equal to the *square of the radius of gyration of the mass distribution.* The standard deviation is thus equal to the radius of gyration, and the variance to its square. As the mass is scattered more widely about the mean, the values of these parameters increase. Small variance, or standard deviation, indicates that the mass is located near its center of mass; in this case the probability that the random variable will take on values in a small interval about the mean value tends to be large.

Some basic properties

Utilizing properties of mathematical expectation, we may derive a number of properties of the variance which are useful. We suppose the variances indicated in the following expressions exist; this means that the squares of the random variables are integrable.

A restatement of the result obtained in Example 5-2-4 shows that

(V1) $\sigma^2[X] = E[X^2] - E^2[X] = E[X^2] - [\mu_X]^2$

Multiplication by a constant produces the following effect on the variance.

(V2) $\sigma^2[aX] = a^2\sigma^2[X]$

PROOF

$$\sigma^2[aX] = E[(aX)^2] - E^2[aX] = a^2 E[X^2] - \{aE[X]\}^2 = a^2\{E[X^2] - E^2[X]\}. \quad \blacksquare$$

This property may be visualized in terms of the mass distribution. If $|a| > 1$, the variable $aX(\cdot)$ spreads the probability mass more widely about the mean than does $X(\cdot)$, with a consequent increase of the moment of inertia. For $|a| < 1$, the opposite condition is true. A change in sign produced by a negative value of a does not alter the spread of the probability mass about the center of mass.

Adding a constant to a random variable does not change the variance. That is,

(V3) $\sigma^2[X + a] = \sigma^2[X]$

PROOF

$$E[(X + a)^2] = E[X^2] + 2aE[X] + a^2$$
$$E^2[X + a] = \{E[X] + a\}^2 = E^2[X] + 2aE[X] + a^2$$

The difference is equal to $\sigma^2[X]$ by (V1). ∎

In terms of the mass distributions, the random variable $X(\cdot) + a$ induces a probability mass distribution which differs from that for $X(\cdot)$ only in being shifted by an amount a. This shifts the center of mass by an amount a. The spread about the center of mass is the same in both cases.

(V4) $\sigma^2[X \pm Y] = \sigma^2[X] + \sigma^2[Y] \pm 2\{E[XY] - E[X]E[Y]\}$

PROOF

$$\sigma^2[X \pm Y] = E[(X \pm Y)^2] - E^2[X \pm Y]$$
$$E[(X \pm Y)^2] = E[X^2] \pm 2E[XY] + E[Y^2]$$
$$E^2[X \pm Y] = \{E[X] \pm E[Y]\}^2 = E^2[X] \pm 2E[X]E[Y] + E^2[Y]$$

Taking the difference and using (V1) gives the desired result. ∎

An important simplification of the formula (V4) results if the random variables are independent, so that $E[XY] = E[X]E[Y]$. More generally, we have

(V5) If $\{X_i(\cdot): 1 \leq i \leq n\}$ is a class of random variables and $X(\cdot) = \displaystyle\sum_{i=1}^{n} \delta_i X_i(\cdot)$, where each δ_i has one of the values $+1$ or -1, then

$$\sigma^2[X] = \sum_{i=1}^{n} \sigma^2[X_i] + \sum_{i \neq j} \delta_i \delta_j \{E[X_i X_j] - E[X_i]E[X_j]\}$$

PROOF

$$\sigma^2[X] = E\left[\left\{\sum_i \delta_i X_i - \sum_i \delta_i \mu_i\right\}^2\right]$$
$$= \sum_i E[(X_i - \mu_i)^2] + \sum_{i \neq j} \delta_i \delta_j E[(X_i - \mu_i)(X_j - \mu_j)]$$

If the product rule $E[X_i X_j] = E[X_i]E[X_j]$ holds for all $i \neq j$, then each of the terms in the final summation in (V5) is zero. In particular, this is the case if the class is independent. Random variables satisfying this product law for expectations are said to be *uncorrelated*. ∎

SOLUTION In Example 5-3-3 it is shown that $E[X] = \mu$.

$$E[X^2] = e^{-\mu} \sum_{k=0}^{\infty} k^2 \frac{\mu^k}{k!} = e^{-\mu} \sum_{k=0}^{\infty} k(k-1) \frac{\mu^k}{k!} + \mu$$

$$= e^{-\mu} \sum_{k=2}^{\infty} \frac{\mu^k}{(k-2)!} + \mu = e^{-\mu} \mu^2 \sum_{r=0}^{\infty} \frac{\mu^r}{r!} + \mu$$

$$= \mu^2 + \mu$$

From this it follows that

$$\sigma^2[X] = \mu \quad \blacksquare$$

The normal distribution

We next obtain the variance for the normal, or gaussian, distribution and obtain some properties of this important distribution.

Example 5-4-5   The Normal Distribution

Find $\sigma^2[X]$ if $X(\cdot)$ has the normal distribution, as defined in Example 5-3-2.

SOLUTION

$$\sigma^2[X] = \frac{1}{\sigma\sqrt{2\pi}} \int (t-\mu)^2 \exp\left[-\frac{1}{2}\left(\frac{t-\mu}{\sigma}\right)^2\right] dt$$

$$= \frac{\sigma^2}{\sqrt{2\pi}} \int \left(\frac{t-\mu}{\sigma}\right)^2 \exp\left[-\frac{1}{2}\left(\frac{t-\mu}{\sigma}\right)^2\right] d\left(\frac{t-\mu}{\sigma}\right)$$

$$= \frac{\sigma^2}{\sqrt{2\pi}} \int x^2 \exp\left(-\frac{1}{2}x^2\right) dx = \sigma^2$$

as may be verified from a table of integrals. ■

Definition 5-4b

A random variable $X(\cdot)$ is said to be *normal* (μ, σ) iffi it is continuous and has the normal density function

$$f_X(t) = \frac{1}{\sigma\sqrt{2\pi}} \exp\left[-\frac{1}{2}\left(\frac{t-\mu}{\sigma}\right)^2\right]$$

If $X(\cdot)$ is normal (μ, σ), we may standardize it in the manner indicated in (V6).   It is of considerable interest and importance that the standardized variable is also normal.

Theorem 5-4A

If $X(\cdot)$ is normal (μ, σ), the standardized random variable

$$Y(\cdot) = \frac{X(\cdot) - \mu}{\sigma}$$

is normal $(0, 1)$.

Example 5-4-2

Let $X(\cdot)$ be a simple random variable given in canonical form by the expression

$$X(\cdot) = \sum_{i=1}^{n} t_i I_{A_i}(\cdot)$$

Let $p_i = P(A_i)$ and $q_i = 1 - p_i$. Determine $\sigma^2[X]$.

SOLUTION

$$X^2(\cdot) = \sum_{i=1}^{n} t_i^2 I_{A_i}(\cdot) \qquad \text{so that} \qquad E[X^2] = \sum_i t_i^2 p_i$$

$$E^2[X] = \left[\sum_i t_i p_i\right]^2 = \sum_i t_i^2 p_i^2 + 2 \sum_{i<j} t_i t_j p_i p_j$$

Using (V1), we get $\sigma^2[X] = \sum_i t_i^2 p_i q_i - 2 \sum_{i<j} t_i t_j p_i p_j$, since $p_i^2 = p_i(1 - q_i)$. ∎

In the last summation, the index notation $i < j$ is intended to indicate the sum of all distinct pairs of integers, with the pair (i, j) considered the same as the pair (j, i). The factor 2 takes account of the two possible ways any pair can appear.

Example 5-4-3

Suppose $\{A_i: 1 \leq i \leq n\}$ is a class of pairwise independent events. Let $X(\cdot) = \sum_{i=1}^{n} a_i I_{A_i}(\cdot)$. Put $P(A_i) = p_i$ and $q_i = 1 - p_i$. Determine $\sigma^2[X]$.

SOLUTION   Since by Example 3-7-1, we may assert $\{a_i I_{A_i}(\cdot): 1 \leq i \leq n\}$ is a class of pairwise independent random variables, we may apply (V5), (V2), and the result of Example 5-4-1 to give

$$\sigma^2[X] = \sum_{i=1}^{n} a_i^2 p_i q_i$$

For the random variable giving the number of successes in Bernoulli trials (Example 5-2-2), $A_i = E_i$, $a_i = 1$, and $p_i = p$. Hence the random variable $N(\cdot)$ with the *Binomial distribution* has $\sigma^2[N] = npq$. ∎

Example 5-4-4   *Poisson Distribution*

The random variable $X(\cdot) = \sum_{k=0}^{\infty} k I_{A_k}(\cdot)$, with the A_k forming a disjoint class and with $P(A_k) = p_k = e^{-\mu} \dfrac{\mu^k}{k!}$, has the Poisson distribution. Determine $\sigma^2[X]$.

It is frequently convenient to make a transformation from a random variable $X(\cdot)$ having a specified mean and variance to a related random variable having zero mean and unit variance. This may be done as indicated in the following:

(**V6**) Consider the random variable $X(\cdot)$ with mean $\mu[X] = \mu$ and standard deviation $\sigma[X] = \sigma$. Then the random variable

$$Y(\cdot) = \frac{X(\cdot) - \mu}{\sigma}$$

has mean $\mu[Y] = 0$ and standard deviation $\sigma[Y] = 1$.

PROOF

$$\mu[Y] = E\left[\frac{X - \mu}{\sigma}\right] = \frac{1}{\sigma}\{E[X] - \mu\} = 0$$

$$\sigma^2[Y] = \frac{1}{\sigma^2}\sigma^2[X - \mu] = \frac{1}{\sigma^2}\sigma^2[X] = 1 \qquad \text{by (V2) and (V3)} \blacksquare$$

Some examples

We now consider a number of simple but important examples.

Example 5-4-1   The Indicator Function $I_A(\cdot)$

$$\sigma^2[I_A] = pq \qquad \text{where } p = P(A) \text{ and } q = P(A^c) = 1 - p$$
$$\sigma^2[I_A I_B] = \sigma^2[I_{AB}] = P(AB)P(A^c \cup B^c)$$
$$\sigma^2[I_A \pm I_B] = P(A)P(A^c) + P(B)P(B^c) \pm 2[P(AB) - P(A)P(B)]$$

PROOF AND DISCUSSION $\sigma^2[I_A] = E[I_A{}^2] - E^2[I_A]$. Since $I_A(\cdot)$ has values only zero or one, $I_A{}^2(\cdot) = I_A(\cdot)$. Using the fact that $E[I_A] = P(A) = p$, we have $\sigma^2[I_A] = p - p^2 = p(1 - p) = pq$. This problem may also be viewed in terms of the mass distribution with the aid of Fig. 5-4-1. Mass $p = P(A)$ is located at $t = 1$, and mass q is located at $t = 0$. The center of mass is at $t = p$. Summing the products of the mass times the square of its distance from the center of mass yields

$$\sigma^2 = qp^2 + pq^2 = pq(p + q) = pq$$

The second case follows from the first and the fact that $(AB)^c = A^c \cup B^c$. The third relation comes from a direct application of (V4). $\blacksquare$

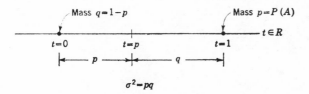

Fig. 5-4-1   Variance for the indicator function $I_A(\cdot)$.

PROOF   This result may be obtained easily from the result of Example 3-9-4. Putting $a = 1/\sigma$, which is positive, and $b = -\mu/\sigma$, we have

$$F_Y(v) = F_X(\sigma v + \mu) \qquad f_Y(v) = \sigma f_X(\sigma v + \mu)$$

Substituting $\sigma v + \mu$ for t in the expression for $f_X(t)$ gives

$$f_Y(v) = \frac{1}{\sqrt{2\pi}} \exp\left(-\frac{1}{2} v^2\right)$$

which is the condition required. ■

It is customary to use special symbols for the distribution and density functions for a random variable $X(\cdot)$ which is normal $(0, 1)$.   In this case it is customary to designate $F_X(\cdot)$ by $\Phi(\cdot)$ and $f_X(\cdot)$ by $\varphi(\cdot)$.   Extensive tables of these functions are found in many places.   The general shape of the graphs for these functions is given in Fig. 3-4-4.   By letting

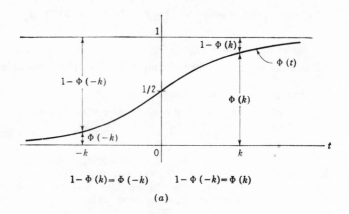

$$1 - \Phi(k) = \Phi(-k) \qquad 1 - \Phi(-k) = \Phi(k)$$

(a)

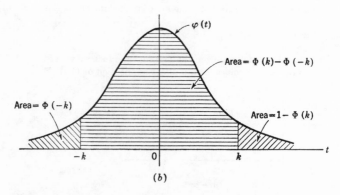

(b)

Fig. 5-4-2   (a) Distribution function $\Phi(\cdot)$ and (b) density function $\varphi(\cdot)$ for a random variable normal $(0, 1)$.

$\mu = 0$ and adjusting the scale for $\sigma = 1$, these become curves for $\Phi(t)$ and $\varphi(t)$ as shown in Fig. 5-4-2.

If $X(\cdot)$ is normal (μ, σ), then

$$F_X(t) = \Phi\left(\frac{t - \mu}{\sigma}\right) \quad \text{and} \quad f_X(t) = \frac{1}{\sigma}\,\varphi\left(\frac{t - \mu}{\sigma}\right)$$

so that tables of $\Phi(\cdot)$ and $\varphi(\cdot)$ suffice for any normally distributed random variable.   As an illustration of this fact, consider the following

Example 5-4-6

A random variable $X(\cdot)$ is normal (μ, σ).   What is the probability that it does not differ from its mean value by more than k standard deviations?

SOLUTION $P(|X - \mu| \leq k\sigma) = P\left(-k \leq \dfrac{X - \mu}{\sigma} \leq k\right)$.   Since $(X - \mu)/\sigma$ is normal $(0, 1)$, the desired probability is $\Phi(k) - \Phi(-k)$.   This may be put into another form which requires only values of $\Phi(\cdot)$ for positive k, by utilizing the symmetries of the distribution or density function.   From the curves in Fig. 5-4-2, it is apparent that $\Phi(-k) = 1 - \Phi(k)$, so that $\Phi(k) - \Phi(-k) = 2\Phi(k) - 1$. ∎

This example merely serves to illustrate how the symmetries of the normal distribution and density functions may be exploited to simplify or otherwise modify expressions for probabilities of normally distributed random variables.   Once these symmetries are carefully noted, the manner in which they may be utilized becomes apparent in a variety of situations. Extensive tables of $\Phi(\cdot)$, $\varphi(\cdot)$, and related functions are to be found in most handbooks, as well as in books on statistics or applied probability. The table below provides some typical values which are helpful in making estimates.

t	$\varphi(t)$	$2\Phi(t) - 1$	$\Phi(t)$
0	0.3989	0	0.5
0.1	0.3970	0.07966	0.5398
0.5	0.3521	0.38292	0.6915
0.67449		0.50000	
1.0	0.2420	0.68268	0.8413
2.0	0.0540	0.95450	0.9772
3.0	0.0044	0.99730	0.9987
4.0	0.0001	0.99994	

The inequality of Chebyshev

We have seen that the standard deviation gives an indication of the spread of possible values of a random variable about the mean.   This fact has been given analytical expression in the following inequality of Chebyshev (spelled variously in the literature as Tchebycheff or Tshebysheff).

Theorem 5-4B   Chebyshev Inequality

Let $X(\cdot)$ be any random variable whose mean μ and standard deviation σ exist. Then

$$P(|X - \mu| \geq k\sigma) \leq \frac{1}{k^2}$$

or, equivalently,

$$P(|X - \mu| \geq k) \leq \frac{\sigma^2}{k^2},$$

PROOF   Since $|X - \mu| \geq k$ iffi $(X - \mu)^2 \geq k^2$, we may use property (E7) with $g(X) = (X - \mu)^2$ to assert

$$k^2 P[(X - \mu)^2 \geq k^2] \leq E[(X - \mu)^2] = \sigma^2 \quad \blacksquare$$

The first form of the inequality is interesting in that it shows the effective spread of probability mass to be measured in multiples of σ. Thus the standard deviation provides an important unit of measure in evaluating the spread of probability mass. This fact is also illustrated in Example 5-4-6. It should be noted that the Chebyshev inequality is quite general in its application. It has played a significant role in providing the estimates needed to prove various limit theorems (Chap. 6). Numerical examples show, however, that it does not provide the best possible estimate of the probability in special cases. For example, in the case of the normal distribution, Chebyshev's estimate for $k = 2$ is $\frac{1}{4}$, whereas from the table above for the normal distribution, it is apparent that

$$P(|X - \mu| \geq 2\sigma) = 1 - 0.9545 = 0.0455$$

Another comparison is made for an important special case in Sec. 6-2. The general inequality is useful, however, for estimates in situations where detailed information concerning the distribution of the random variable is not available.

A number of other special mathematical expectations are used extensively in mathematical statistics to describe important characteristics of a probability mass distribution. For a careful treatment of some of the more important of these, one may consult any of a number of works in mathematical statistics (e.g., Brunk [1964], Cramér [1946], or Fisz [1963]).

It may be well at this point to make a new comparison of the situation in the real world, in the mathematical model, and in the auxiliary model, as we have done previously.

Real world	Mathematical model	Auxiliary model
Empirical relative frequencies	Mathematical probabilities	Probability masses
Observed numerical phenomena	Random variables	Mappings and mass transfer
Averages	Sums and integrals; expectations	Moments of mass distributions

5-5   Random samples and random variables

The process of sampling is fundamental in statistics. In Examples 2-8-9 and 2-8-10, the idea of random sampling is touched upon briefly. In this section, we examine the idea in terms of random variables.

Suppose a physical quantity is under observation in some sort of experiment. A sequence of n observations $(t_1, t_2, \ldots, t_n)$ is recorded. Under appropriate conditions, these observations may be considered to be n observations, or "samples," of a single random variable $X(\cdot)$, having a given probability distribution. A more natural way to consider these observations, from the viewpoint of our mathematical model, is to think of them as a single observation of the values of a class of random variables $\{X_1(\cdot), X_2(\cdot), \ldots, X_n(\cdot)\}$, where $t_1 = X_1(\xi)$, $t_2 = X_2(\xi)$, etc.

Definition 5-5a

If the conditions of the physical sampling or observation process described above are such that

　　1. The class $\{X_1(\cdot), X_2(\cdot), \ldots, X_n(\cdot)\}$ is an independent class, and

　　2. Each random variable $X_i(\cdot)$ has the same distribution as does $X(\cdot)$,

the set of observations is called a *random sample of size n of the variable* $X(\cdot)$.

This model seems to be a satisfactory one in many physical sampling processes. The concept of stochastic independence is the mathematical counterpart of an assumed "physical independence" between the various members of the sample. The condition that the $X_i(\cdot)$ have the *same distribution* seems to be met when the sampling is from an *infinite population* (that is, one so large that the removal of a unit does not appreciably affect the distribution in the population) or from a *finite population with replacement*.

When a sample is taken, it is often desirable to average the set of values observed. The significance of this operation from the point of view of the probability model may be obtained as follows. First we make the

Definition 5-5b

The random variable $A(\cdot)$ given by the expression

$$A(\cdot) = \frac{1}{n} \sum_{i=1}^{n} X_i(\cdot)$$

is known as the *sample mean*.

Each choice of an elementary outcome ξ corresponds to the choice of a set of values $(t_1, t_2, \ldots, t_n)$. But this also corresponds to the choice of the number

$$A(\xi) = \frac{1}{n}(t_1 + t_2 + \cdots + t_n)$$

which is the sample mean. The following simple but important theorem shows the significance of the sample mean in statistical analysis.

Theorem 5-5A

If $A(\cdot)$ is the sample mean for a sample of size n of the random variable $X(\cdot)$, then

$$\mu[A] = \mu[X] \qquad \text{and} \qquad \sigma^2[A] = \frac{1}{n}\sigma^2[X]$$

provided $\mu[X]$ and $\sigma^2[X]$ exist

PROOF   The first relation is a direct result of the linearity property for expectations, and the second is a direct result of the additivity property (V5) for variances of independent random variables. Thus

$$\mu[A] = E[A] = \frac{1}{n}\sum_{i=1}^{n} E[X_i] = \mu[X]$$

$$\sigma^2[A] = \frac{1}{n^2}\sum_{i=1}^{n} \sigma^2[X_i] = \frac{1}{n}\sigma^2[X] \quad \blacksquare$$

This theorem says that the mean of the sample mean is the same as the mean of the random variable being sampled. The variance of the sample mean, however, is only $1/n$ times the variance of the random variable being sampled. In view of the role of the variance in measuring the spread of possible values of the random variable about its mean, this indicates that the values of the sample mean can be concentrated about the mean by taking large enough samples. This may be given analytical formulation with the help of the Chebyshev inequality (Theorem 5-4B). If we put $\mu[X] = \mu$ and $\sigma^2[X] = \sigma^2$, we have

$$P(|A - \mu| \geq k\sigma) \leq \frac{\sigma^2/n}{k^2\sigma^2} = \frac{1}{nk^2}$$

For any given k we can make the probability that $A(\cdot)$ differs from μ by more than $k\sigma$ as small as we please by making the sample size n sufficiently large.

In various situations of practical interest, much better estimates are available than that provided by Chebyshev inequality. One of the central problems of mathematical statistics is to study such estimates. For an introduction to such studies, one may refer to any of a number of standard works on statistics (e.g., Brunk [1964] or Fisz [1963]).

5-6   Probability and information

Probability ideas are being applied to problems of communication, control, and data processing in a variety of ways. One of the important aspects of this statistical communication theory is the so-called *information theory*, whose foundations were laid by Claude Shannon in a series of brilliant researches carried out in the 1940s at Bell Telephone Laboratories. In this section, we sketch briefly some of the central ideas of information theory and see how they may be given expression in terms of the concepts of probability theory.

We suppose the communication system under study transmits signals in terms of some known, finite set of symbols. These may be the symbols of an alphabet, words chosen from a given code-word list, or messages from a prescribed set of possible messages. For convenience, we refer to the unit of signal under consideration as a *message*. The situation assumed is this: when a signal conveying a message is received, it is known that *one of the given set of possible messages has been sent*. The usual communication channel is characterized by noise or disturbances of a "random" character which perturb the transmitted signal. The actual physical process may be quite complicated.

The behavior of such a system is characterized by *uncertainties*. The problem of communication is to remove these uncertainties. If there were no uncertainty about the message transmitted over a communication system, there would be no real reason for the system. All that is needed in this case is a device to generate the appropriate signal at the receiving station. There are two sources of uncertainty in the communication system:

1. *The message source* (whether a person or a physical device under observation). It is not known with certainty which of the possible messages will be sent at any given transmission time.

2. *The transmission channel*. Because of the effects of noise, it is not known with certainty which signal was transmitted even when the received signal is known.

This characterization of a communication system points to two fundamental ideas underlying the development of a model of information theory.

1. The signal transmitted (or received) is one of a *set* of conceptually possible signals. The signal actually sent (or received) is chosen from this set.

2. The receipt of information is equivalent to the removal of uncertainty.

The problem of establishing a model for information theory requires first the identification of a suitable measure of information or uncertainty.

Shannon, in a remarkable work [1948], identified such a concept and succeeded in establishing the foundations of a mathematical theory of information. In this theory, uncertainty is related to probabilities. Shannon showed that a reasonable model for a communication source describes the source in terms of probabilities assigned to the various messages and subsets of messages in the basic set. The effects of noise in the communication channel are characterized by conditional probabilities. A given signal is transmitted. The signal is perturbed in such a manner that one can assign, in principle at least, conditional probabilities to each of the possible received signals, given the particular signal transmitted.

In this theory, the receipt of information is equivalent to the removal of uncertainty about which of the possible signals is transmitted. The concept of information employed has nothing to do with the meaning of the message or with its value to the recipient. It simply deals with the problem of getting knowledge of the correct message to the person at the receiving end.

For a review of the ideas which led up to the formulation of the mathematical concepts, Shannon's original papers are probably the best source available. We simply begin with some definitions, due principally to Shannon, but bearing the mark of considerable development by various workers, notably R. M. Fano and a group of associates at the Massachusetts Institute of Technology (Fano [1961] or Abramson [1963]). The importance of the concepts defined rests upon certain properties and the interpretation of these properties in terms of the communication problem.

Probability schemes and uncertainty

The output of an information source or of a communication channel at any given time can be considered to be a random variable. This may be a random variable whose "values" are nonnumerical, but this causes no difficulty. The random variable indicates in some way which of the possible messages is chosen (transmitted or received). We limit ourselves to the case where the number of possible messages is finite.   Information theory depends upon the identification of which of the possible outcomes has been realized.   Once this is known, it is assumed that the proper signal can be generated (i.e., that the "value" is known).   If the random variable is

$$X(\cdot) = \sum_{i \in J} m_i I_{A_i}(\cdot)$$

where m_i represents the ith message, and A_i is the event that the message m_i is sent, there is uncertainty about which message m_i will be sent. This means that there is uncertainty about which event in the partition $\{A_i \colon i \in J\}$ will occur.   This uncertainty must be related in some manner to the set of probabilities $\{P(A_i) \colon i \in J\}$ for the events in the partition.   For example, consider three possible partitions of the basic space into two events (corresponding to two possible messages):

$$\begin{bmatrix} A_1 & A_2 \\ 0.5 & 0.5 \end{bmatrix} \qquad \begin{bmatrix} B_1 & B_2 \\ 0.3 & 0.7 \end{bmatrix} \qquad \begin{bmatrix} C_1 & C_2 \\ 0.01 & 0.99 \end{bmatrix}$$

It is intuitively obvious that the first system, or scheme, indicates the greatest uncertainty as to the outcome; the next two schemes indicate decreasing amounts of uncertainty as to which event may be expected to occur.   One may test his intuition by asking himself how he would bet on the outcome in each case.   There is little doubt how one would bet in the third case; one would not bet with as much confidence in the second case; and in the first case the outcome would be so uncertain that he may reasonably bet either way.   In evaluating more complicated schemes, or in comparing schemes with different numbers of events, the amounts of uncertainty may not be obvious.   The problem is to find a measure which assigns a number to the uncertainty of the outcome. Actually, Shannon's measure is a measure of the *average* uncertainty for the various events in the partition; and as the examples above suggest, this measure must be a function of the set of probabilities of the events in the partition.   A number of intuitive and theoretical considerations led to the introduction of a measure of uncertainty which depends upon the logarithms of the probabilities.   We shall not attempt to trace the development that led to this choice, but shall simply introduce the measure discovered by Shannon (and anticipated by others) and see that the properties of such a measure admit of an appropriate interpretation in terms of uncertainty.

Mutual and self-information of individual events

Shannon's measure refers to probability schemes consisting of partitions and the associated set of probabilities.   It is helpful, however, to begin with the idea of the mutual information in any two events; these events may or may not be given as members of a partition. At this point we simply make a mathematical definition. The justification of this definition must rest on its mathematical consequences and on the interpretations which can be made in terms of uncertainties among events in the real world.

Definition 5-6a

The *mutual information* in two events A and B is given by the expression

$$\mathcal{I}(A:B) = \log_2 \frac{P(AB)}{P(A)P(B)} \qquad \text{(in bits)}$$

It should be noted that the logarithm in the definition is to the base 2. If any other base had been chosen, the result would be to multiply the expression above by an appropriate constant. This is equivalent to a scale change. *Unless otherwise indicated, logarithms in this section are taken to base* 2, in keeping with the custom in works on information theory. With this choice of base, the unit of information is the *bit;* the term was coined as a contraction of the term *binary digit*. Reasons for the choice of this term are discussed in practically all the introductions to information theory.

A number of facts about the properties of mutual information are easily derived. First we note a symmetry property which helps justify the term *mutual*.

$$\mathcal{I}(A:B) = \mathcal{I}(B:A) \qquad \text{(symmetry property)}$$

The mutual information in events A and B is the same as the mutual information in events B and A. The order of considering or listing is not important. The function may also be written in a form which hides the symmetry but which leads to some relationships that are important in the development of the topic.

$$\mathcal{I}(A:B) = \log \frac{P(A|B)}{P(A)} = \log \frac{P(B|A)}{P(B)}$$

An examination of the possibilities shows that mutual information may take on any real value, positive, negative, or zero. That is,

$$-\infty < \mathcal{I}(A:B) < \infty$$

The phenomenon of negative mutual information usually surprises the beginner. The following argument shows the significance of this condition. From the form above, which uses conditional probabilities, it is apparent that

$$\mathcal{J}(A:B) < 0 \qquad \text{iffi } P(A|B) < P(A) \qquad \text{iffi } P(B|A) < P(B)$$

Use of elementary properties of conditional probabilities verifies the equivalence of the last two inequalities as well as the fact that

$$P(A|B) < P(A) \qquad \text{iffi } P(A^c|B) > P(A^c) \qquad \text{iffi } P(B^c|A) > P(B^c)$$

so that

$$\mathcal{J}(A:B) < 0 \qquad \text{iffi } \mathcal{J}(A^c:B) > 0 \qquad \text{iffi } \mathcal{J}(A:B^c) > 0$$

We may thus say that the mutual information between A and B is negative under the condition that the occurrence of B makes the non-occurrence of A more likely or the occurrence of A makes the non-occurrence of B more likely. It is also easy to see from the definition that

$$\mathcal{J}(A:B) = 0 \qquad \text{iffi } A, B \text{ are independent}$$

The mutual information is zero if the occurrence of one event does not influence (or condition) the probability of the occurrence of the other.

We may rewrite the expression, using conditional probabilities in the following manner to exhibit an important property.

$$\mathcal{J}(A:B) = - \log P(A) - [- \log P(A|B)] \leq - \log P(A)$$
$$\mathcal{J}(A:B) = - \log P(B) - [- \log P(B|A)] \leq - \log P(B)$$

It is apparent that

$$\mathcal{J}(A:B) \leq - \log P(A) \qquad \text{with equality iffi } P(A|B) = 1$$

In particular, $\mathcal{J}(A:A) = - \log P(A)$ and $\mathcal{J}(B:B) = - \log P(B)$. The mutual information of an event with itself is naturally interpreted as an information property of that event. It is thus natural to make the

Definition 5-6b

The *self-information* in an event A is given by

$$\mathcal{J}(A) = - \log_2 P(A) \qquad \text{(in bits)}$$

The self-information is nonnegative; it has value zero iffi $P(A) = 1$, which means there is no uncertainty about the occurrence of A. In this case, its occurrence removes no uncertainty, hence conveys no information.

Parallel definitions may be made when the probabilities involved are conditional probabilities.

Definition 5-6c

The *conditional mutual information in A and B, given C*, is

$$\mathcal{I}(A:B|C) = \log_2 \frac{P(AB|C)}{P(A|C)P(B|C)} \qquad \text{(in bits)}$$

The *conditional self-information in A, given C*, is

$$\mathcal{I}(A|C) = -\log_2 P(A|C) \qquad \text{(in bits)}$$

Properties of the conditional-information measures are similar to those for the information, since conditional probability is a probability measure. Some interpretations are of interest. The conditional self-information is nonnegative. $\mathcal{I}(A|C) = 0$ iff $P(A|C) = 1$. This means that there is no uncertainty about the occurrence of A, given that C is known to have occurred. $\mathcal{I}(A|C) = \mathcal{I}(A)$ iff A and C are independent. Independence of the events A and C implies that knowledge of the occurrence of C gives no knowledge of the occurrence of A.

Some simple but important relations between the concepts defined above are of interest.

$$\begin{aligned}\mathcal{I}(A:B) &= \mathcal{I}(A) - \mathcal{I}(A|B) = \mathcal{I}(B) - \mathcal{I}(B|A) \\ &= \mathcal{I}(A) + \mathcal{I}(B) - \mathcal{I}(AB)\end{aligned}$$

The last relation may be rewritten

$$\mathcal{I}(AB) = \mathcal{I}(A) + \mathcal{I}(B) - \mathcal{I}(A:B)$$

These relations may be given interpretations in terms of uncertainties. We suppose $\mathcal{I}(A)$ is the uncertainty about the occurrence of event A (i.e., the information received when event A occurs). If event B occurs, it may or may not change the uncertainty about event A. We suppose $\mathcal{I}(A|B)$ is the residual uncertainty about A when one has knowledge of the occurrence of B. We may interpret $\mathcal{I}(B)$ and $\mathcal{I}(B|A)$ similarly. Then, according to the first relation, $\mathcal{I}(A:B) = \mathcal{I}(B:A)$ is the information about A provided by the occurrence of the event B; similarly, the mutual information is the information about B provided by the occurrence of A. The last relation above may then be interpreted to mean that the information in the joint occurrence of events A and B is the information in the occurrence of A plus that in the occurrence of B minus the mutual information. Since $\mathcal{I}(A:B) = 0$ iff A and B are independent, the information in the joint occurrence AB is the sum of the information in A and in B iff the two events are independent.

Probability schemes

We turn next to the problem of measuring the average uncertainties in systems or schemes in which any one of several outcomes is possible. In order to be able to speak with precision and economy of statement, we formalize some notions and introduce some terminology.

The concept of a partition has played an important role in the development of the theory of probability. For a given probability measure defined on a suitable class of events, there is associated with each partition a corresponding set of probabilities. If the partition is

$$\alpha = \{A_i : i \in J\}$$

there is a corresponding set of probabilities $\{p_i = P(A_i) : i \in J\}$ with the properties

(1) $p_i \geq 0 \qquad i \in J \qquad$ and $\qquad$ (2) $\sum_{i \in J} p_i = 1$

It should be apparent that, if we were given any set of numbers p_i satisfying these two conditions, we could define a probability space and a partition such that these are the probabilities of the members of the partition. It is frequently convenient to use the following terminology in dealing with such a set of numbers:

Definition 5-6d

If $\{p_i : i \in J\}$ is a finite or countably infinite set of nonnegative numbers whose sum is 1, we refer to this indexed or ordered set as a *probability vector*. If there is a finite number n of elements, it is sometimes useful to refer to the set as a *probability n-vector*.

The combination of a partition and its associated probability vector we refer to according to the following

Definition 5-6e

If $\alpha = \{A_i : i \in J\}$ is a finite or countably infinite partition with associated probability vector $\{P(A_i) : i \in J\}$, we speak of the collection $\mathbf{\alpha} = \{[A_i, P(A_i)] : i \in J\}$ as a *probability scheme*. If the partition has a finite number of members, we speak of a *finite (probability) scheme*.

The simple schemes used earlier in this section to illustrate the uncertainty associated with such schemes are special cases having two member events each. It is easy to extend this idea to joint probability schemes in which the partition is a joint partition $\alpha\mathcal{B}$ of the type discussed in Sec. 2-7.

Definition 5-6f

If $\mathcal{A}\mathcal{B} = \{A_iB_j : i \in I, j \in J\}$ is a joint partition, the collection $\mathbf{\mathcal{A}\mathcal{B}} = \{[A_iB_j, P(A_iB_j)] : i \in I, j \in J\}$ is called a *joint probability scheme.*

This concept may be extended in an obvious way to the joint scheme for three or more schemes. In the case of the joint scheme $\mathbf{\mathcal{A}\mathcal{B}}$, we must have the following relationships between probability vectors. We let

$$P(A_iB_j) = p(i, j) \qquad P(A_i) = p(i, *) \qquad P(B_j) = p(*, j)$$

Then $\{p(i, j) : i \in I, j \in J\}$, $\{p(i, *) : i \in I\}$, and $\{p(*, j) : j \in J\}$ are all probability vectors. They are related by the fact that

$$p(i, *) = \sum_{j \in J} p(i, j)$$
$$p(*, j) = \sum_{i \in I} p(i, j)$$

Definition 5-6g

Two schemes $\mathcal{A}$ *and* $\mathcal{B}$ are said to be *independent* iffi

$$P(A_iB_j) = P(A_i)P(B_j)$$

for all $i \in I$ and $j \in J$.

Average uncertainty for probability schemes

We have already illustrated the fact that the uncertainty as to the outcome of a probability scheme is dependent in some manner on the probability vector for the scheme. Also, we have noted that in a communication system the probability scheme of interest is that determined by the choice of a message from among a set of possible messages. Shannon developed his very powerful mathematical theory of information upon the following basic measure for the average uncertainty of the scheme associated with the choice of a message.

Definition 5-6h

For schemes $\mathcal{A}$, $\mathcal{B}$, and $\mathbf{\mathcal{A}\mathcal{B}}$, described above, we define the *entropies*

$$H(\mathcal{A}) = - \sum_{i \in I} p(i, *) \log_2 p(i, *)$$
$$H(\mathcal{B}) = - \sum_{j \in J} p(*, j) \log_2 p(*, j)$$
$$H(\mathcal{A}\mathcal{B}) = - \sum_{i,j} p(i, j) \log_2 p(i, j)$$

The term *entropy* is used because the mathematical functions are the same as the entropy functions which play a central role in statistical mechanics. Numerous attempts have been made to relate the entropy concept in communication theory to the entropy concept in statistical mechanics. For an interesting discussion written for the nonspecialist, one may consult Pierce [1961, chaps. 1 and 10]. For our purposes, we shall be concerned only to see that the functions introduced do have the properties that one would reasonably demand of a measure of uncertainty or information.

Since the partitions of interest in the probability schemes are related to random variables $X(\cdot)$ and $Y(\cdot)$ whose values are messages (or numbers representing the messages), the corresponding entropies are often designated $H(X), H(Y),$ and $H(XY)$. The random variables rather than the schemes are displayed by this notation. For our purposes, it seems desirable to continue with the notation designating the scheme.

We have alluded to the fact that the value of the entropy in some sense measures the *average* uncertainty for the scheme. We wish to see that this is the case, provided the self- and mutual-information concepts introduced earlier measure the uncertainty associated with any single event or pair of events.

First let us note that if we have a sum of the form

$$\sum_{i \in I} p_i a_i$$

where the p_i are components of the probability vector for a scheme, *this sum is in fact the mathematical expectation of a random variable.* We suppose $\{A_i : i \in I\}$ is a partition and that $p_i = P(A_i)$ for each i. Then if we form the random variable (in canonical form)

$$X(\cdot) = \sum_{i \in I} a_i I_{A_i}(\cdot)$$

it follows immediately from the definition of mathematical expectation that

$$\sum_{i \in I} p_i a_i = \sum_{i \in I} a_i P(A_i) = E[X]$$

Now suppose we let $a_i = -\log_2 p_i = -\log_2 P(A_i) = \mathcal{I}(A_i)$. That is, a_i is the self-information of the event A_i. We define the random variable $\mathcal{I}_a(\cdot)$ by

$$\mathcal{I}_a(\cdot) = \sum_{i \in I} \mathcal{I}(A_i) I_{A_i}(\cdot)$$

This is the random variable whose value for any ξ is the self-information $\mathcal{I}(A_i) = -\log_2 P(A_i)$ for the event A_i which occurs when this ξ is chosen. The mathematical expectation or probability-weighted average for this random variable is

$$E[\mathcal{I}_{\mathcal{C}}] = \sum_{i \in I} \mathcal{I}(A_i)P(A_i) = - \sum_{i \in I} p_i \log_2 p_i = H(\mathcal{C})$$

We may define similar random variables $\mathcal{I}_{\mathcal{B}}(\cdot)$ and $\mathcal{I}_{\mathcal{C}\mathcal{B}}(\cdot)$ for the schemes $\mathcal{B}$ and $\mathcal{C}\mathcal{B}$ and have as a result

$$E[\mathcal{I}_{\mathcal{B}}] = H(\mathcal{B}) \qquad \text{and} \qquad E[\mathcal{I}_{\mathcal{C}\mathcal{B}}] = H(\mathcal{C}\mathcal{B})$$

We may also define the random variable $\mathcal{I}_{\mathcal{C}|\mathcal{B}}(\cdot)$ whose value for any $\xi \in A_i B_j$ is the conditional self-information in A_i, given B_j. That is,

$$\mathcal{I}_{\mathcal{C}|\mathcal{B}}(\cdot) = \sum_{i,j} \mathcal{I}(A_i|B_j)I_{A_i B_j}(\cdot) = - \sum_{i,j} \log_2 P(A_i|B_j)I_{A_i B_j}(\cdot)$$

In a similar manner the variable $\mathcal{I}_{\mathcal{B}|\mathcal{C}}(\cdot)$ may be defined. It is also convenient to define the random variable

$$\mathcal{I}_{\mathcal{C}:\mathcal{B}}(\cdot) = \mathcal{I}_{\mathcal{C}}(\cdot) + \mathcal{I}_{\mathcal{B}}(\cdot) - \mathcal{I}_{\mathcal{C}\mathcal{B}}(\cdot)$$

Conditions for independence of the schemes $\mathcal{C}$ and $\mathcal{B}$ show that the following relations hold iffi $\mathcal{C}$ and $\mathcal{B}$ are independent:

$$\mathcal{I}_{\mathcal{C}\mathcal{B}}(\cdot) = \mathcal{I}_{\mathcal{C}}(\cdot) + \mathcal{I}_{\mathcal{B}}(\cdot)$$
$$\mathcal{I}_{\mathcal{C}|\mathcal{B}}(\cdot) = \mathcal{I}_{\mathcal{C}}(\cdot) \qquad \mathcal{I}_{\mathcal{B}|\mathcal{C}}(\cdot) = \mathcal{I}_{\mathcal{B}}(\cdot)$$

In terms of these functions we may define two further entropies:

$$H(\mathcal{C}:\mathcal{B}) = E[\mathcal{I}_{\mathcal{C}:\mathcal{B}}] \qquad \text{and} \qquad H(\mathcal{C}|\mathcal{B}) = E[\mathcal{I}_{\mathcal{C}|\mathcal{B}}]$$

These entropies may be expanded in terms of the probabilities as follows:

$$H(\mathcal{C}:\mathcal{B}) = \sum_{i,j} P(A_i B_j) \log_2 \frac{P(A_i B_j)}{P(A_i)P(B_j)}$$
$$H(\mathcal{C}|\mathcal{B}) = - \sum_{i,j} P(A_i B_j) \log_2 P(A_i|B_j)$$

Linearity properties of the mathematical expectation show that various relations developed between information expressions for events can be extended to the entropies. Thus

$$H(\mathcal{C}:\mathcal{B}) = H(\mathcal{C}) + H(\mathcal{B}) - H(\mathcal{C}\mathcal{B})$$
$$= H(\mathcal{C}) - H(\mathcal{C}|\mathcal{B})$$
$$= H(\mathcal{B}) - H(\mathcal{B}|\mathcal{C})$$

Interpretations for these expressions parallel those for the information expressions, with the addition of the term *average*.

The following very simple example deals with a system studied extensively in information theory. It shows how the entropy concepts appear in this theory.

Example 5-6-1   Binary Symmetric Channel

In information theory, a binary channel is one which transmits a succession of binary symbols, which may be represented by 0 and 1 (see also Example 2-5-7). We suppose the system deals independently with each successive symbol, so that we may study the behavior in terms of one symbol at a time. Let S_0 be the event a 0 is sent and S_1 be the event a 1 is sent; let R_0 and R_1 be the events of receiving a 0 or a 1, respectively. If the probabilities of the two input events are $P(S_0) = \pi$ and $P(S_1) = \bar{\pi} = 1 - \pi$, then the *input scheme* (which characterizes the message source) consists of the collection $\mathbf{S} = \{(S_0, \pi), (S_1, \bar{\pi})\}$. The channel is characterized by the conditional probabilities of the outputs, given the inputs. The channel is referred to as a *binary symmetric channel* (BSC) if $P(R_1|S_0) = P(R_0|S_1) = p$ and $P(R_0|S_0) = P(R_1|S_1) = q$, where $q = 1 - p$; that is, the probability that an input digit is changed into its opposite is p and the probability that an input digit is transmitted unchanged is q. These probability relations are frequently represented schematically as in Fig. 5-6-1.

The *binary entropy function* $H(p) = -p \log_2 p - q \log_2 q$ plays an important role in the theory of binary systems. [Note that we have used $H(\cdot)$ in a different sense here. This is in keeping with a custom that need cause no real difficulty]. A sketch of this function against the value of the transition probability p is shown in Fig. 5-6-2. Two characteristics important for our purposes in this example are evident in that figure. These are easily checked analytically.

$$H(p) = H(1 - p) = H(q) \qquad H(p) \leq H(\tfrac{1}{2}) = 1$$

If we let $\mathbf{\mathcal{R}}$ be the output scheme consisting of the partition $\{R_0, R_1\}$ and the associated probability vector, we may calculate the average mutual information between the input and output schemes and show it to be

$$H(\mathbf{S}:\mathbf{\mathcal{R}}) = H(\pi p + \bar{\pi}q) - H(p)$$

To see this, we write

$$H(\mathbf{S}:\mathbf{\mathcal{R}}) = \sum_{i,j} P(S_j)P(R_i|S_j) \log P(R_i|S_j) - \sum_j P(R_j) \log P(R_j)$$

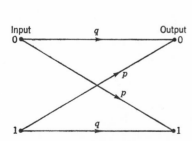

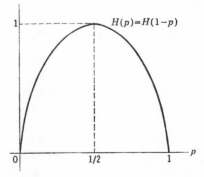

Fig. 5-6-1 Schematic representation of the binary symmetric channel.

Fig. 5-6-2 The binary entropy function.

We use the values

$$P(R_0) = \pi q + \bar{\pi} p \qquad P(R_1) = \pi p + \bar{\pi} q$$
$$P(R_0 S_0) = \pi q \qquad\quad P(R_1 S_0) = \pi p$$
$$P(R_0 S_1) = \bar{\pi} p \qquad\quad P(R_1 S_1) = \bar{\pi} q$$

Substituting and recognizing the form for the binary entropy function, we obtain

$$H(\mathbf{S}:\mathbf{R}) = \pi[-H(p)] + \bar{\pi}[-H(p)] + H(\pi p + \bar{\pi} q)$$

from which the desired result follows because $\pi + \bar{\pi} = 1$. This function gives the average information about the input provided by the receipt of symbols at the output. If we set the input probabilities equal to $\frac{1}{2}$, so that either input symbol is equally likely, the resulting value is

$$H(\mathbf{S}:\mathbf{R}) = 1 - H(p)$$

which is a maximum, since the maximum value of $H(\cdot)$ is 1 and it occurs when the argument is $\frac{1}{2}$. ∎

The entropy for an input-message scheme plays a central role in the theory of information. If a source produces one of n messages $\{m_i: 1 \leq i \leq n\}$ with corresponding probabilities $\{p_i: 1 \leq i \leq n\}$, the *source entropy* $H(\mathbf{S})$ is given by

$$H(\mathbf{S}) = - \sum_{i=1}^{n} p_i \log_2 p_i$$

A basic coding theorem shows that these messages may be represented by a binary code in which the probability-weighted average L of the word lengths satisfies

$$H(\mathbf{S}) \leq L < H(\mathbf{S}) + 1$$

By considering coding schemes in which several messages are coded in groups, it is possible to make the average code-word length as close to the source entropy as desired. This is done at the expense of complications in coding, of course. Roughly speaking, then, the source entropy in bits (i.e., when logarithms are to base 2) measures the minimum number of binary digits per code word required to encode the source. It is interesting to note that the codes which achieve this minimum length do so by assigning to each message m_i a binary code word whose length is the integer next larger than the self-information $- \log_2 p_i$ for that message. In the more sophisticated theory of coding for efficient transmission in the presence of noise, the source entropy and the mutual and conditional entropies for channel input and output play similar roles. It is outside the scope of this book to deal with these topics, although

some groundwork has been laid.   For a clear, concise, and easily readable
treatment, one may consult the very fine work by Abramson [1963], who
deals with simple cases in a way that illuminates the content and essential
character of the theory.

We shall conclude this section by deriving several properties of the
entropy functions which strengthen the interpretation in terms of uncertainty or information.   In doing so, we shall find the following inequality
and some inequalities derived from it very useful.

$$\log_e x \leq x - 1 \qquad \text{or} \qquad \log_2 x \leq (x - 1) \log_2 e$$

with equality iffi $x = 1$.   This inequality follows from the fact that the
curve $y = \log_e x$ is tangent to the line $y = x - 1$ at $x = 1$ while the
slope of $y = \log_e x$ is a decreasing function of x.

In calculations with entropy expressions, we note that $x \log x$ is 0
when x has the value 1.   Since $x \log x$ approaches 0 as x goes to 0, we
define $0 \log 0$ to be 0.

Consider two probability vectors $\{p_i : 1 \leq i \leq n\}$ and $\{q_i : 1 \leq i \leq n\}$.
Then we show that

$$-\sum_i p_i \log p_i \leq -\sum_i p_i \log q_i \qquad \text{or} \qquad \sum_i p_i \log \frac{q_i}{p_i} \leq 0$$

with equality iffi $p_i = q_i$ for each i.   To show this, we note

$$0 = \sum_i p_i \left(\frac{q_i}{p_i} - 1 \right)$$

$$\geq \sum_i p_i \log_e \frac{q_i}{p_i} = \sum_i p_i \log_e q_i - \sum_i p_i \log_e p_i$$

Equality holds iffi $q_i/p_i = 1$ for each i.

We now collect some properties and relations for entropies which are
important in information theory.   Some of these properties have already
been considered above.

(**H1**)   If, for some $i = i_0$, $P(A_{i_0}) = 1$, then $H(\mathfrak{C}) = 0$.

(**H2**) $H(\mathfrak{C}) = h[p(1, *), p(2, *), \ldots, p(n, *)]$

$$\leq h \left[\frac{1}{n}, \frac{1}{n}, \ldots, \frac{1}{n} \right] = \log_2 n$$

with equality iffi $p(i, *) = 1/n$ for each i.

(**H3**) $H(\mathfrak{C}) = h(p_1, p_2, \ldots, p_n) = H(\mathfrak{C}') = h(p_1, p_2, \ldots, p_n, 0)$

(**H4**) $H(\mathfrak{C}\mathfrak{B}) = H(\mathfrak{C}) + H(\mathfrak{B}) - H(\mathfrak{C} : \mathfrak{B})$

$$= H(\mathfrak{C}) + H(\mathfrak{B}|\mathfrak{C}) = H(\mathfrak{B}) + H(\mathfrak{C}|\mathfrak{B})$$

(**H5**) $H(\mathfrak{C}|\mathfrak{B}) \leq H(\mathfrak{C})$ and $H(\mathfrak{B}|\mathfrak{C}) \leq H(\mathfrak{B})$ with equality iffi schemes $\mathfrak{C}$ and $\mathfrak{B}$
are independent schemes.

PROOF

$$H(\mathfrak{A}|\mathfrak{B}) - H(\mathfrak{A}) = \sum_{i,j} P(A_i B_j)[\log P(A_i) - \log P(A_i|B_j)]$$

$$= \sum_{i,j} P(A_i B_j) \log \frac{P(A_i)P(B_j)}{P(A_i B_j)}$$

$$\leq \sum_{i,j} P(A_i B_j) \left[\frac{P(A_i)P(B_j)}{P(A_i B_j)} - 1\right] \log_2 e = 0$$

Equality holds iffi $P(A_i B_j) = P(A_i)P(B_j)$ for each i, j. ∎

(H6) $H(\mathfrak{A}|\mathfrak{B}\mathfrak{C}) \leq H(\mathfrak{A}|\mathfrak{B})$

PROOF   The proof is similar to that in (H5), but the details are somewhat more complicated.   We omit the proof. ∎

Property (H1) indicates that the average uncertainty in a scheme is zero if there is probability 1 that one of the events A_{i_0} will occur.   Property (H2) says that the average uncertainty is a maximum iffi the events of the scheme are all equally likely.   Property (H3) indicates that no change in the uncertainty of a scheme is made if an event of zero probability is split off as a separate event.   Property (H4) has been discussed above. Property (H5) says that conditioning (i.e., giving information about one scheme) cannot increase the average uncertainty about the other.   Property (H6) extends this result to conditioning by several schemes.

A number of investigators have shown that entropies defined as functions of the probabilities in schemes must have the functional form given in the definition, if the entropy function is to have the properties listed above.   Shannon showed this in his original investigation [1948]. Investigators have differed in the list of properties taken as axiomatic, although they have usually taken an appropriate subset of the properties listed above.   Their lists have been equivalent in the sense that any one implies the others.   As Shannon has pointed out, however, the importance of the entropy concept lies not in its uniqueness, but in the fact that theorems of far-reaching consequence for communication systems may be proved about it.   For a discussion of these developments, one may consult any of several references (cf. Fano [1961] or Abramson [1963]).

5-7   Moment-generating and characteristic functions

In the analytical theory of probability distributions, two special functions play a very powerful role.   Although these two functions are usually treated somewhat independently, we shall emphasize the fundamental connection made evident in the following definition.

Definition 5-7a

If $X(\cdot)$ is a random variable and s is a parameter, real or complex, the function of s defined by

$$M_X(s) = E[e^{sX}]$$

is called the *moment-generating function for* $X(\cdot)$. If $s = iu$, where i is the imaginary unit having the formal properties of the square root of -1, the function $\varphi_X(\cdot)$ defined by

$$\varphi_X(u) = M_X(iu) = E[e^{iuX}]$$

is called the *characteristic function for* $X(\cdot)$

Many works on probability theory limit the parameter s in the definition of the moment-generating function to real values. It seems desirable, however, to treat the more general case, in order to exploit an extensive theory developed in connection with classical integral transforms. Examination of the following integral expressions reveals the connection of the functions defined above with the classical theory.

$$M_X(s) = \int e^{st} \, dF_X(t) \qquad \text{(Laplace-Stieltjes transform)}$$
$$\varphi_X(u) = \int e^{iut} \, dF_X(t) \qquad \text{(Fourier-Stieltjes transform)}$$

Absolutely continuous case:

$$M_X(s) = \int e^{st} f_X(t) \, dt \qquad \text{(Laplace transform)}$$
$$\varphi_X(u) = \int e^{iut} f_X(t) \, dt \qquad \text{(Fourier transform)}$$

Discrete case:

$$M_X(s) = \sum_i p_i e^{st_i} \qquad \text{(Dirichlet series)}$$

The names in parentheses indicate that the functions in the various cases are well-known forms which have been studied intensively. The results of this nonprobabilistic theory are available for use in probability theory.

Before considering examples, we shall list some important properties of the moment-generating and characteristic functions. We shall state the results in terms of the moment-generating function. Where it is necessary to limit statements essentially to characteristic functions, this may be done by requiring that $s = iu$, with u real.

For a number of reasons, most probabilists prefer to use the characteristic function. One of these is the fact that the integral defining $\varphi_X(u) = M_X(iu)$ exists for all real u. The integral for $M_X(s)$, where s is complex, may exist only for s in a particular region of the complex

plane, the *region of convergence*.   In general, the region of convergence
is a vertical strip in the complex plane; in special cases this may be a
half plane, the whole plane, or the degenerate strip consisting of the
imaginary axis.   Two cases are frequently encountered in practice: (1)
the probability mass is all located in a finite interval on the real line,
in which case the region of convergence for the moment-generating function is the whole complex plane; (2) the probability mass is distributed
on the line in such a manner that there are positive numbers A and γ
such that, for large t, $P(|X| \geq t) \leq Ae^{-\gamma t}$, in which case the region of
convergence is the strip $-\gamma < \operatorname{Re} s < \gamma$.   When the region of convergence for the moment-generating function is a proper strip, the moment-
generating function can be expanded in a Taylor's series about the origin
(i.e., it is an analytic function).   The importance of this fact appears in
the development below.

(**M1**)   Consider two random variables $X(\cdot)$ and $Y(\cdot)$ with distribution functions
$F_X(\cdot)$ and $F_Y(\cdot)$, respectively.   Let $M_X(\cdot)$ and $M_Y(\cdot)$ be the corresponding moment-
generating functions for the two variables.   Then $M_X(iu) = M_Y(iu)$ for all real u
iffi $F_X(t) = F_Y(t)$ for all real t.

In the case of a proper strip of convergence, the equality of the moment-
generating functions extends (by analytic continuation) throughout the
interior of the region of convergence.   The property stated is a special
case of a uniqueness theorem on the Laplace-Stieltjes transform or the
Laplace-Fourier transform (cf. Widder [1941, p. 243, theorem 6A] or
Loève [1963, sec. 12.1]).

(**M2**)   If $E[|X|^n]$ exists, the nth-order derivative of the characteristic function
exists and

$$\varphi_X{}^{(n)}(0) = \frac{d^n}{du^n} M_X(iu) \Big|_{u=0} = i^n E[X^n]$$

In important special cases we can be assured that the derivatives and the
nth-order moments exist, as indicated in the following property.

(**M3**)   If the region of convergence for $M_X(\cdot)$ is a proper strip in the s plane
(which will include the imaginary axis), derivatives of all orders exist and

$$M_X{}^{(n)}(s) = \frac{d^n}{ds^n} M_X(s) = \int t^n e^{st} \, dF_X(t)$$
$$M_X{}^{(n)}(0) = E[X^n]$$

This theorem is based on the analytic property of the bilateral Laplace-
Stieltjes integral (cf. Widder [1941]).   Two important cases of mass distribution ensuring the condition of convergence in a proper strip are noted
above—all probability mass in a finite interval or an exponential decay
of the mass distribution.

Under the conditions of property (M3), the moment-generating function may be expanded in a power series in s about $s = 0$ to give

$$M_X(s) = \sum_{n=0}^{\infty} \frac{M_X^{(n)}(0)}{n!} s^n = \sum_{n=0}^{\infty} \frac{E[X^n]}{n!} s^n$$

Thus, if the moment-generating function is known in explicit form and can be expanded in a power series, the various moments for the mass distribution can be determined from the coefficients of the expansion.

(**M4**) If $Z(\cdot) = aX(\cdot) + b$, then $M_Z(s) = e^{bs}M_X(as)$.

PROOF

$$M_Z(s) = E[e^{asX+bs}] = E[e^{asX}e^{sb}] = e^{sb}E[e^{asX}] \quad \blacksquare$$

The moment-generating function is particularly useful in dealing with sums of independent random variables because of the following property.

(**M5**) If $X(\cdot)$ and $Y(\cdot)$ are *independent* random variables and if $Z(\cdot) = X(\cdot) + Y(\cdot)$, then $M_Z(s) = M_X(s)M_Y(s)$ for all s.

PROOF It is known that $g(t) = e^{st}$ is a Borel function. Hence the random variables $e^{sX(\cdot)}$ and $e^{sY(\cdot)}$ are independent when $X(\cdot)$ and $Y(\cdot)$ are. Now $E[e^{s(X+Y)}] = E[e^{sX}e^{sY}] = E[e^{sX}]E[e^{sY}]$ by property (E6). $\blacksquare$

The result can be extended immediately by induction to any finite number of random variables.

We consider now several examples.

Example 5-7-1

Uniformly distributed random variable. Let $X(\cdot)$ be distributed uniformly on the interval $a \leq t \leq b$.

SOLUTION

$$f_X(t) = \frac{1}{b-a}\left[u(t-a) - u(t-b)\right]$$
$$M_X(s) = \frac{1}{b-a}\int_a^b e^{st}\, dt = \frac{1}{s(b-a)}\left[e^{sb} - e^{sa}\right]$$

Using the well-known expansion of e^{bs} and e^{as} in a power series about $s = 0$, we obtain

$$M_X(s) = \sum_{n=0}^{\infty} \frac{b^{n+1} - a^{n+1}}{(n+1)(b-a)}\frac{s^n}{n!}$$

Thus

$$E[X^n] = \frac{b^{n+1} - a^{n+1}}{(n+1)(b-a)}$$

which may be checked by straightforward evaluation of the integral

$$\frac{1}{b-a}\int_a^b t^n\, dt$$

As special cases we have:

$$\mu[X] = E[X] = \frac{b^2 - a^2}{2(b-a)} = \frac{a+b}{2}$$

$$\sigma^2[X] = E[X^2] - E^2[X] = \frac{b^3 - a^3}{3(b-a)} - \frac{(a+b)^2}{4} \quad \blacksquare$$

The next result is so important that we state it as a theorem.

Theorem 5-7A

The random variable $X(\cdot)$ is normal (μ, σ) iffi

$$M_X(s) = e^{\mu s + \sigma^2 s^2/2} \qquad \text{for each } s$$

PROOF   First, consider the random variable $Y(\cdot)$, which is normal $(0, 1)$.

$$M_Y(s) = \frac{1}{\sqrt{2\pi}} \int e^{st} e^{-t^2/2}\, dt$$

Evaluation can be carried out by noting that the combined exponent $st - t^2/2 = s^2/2 - s^2/2 + st - t^2/2 = s^2/2 - \frac{1}{2}(t-s)^2$.   Thus

$$M_Y(s) = \frac{e^{s^2/2}}{\sqrt{2\pi}} \int e^{-(t-s)^2/2}\, dt$$

Making the change of variable $x = t - s$, we have, for each fixed s,

$$M_Y(s) = \frac{e^{s^2/2}}{\sqrt{2\pi}} \int e^{-x^2/2}\, dx = e^{s^2/2}$$

The random variable $X(\cdot) = \sigma Y(\cdot) + \mu$ is normal (μ, σ), as may be seen from Example 3-9-4.   By properties (M4) and (M1) the theorem follows.   ∎

In many investigations, it is simpler to identify the moment-generating function or the characteristic function than to identify the density function directly.

Example 5-7-2

We may use the series expansion for the moment-generating function for a normally distributed random variable to obtain its mean and variance.

SOLUTION   Using the first three terms of the expansion $e^z = 1 + z + z^2/2 + \cdots$, we obtain

$$M_X(s) = 1 + \left(\mu s + \frac{\sigma^2 s^2}{2}\right) + \frac{1}{2}\left(\mu s + \frac{\sigma^2 s^2}{2}\right)^2 + \cdots$$

$$= 1 + \mu s + (\sigma^2 + \mu^2)\frac{s^2}{2} + \cdots$$

Hence

$$\mu[X] = E[X] = \mu \qquad \text{(coefficient of } s)$$

$$E[X^2] = \sigma^2 + \mu^2 \qquad \left(\text{coefficient of } \frac{s^2}{2}\right)$$

so that $\sigma^2[X] = \sigma^2 + \mu^2 - \mu^2 = \sigma^2$.   ∎

Our next result, also stated as a theorem, displays an important property of normally distributed random variables.

Theorem 5-7B

The sum of a finite number of independent, normally distributed random variables is normally distributed.

PROOF   Let $X_i(\cdot)$ be normal (μ_i, σ_i), $1 \leq i \leq n$.   The $X_i(\cdot)$ form an independent class.   Let $M_i(\cdot)$ be the moment-generating function for X_i, and let $M_X(\cdot)$ be the moment-generating function for sum $X(\cdot)$.   By property (M5),

$$M_X(s) = \prod_{i=1}^{n} M_i(s)$$

By Theorem 5-7A, $M_X(s) = e^{\mu s + \sigma^2 s^2/2}$, where $\mu = \sum_{i=1}^{n} \mu_i$ and $\sigma^2 = \sum_{i=1}^{n} \sigma_i^2$. ∎

One important consequence of this theorem for sampling theory is that, if the sampling is from a normal distribution, the sample average $A(\cdot)$ defined in Sec. 5-5 (Theorem 5-5A) is a normally distributed random variable.   This is important in determining the probability that the sample average differs from the mean by any given amount.

The importance of the moment-generating function is greatly enhanced by certain limit theorems.   We state, without proof, the following general theorem as one of the fundamental properties of the moment-generating function (Loève [1963, sec. 12.2]).

(M6)   Suppose $\{F_n(\cdot): 1 \leq n < \infty\}$ is a sequence of probability distribution functions and, for each n, $M_n(\cdot)$ is the moment-generating function corresponding to $F_n(\cdot)$.

(A)   If $F(\cdot)$ is a probability distribution function such that $F_n(t) \to F(t)$ at every point of continuity of $F(\cdot)$, then

$M_n(iu) \to M(iu)$ for each real u

(B)   If $M_n(iu) \to M(iu)$ for each real u and $M(iu)$ is continuous at $u = 0$, then $F_n(t) \to F(t)$ at every point of continuity of $F(\cdot)$.

One of the historically important applications of this theorem has been its role in establishing the central limit theorem (discussed very briefly in Chap. 6).   Its importance is not limited to this application, however, for the moment-generating function or the characteristic function is an important analytical tool in many phases of the development of probability theory.   For the mathematical foundations of this theory, one should consult standard works such as Loève [1963, chap. 4], Cramér [1946, chap. 10], or Gnedenko [1962, chap. 7].   Applications are found in most books on statistics or on random processes.

Problems

5-1. A discrete random variable $X(\cdot)$ has values t_k, $k = 1, 2, 3, 4, 5$, with corresponding probabilities $p_k = P(X = t_k)$ given below:

k	t_k	p_k
1	-1	0.3
2	0	0.1
3	3	0.2
4	5	0.3
5	7	0.1

(a) Calculate $E[X]$ and $E[X - 2.5]$.   Answer: 2.5, 0

(b) Calculate $E[(X - 2.5)^2]$.   Answer: 8.25

5-2. A man stands in a certain position (which we may call the origin). He tosses a fair coin. If a head appears, he moves *three* units to the right (which we designate as plus three units); if a tail appears, he moves *two* units to the left (in the negative direction). Make the usual assumptions on coin flipping; i.e., let A_i be the event of a head on the ith throw. Assume the A_i form an independent class, with $P(A_i) = \frac{1}{2}$ for each i. Let $X_n(\cdot)$ be the random variable whose value is the position of the man after n successive flips of the coin (the sequence of n flips is considered one composite trial).

(a) What are the possible values of $X_4(\cdot)$, and what are the probabilities that each of these values is taken on? Determine from these numbers the mathematical expectation $E[X_4]$.

(b) Show that, for any $n \geq 1$, $X_n(\cdot)$ may be written

$$X_n(\cdot) = \sum_{i=1}^{n} [5I_{A_i}(\cdot) - 2]$$

Determine $E[X_n]$ from this expression, and compare the result for $n = 4$ with that obtained in part a.

5-3. A man bets on three horse races. He places a $2 bet on one horse in each race. Let H_1, H_2, and H_3 be the events that the first, second, and third horses, respectively, win their races. Suppose the H_i form an independent class and that

$$p_1 = P(H_1) = 0.2 \quad \text{and} \quad H_1 \text{ pays } \$10$$
$$p_2 = P(H_2) = 0.3 \quad \text{and} \quad H_2 \text{ pays } \$5$$
$$p_3 = P(H_3) = 0.6 \quad \text{and} \quad H_3 \text{ pays } \$3$$

(a) Let $W(\cdot)$ be the random variable which represents the man's net winnings in dollars (a negative win is a loss). Express $W(\cdot)$ in terms of the indicator functions for H_1, H_2, and H_3 and appropriate constants.

(b) Calculate the possible values for $W(\cdot)$ and the corresponding probabilities.

(c) What is the probability of coming out ahead on the betting (i.e., of having positive "winnings")?   Answer: $P(W > 0) = 0.344$

(d) Determine the "expected" value of the winnings, $E[W]$.   Answer: -0.7

5-4. A random variable $X(\cdot)$ has a distribution function $F_X(\cdot)$ which rises linearly from zero at $t = a$ to the value unity at $t = b$. Determine $E[X]$, $E[X^2]$, and $E[X^3]$.

ANSWER: $E[X] = (b + a)/2$, $E[X^2] = (b^3 - a^3)/[3(b - a)]$

5-5. A random variable $X(\cdot)$ produces a probability mass distribution which may be described as follows. A point mass of 0.2 is located at each of the points $t = -1$ and $t = 2$. The remaining mass is distributed smoothly between $t = -2$ and $t = 2$, with a density that begins with zero value at $t = -2$ and increases linearly to a maximum density at $t = 2$. Find $E[X]$ and $E[(X - 1)^2]$. ANSWER: 0.6, 1.6

5-6. The distribution function $F_X(\cdot)$ for a random variable $X(\cdot)$ has jumps of 0.1 at $t = -2$ and $t = 1$; otherwise it increases linearly at the same rate between $t = -3$ and $t = 2$. Determine $E[X]$ and $E[2\sqrt{X + 3}]$. ANSWER: -0.5, 2.99

5-7. Suppose the sales demand for a given item sold by a store is represented by a random variable $X(\cdot)$. The item is stocked to capacity at the beginning of the period. There are fixed costs associated with storage which are approximately proportional to the amount of storage N. A fixed gross profit is made on sales. If the demand exceeds the supply, however, certain losses (due to the decrease of sales of other items merchandised, etc.) occur; this loss may be approximated by a constant times the excess of the demand over the supply. We may thus express the profit or gain $G(\cdot)$ in terms of the demand $X(\cdot)$ as follows:

$$G(\cdot) = \begin{cases} gX(\cdot) - cN & \text{for } 0 \leq X(\cdot) \leq N \\ (g - c)N - L[X(\cdot) - N] & \text{for } X(\cdot) \geq N \end{cases}$$

(a) If $f_X(t) = \alpha e^{-\alpha t}u(t)$, determine the expected profit in terms of α, g, c, L, and N.

(b) For any given storage capacity N, what is the probability that demand will exceed the supply? Express $E[G]$ in terms of μ_X, $P(X > N)$, g, c, L, and N.

5-8. A random variable $X(\cdot)$ has values $t_i = (i - 1)^2$, $i = 1, 2, 3, 4$, and random variable $Y(\cdot)$ has range $u_j = 2j$, $j = 1, 2, 3$. The values of $p(i, j) = P(X = t_i, Y = u_j)$ are as follows:

	$i = 1$	$i = 2$	$i = 3$	$i = 4$
$j = 1$	0.05	0.10	0.05	0.10
$j = 2$	0.05	0.09	0.01	0.05
$j = 3$	0.10	0.21	0.04	0.15

Determine the mathematical expectations $E[X]$, $E[Y]$, and $E[2X - 3Y]$.

ANSWER: 3.50, 4.40, -6.20

5-9. A random variable $X(\cdot)$ has range $(\frac{1}{2}, \frac{3}{2}, \frac{5}{2}) = (t_1, t_2, t_3)$. A random variable $Y(\cdot)$ has range $(\frac{1}{2}, \frac{3}{2}) = (u_1, u_2)$. The values of $p(i, j) = P(X = t_i, Y = u_j)$ are as follows:

$p(1, 1) = \frac{1}{8}$ $p(2, 1) = \frac{1}{8}$ $p(3, 1) = \frac{1}{8}$
$p(1, 2) = \frac{1}{8}$ $p(2, 2) = \frac{1}{4}$ $p(3, 2) = \frac{1}{4}$

(a) Let the random variable $Z(\cdot)$ be defined by $Z(\xi) = X(\xi) + Y(\xi)$. Determine the distribution function $F_Z(\cdot)$.

(b) Determine $E[X]$, $E[Y]$, and $E[Z]$, and verify in this case that

$$E[Z] = E[X] + E[Y]$$

5-10. If $g(\cdot)$ is an odd Borel function and if the density function $f_X(\cdot)$ for the random variable $X(\cdot)$ is an even function, show that $E[g(X)] = 0$.

5-11. Two real-valued functions $g(\cdot)$ and $h(\cdot)$ are said to be *orthogonal* on the interval (a, b) iffi

$$\int_a^b g(t)h(t)\, dt = 0$$

Two real-valued functions $g(\cdot)$ and $h(\cdot)$ are said to be *orthogonal* on the interval (a, b) *with respect to the weighting function* $r(\cdot)$ iffi

$$\int_a^b g(t)h(t)r(t)\, dt = 0$$

(a) If $g(\cdot)$ is a Borel function orthogonal to the density function $f_X(\cdot)$ for the random variable $X(\cdot)$ on the interval $(-\infty, \infty)$, show that $E[g(X)] = 0$.

(b) If $g(\cdot)$ and $h(\cdot)$ are two Borel functions orthogonal on the interval $(-\infty, \infty)$ with respect to the density function $f_X(\cdot)$, show that $E[g(X)h(X)] = 0$.

(c) If the random variable $X(\cdot)$ is uniformly distributed on the interval $[0, 2\pi]$, if n and m are nonzero integers, and if a and b are arbitrary real constants, show that the following hold:

$$E[\sin (nX + a)] = E[\cos (mX + b)] = 0$$
$$E[\sin (nX + a) \cos (mX + b)] = 0$$
$$E[\sin (nX + a) \sin (mX + b)] = 0 \qquad \text{for } n \neq m$$
$$E[\cos (nX + a) \cos (mX + b)] = 0 \qquad \text{for } n \neq m$$

(*Suggestion:* Apply the well-known orthogonality relations on sinusoidal functions used in developing Fourier series.)

5-12. Suppose $X(\cdot)$ and $\theta(\cdot)$ are independent random variables with $\theta(\cdot)$ uniformly distributed on the interval $[0, 2\pi]$. Show that, for any nonzero integer n,

$$E\{\cos [X(\cdot) + n\theta(\cdot)]\} = E\{\sin [X(\cdot) + n\theta(\cdot)]\} = 0$$

Note that this result includes the special case $X(\cdot) = c$ $[P]$, in which case $E[X] = c$. [*Suggestion:* Use formulas for $\cos (x + y)$ and $\sin (x + y)$ and the results of Prob. 5-11.]

5-13. A discrete random variable has the following set of values and probabilities. Calculate the mean and standard deviation.

$t_i =$	-2	-1	0	1	5	6	
$p_i =$	0.20	0.10	0.25	0.05	0.20	0.20	ANSWER: $\mu = 1.75$, $\sigma = 3.17$

5-14. A random variable $X(\cdot)$ has range $(-30, -10, 0, 20, 40)$ and corresponding probabilities $(0.2, 0.1, 0.3, 0.3, 0.1)$; for example, $P(X = -30) = 0.2$, $P(X = 40) = 0.1$, etc. Determine the mean value $\mu[X]$ and the variance $\sigma^2[X]$.

ANSWER: 3, 461

5-15. A random variable $X(\cdot)$ is distributed according to the binomial distribution with $n = 10$, $p = 0.1$. Consider the random variable $Y(\cdot) = X^2(\cdot) - 1$.

(a) Calculate the mean value of $Y(\cdot)$.   ANSWER: 0.9

(b) Calculate the standard deviation of $Y(\cdot)$. (Simplify as far as possible, and indicate how tables could be used to determine numerical results.)

5-16. A man stands in a certain position (which we may call the origin). He makes a sequence of n moves or steps. At each move, with probability p, he will move a units to the right, and with probability $q = 1 - p$, he will move b units to the left. The various moves are stochastically independent events. Let $X_n(\cdot)$ be the random variable which gives the distance moved after n steps (cf. Prob. 5-2).

(a) Obtain an expression for $X_n(\cdot)$ if A_i is the event of a move to the right at the ith step.

(b) Determine expressions for $\mu_n = E[X_n]$ and $\sigma_n{}^2 = \sigma^2[X_n]$.

ANSWER: $n(ap - bq)$, $n(a + b)^2 pq$

(c) Determine numerical values for $n = 4$, $p = 0.3$, $a = 2$, and $b = 1$.

Aids to computation

From the geometric series

$$\sum_{k=0}^{\infty} z^k = \frac{1}{1 - z} \qquad \text{for } |z| < 1$$

the following may be derived by differentiation and algebraic manipulation:

(1) $\displaystyle\sum_{k=1}^{\infty} z^k = \frac{z}{1 - z}$ 　　　　　(2) $\displaystyle\sum_{k=1}^{\infty} kz^k = \frac{z}{(1 - z)^2}$

(3) $\displaystyle\sum_{k=1}^{\infty} k^2 z^k = \frac{z(1 + z)}{(1 - z)^3}$ 　　$|z| < 1$

5-17. An experiment consists of a sequence of tosses of an honest coin (i.e., to each elementary event corresponds an infinite sequence of heads and tails). Let A_k be the event that a head appears for the first time at the kth toss in a sequence, and let H_k be the event of a head at the kth toss. Suppose the H_k form an independent class with $P(H_k) = \frac{1}{2}$ for each k. For a given sequence corresponding to the elementary outcome ξ, let $X(\xi)$ be the number of the toss in the sequence for which the first head appears.

(a) Express $X(\cdot)$ in terms of indicator functions for the A_k.

(b) Determine $\mu[X]$ and $\sigma^2[X]$. (*Suggestion:* Use the appropriate formulas above for summing the series.) 　　ANSWER: $\mu[X] = 2$, $\sigma^2[X] = 2$

5-18. Consider two random variables $X(\cdot)$ and $Y(\cdot)$ with joint probability distribution which is uniform over the triangular region having vertices at points $(0, 0)$, $(0, 1)$, and $(1, 0)$. Let $Z(\cdot) = X(\cdot) + Y(\cdot)$.

(a) Determine the distribution functions $F_X(\cdot)$ and $F_Z(\cdot)$ and the density functions $f_X(\cdot)$ and $f_Z(\cdot)$.

(b) Calculate μ_X, μ_Y, μ_Z, $\sigma_X{}^2$, $\sigma_Y{}^2$, and $\sigma_Z{}^2$. 　　ANSWER: $\sigma_X{}^2 = \sigma_Y{}^2 = \sigma_Z{}^2 = \frac{1}{18}$

5-19. Consider two random variables $X(\cdot)$ and $Y(\cdot)$ with joint probability distribution which is uniform over the triangular region having vertices at points $(0, 0)$, $(1, 0)$, and $(0, -1)$. Let $W(\cdot) = X(\cdot) - Y(\cdot)$.

(a) Determine the distribution functions $F_X(\cdot)$ and $F_W(\cdot)$ and the density functions $f_X(\cdot)$ and $f_W(\cdot)$.

(b) Calculate μ_X, μ_Y, μ_W, $\sigma_X{}^2$, $\sigma_Y{}^2$, and $\sigma_W{}^2$.

5-20. If $X(\cdot)$ and $Y(\cdot)$ are real-valued random variables whose variances exist, show that $(\sigma[X] - \sigma[Y])^2 \leq \sigma^2[X - Y]$. (*Suggestion:* Use the Schwarz inequality and the fact that the variance does not depend upon an additive constant.)

5-21. Scores on college board examinations vary between 200 and 800. To a good approximation, these may be considered to be values of a random variable which is normally distributed with mean 500 and standard deviation 100. What is the probability of selecting a person at random with a score of 700 or more?

ANSWER: 0.023

5-22. Suppose that $X(\cdot)$ is normal (15, 2). Use the table of normal probability functions (Sec. 5-4) to determine:

(a) $P(X < 13)$

ANSWER: 0.1587

(b) $P(X \leq 15)$

(c) $P(|X - 15| \leq 4)$

ANSWER: 0.9545

5-23. A random variable $X(\cdot)$ is distributed normally (2, 0.25); that is, $\mu = 2$, $\sigma = 0.25$

(a) Determine the mean and standard deviation for the variable

$$Y(\cdot) = 2X(\cdot) - 4$$

(b) What is the probability $1.5 \leq X(\cdot) \leq 2.25$?

ANSWER: 0.8185

(c) What is the probability $|X(\cdot) - 2| \leq 0.5$?

5-24. A lot of manufactured items have weights, when packaged, which are distributed approximately normally, with mean $\mu = 10$ pounds and standard deviation $\sigma = 0.25$ pound. A load of 100 items is gathered randomly from the lot for shipment. What is the probability that the weight of the load is no greater than 1,005 pounds? State your assumptions.

5-25. Readings of a certain physical quantity may be considered to be values of a random variable which has the normal distribution with mean μ and standard deviation $\sigma = 0.20$ unit. A sequence of n independent readings may be represented by the respective values of n independent random variables, each with the same distribution.

(a) Determine the probability r that a single reading will lie within 0.1 unit of the mean value μ.

ANSWER: $r = 0.383$

(b) Determine, in terms of r as defined in part a, the probability that at least one reading in four lies within 0.1 unit of μ.

ANSWER: $1 - (1 - r)^4$

(c) Determine, in terms of r, the probability that exactly one reading in four lies within 0.1 unit of μ.

(d) Determine the probability that the average of four readings lies within 0.1 unit of μ.

ANSWER: 0.683

5-26. Suppose $X(\cdot)$ is a random variable and $g(\cdot)$ is a Borel function which may be expanded in a Taylor's series about the point $t = \mu = \mu[X]$. Thus $g(t) = g(\mu) + g'(\mu)(t - \mu) + r(t)$. If the remainder $r(t)$ is negligible throughout the set of t on which most of the probability mass is located, determine expressions for $\mu[g(X)]$ and $\sigma^2[g(X)]$.

ANSWER: $g(\mu)$, $[g'(\mu)]^2\sigma^2[X]$

5-27. A sample of size n is taken (with or without replacement). From this sample, a single value is taken at random.

Let A be the event a certain result is achieved on the final choice from the sample of size n; A_i be the event of a success on the ith trial in taking the original sample;

and B_k be the event of exactly k successes in the n trials making up the sample. Consider $N(\cdot) = \sum_{i=1}^{\infty} I_{A_i}(\cdot)$, the number of successes in the n trials.

(a) Write $N(\cdot)$ in terms of the indicator functions $I_{B_k}(\cdot)$.

(b) Assume $P(A|B_k) = k/n$. Show that $P(A)$ can be expressed simply in terms of $E[N]$. Give the formula. ANSWER: $P(A) = \dfrac{1}{n} E[N]$

(c) Use the results of parts a and b to express $P(A)$ in terms of the $P(A_i)$. What does this reduce to if $P(A_i) = p$ for each i?

5-28. In information theory, frequent attention is given to binary schemes having two events (often designated by 0 and 1) with probabilities p and $q = 1 - p$. In this case, the entropy function is usually indicated by $H(p)$ and is given by $H(p) = -p \log_2 p - q \log_2 q$.

(a) Show that the derivative $H'(p) = \log_2 (q/p)$. (Note that, in differentiating the logarithm, it is desirable to convert to base e.)

(b) Show that the graph for $H(p)$ versus p has a maximum at $p = \frac{1}{2}$ and is symmetrical about $p = \frac{1}{2}$ (Fig. 5-6-2).

5-29. Consider a set of N boxes, each of which is occupied by an object with probability p and is empty with probability $1 - p$. If A_i is the event the ith box is occupied, the A_i are assumed to form an independent class. What is the information about the occupancy of the ith box provided by the knowledge that k of the N boxes are occupied. Let B_k be the event k boxes are occupied. ANSWER: $\log_2 (k/Np)$

5-30. Each of two objects, a and b, is to be put into one of N cells. Either a or b may occupy any cell with probability $1/N$, but they cannot both occupy the same cell. Let A_i be the event that object a is in the cell numbered i, and B_j be the event that object b is in the cell numbered j. Let $O_i = A_i \uplus B_i$ be the event that cell i is occupied, and let $D_{ij} = O_i O_j = A_i B_j \uplus A_j B_i$ be the event that cells i and j are occupied. Assume $P(A_i) = P(B_j) = 1/N$ for each i and j, and $P(A_i|B_j) = P(B_j|A_i) = 1/(N - 1)$ for $i \neq j$. Find and interpret the following quantities:

(a) $\jmath(A_i:B_j)$ ANSWER: $\log \dfrac{N}{N - 1}$ for $i \neq j$, and $-\infty$ for $i = j$

(b) $\jmath(O_i:B_j)$ ANSWER: $\log \dfrac{N}{N - 1} - 1$ for $i \neq j$, and $\log N - 1$ for $i = j$

(c) $\jmath(D_{ij}:B_k)$ for $i \neq j$
(d) $\jmath(D_{ij}:O_k)$ for $i \neq j$
(Note that the latter two quantities are meaningless for $i = j$.)

5-31. Show that $\jmath(D:ABC) = \jmath(D:A) + \jmath(D:B|A) + \jmath(D:C|AB)$. Interpret this expression.

5-32. Consider a set of messages m_0, m_1, m_2, m_3, m_4, m_5, m_6, m_7 which have probabilities $\frac{1}{4}$, $\frac{1}{4}$, $\frac{1}{8}$, $\frac{1}{8}$, $\frac{1}{16}$, $\frac{1}{16}$, $\frac{1}{16}$, $\frac{1}{16}$ and code words 000, 001, 010, 011, 100, 101, 110, 111, respectively. Message m_5 is received digit by digit. What information is provided about the message m_5 by each successive digit received? (*Suggestion:* Let W_i be the event the ith code word is sent. Let A_i be the event that the first letter in the code word received is the first letter in the ith code word. Similarly for B_i and C_i and the second and third letters of the code word received. Calculate the conditional probabilities needed to apply the results of Prob. 5-31.)

ANSWER: 2, 1, 1

5-33. A binary communication system is disturbed by noise in such a way that each input digit has a probability p of being changed by the channel noise. Eight messages may be transmitted with equal probabilities; they are represented by the following code words:

$$u_1 \equiv 0\ 0\ 0\ 0 \qquad u_5 \equiv 1\ 0\ 0\ 1$$
$$u_2 \equiv 0\ 0\ 1\ 1 \qquad u_6 \equiv 1\ 0\ 1\ 0$$
$$u_3 \equiv 0\ 1\ 0\ 1 \qquad u_7 \equiv 1\ 1\ 0\ 0$$
$$u_4 \equiv 0\ 1\ 1\ 0 \qquad u_8 \equiv 1\ 1\ 1\ 1$$

If the sequence of digits $v = 0100$ is received, determine the amount of information provided about message u_3 by each digit as it is received. (*Suggestion:* Let U_i be the event the code word u_i is sent; A be the event the first letter received is a zero; B be the event the second letter received is a zero; etc. Calculate the conditional probabilities needed to use the results of Prob. 5-31, extended to the case of five events.)

ANSWER: $1 + \log q,\ 1 + \log q,\ 1 + \log q,\ -\log[4q(p^2 + q^2)]$

5-34. Consider a scheme $\mathfrak{A} = \{(A_i, p_i): 1 \leq i \leq n\}$.

(a) Suppose, for a given pair of integers r and s, it is possible to modify p_r and p_s but all other p_i are to remain unchanged. Show that if the smaller of the two values p_r, p_s is increased and the larger is decreased accordingly, the value of $H(\mathfrak{A})$ is increased, with a maximum value occurring when the two probabilities are equal. (*Suggestion:* Let $p_r + p_s = 2a$, $p_r = a - \delta$, and $p_s = a + \delta$.)

(b) Suppose $\displaystyle\sum_{i=1}^{m} p_i = a$. Show that $-\displaystyle\sum_{i=1}^{m} p_i \log p_i \leq a \log (m/a)$ with equality iffi $p_i = a/m$ for each i. What can be concluded about maximizing $H(\mathfrak{A})$ if any subset of the probabilities may be modified while the others are held constant?

Aid to computation

x	$-\log_2 x$	$-x \log_2 x$	x	$-\log_2 x$	$-x \log_2 x$
0.1	3.3219	0.3322	0.6	0.7370	0.4422
0.2	2.3219	0.4644	0.7	0.5146	0.3602
0.3	1.7370	0.5211	0.8	0.3219	0.2575
0.4	1.3219	0.5288	0.9	0.1520	0.1368
0.5	1.0000	0.5000			

5-35. Joint probabilities $p(i, j)$ for a joint probability scheme $\mathfrak{A}\mathfrak{B}$ are shown below:

	$j = 1$	$j = 2$	$j = 3$
$i = 1$	0.1	0.1	0.1
$i = 2$	0.1	0.2	0.1
$i = 3$	0.1	0.1	0.1

Calculate $H(\mathfrak{A})$, $H(\mathfrak{B})$, and $H(\mathfrak{A}\mathfrak{B})$, and show for this case that $H(\mathfrak{A}\mathfrak{B}) \neq H(\mathfrak{A})H(\mathfrak{B})$ and $H(\mathfrak{A}\mathfrak{B}) < H(\mathfrak{A}) + H(\mathfrak{B})$.

5-36. Consider N schemes $\mathfrak{A}_k\mathfrak{B}_k$ with joint probabilities $p_k(i, j)$, $1 \leq i \leq n$, $1 \leq j \leq m$, $1 \leq k \leq N$. Let λ_1, λ_2, . . . , λ_N be nonnegative numbers whose sum is unity, and put $p_0(i, j) = \sum\limits_{k=1}^{N} \lambda_k p_k(i, j)$. The $p_0(i, j)$ are joint probabilities of a scheme which we call $\mathfrak{A}_0\mathfrak{B}_0$.

(a) Show $H(\mathfrak{A}_0\mathfrak{B}_0) \geq \sum\limits_{k=1}^{N} \lambda_k H(\mathfrak{A}_k\mathfrak{B}_k)$.

(b) It is possible to construct examples which show that, in general,

$$H(\mathfrak{A}_0 : \mathfrak{B}_0) \not\geq \sum_{k=1}^{N} \lambda_k H(\mathfrak{A}_k : \mathfrak{B}_k)$$

However, in the special case (of importance in information theory) that

$$\frac{p_k(i, j)}{p_k(i, *)} = \frac{p_0(i, j)}{p_0(i, *)} \qquad \text{for all } k = 1, 2, . . . , N$$

the inequality does hold. Show that this is so.

5-37. Show that the average probability discussed in Sec. 2-5, just prior to and in Example 2-5-5, may be expressed as the expectation of a random variable. Show that the result in Example 2-5-5 is then a consequence of property (E7).

5-38. For the binary symmetric channel, use the notation of Example 5-6-1. Show that

$$H(\mathfrak{S}\mathfrak{R}) = H(p) + H(\pi)$$
$$H(\mathfrak{R}) = H(p\pi + q\bar{\pi})$$
$$H(\mathfrak{S}) = H(\pi)$$

Obtain the result of Example 5-6-1 by use of the appropriate relationship between the entropy functions.

5-39. The simple random variable $X(\cdot)$ takes on values -2, -1, 0, 2, 4 with probabilities 0.1, 0.3, 0.2, 0.3, 0.1, respectively.

(a) Write an expression for the moment-generating function $M_X(\cdot)$.

(b) Show by direct calculation that $M'_X(0) = E[X]$ and $M''_X(0) = E[X^2]$.

5-40. If $M_X(\cdot)$ is the moment-generating function for a random variable $X(\cdot)$, show that $\sigma^2[X]$ is the second derivative of $e^{-s\mu}M_X(s)$ with respect to s, evaluated at $s = 0$. [*Suggestion:* Use properties (M3) and (M4).]

5-41. Simple random variables $X(\cdot)$ and $Y(\cdot)$ are independent. The variable $X(\cdot)$ takes on values -2, 0, 1, 2 with probabilities 0.2, 0.3, 0.3, 0.2, respectively. Random variable $Y(\cdot)$ takes on values 0, 1, 2 with probabilities 0.3, 0.4, 0.3, respectively.

(a) Write the moment-generating functions $M_X(\cdot)$ and $M_Y(\cdot)$.

(b) Use the moment-generating functions to determine the range and the probabilities for each value of the random variable $Z(\cdot) = X(\cdot) + Y(\cdot)$.

5-42. Suppose $X(\cdot)$ is *binomially distributed* (Example 3-4-5), with basic probability p and range $t = 0, 1, 2, . . . , n$.

(a) Show that $M_X(s) = (pe^s + q)^n$.

(b) Use this expression to determine $\mu[X]$ and $\sigma^2[X]$.

5-43. An *exponentially distributed* random variable $X(\cdot)$ has probability density function $f_X(\cdot)$ defined by

$$f_X(t) = \alpha e^{-\alpha t} u(t)$$

(a) Determine the moment-generating function $M_X(s)$.

(b) Determine μ_X and $\sigma_X{}^2$ with the aid of the moment-generating function.

Selected references

General ideas about expectations

GNEDENKO [1962], chap. 5. Cited at the end of our Chap. 3.

PARZEN [1960], chap. 8. Cited at the end of our Chap. 2.

WADSWORTH AND BRYAN [1960], chap. 7. Cited at the end of our Chap. 3.

Normal distribution

ARLEY AND BUCH [1950]: "Introduction to the Theory of Probability and Statistics," chap. 7 (transl. from the Danish). A very fine treatment of applications of probability theory, particularly to statistics. Does not emphasize the measure-theoretic foundation of probability theory, but is generally a sound, careful treatment, designed for the engineer and physicist.

CRAMÉR [1946]: "Mathematical Methods of Statistics," chap. 17. A standard reference for the serious worker in statistics. Lays the foundations of topics in probability theory necessary for statistical applications.

Random samples and random variables

BRUNK [1964], chap. 6. Cited at the end of our Chap. 2.

FISZ [1963]. Cited at the end of our Chap. 2.

Probability and information

SHANNON [1948]: A Mathematical Theory of Communication, *Bell System Technical Journal*. Original papers reporting the fundamental work in establishing the field now known as information theory. For insight into some of the central ideas, these papers are still unsurpassed.

ABRAMSON [1963]: "Information Theory and Coding," chaps. 2, 5. A very readable yet precise account of fundamental concepts of information theory and their application to communication systems.

FANO [1961]: "Transmission of Information." A book by one of the fundamental contributors to information theory. Although designed as a text, it has something of the character of a research monograph. For fundamental measures of information, see chap. 2.

Moment-generating and characteristic functions

BRUNK [1964], sec. 81. Cited at the end of our Chap. 2.

CRAMÉR [1946], chap. 10. Cited above.

GNEDENKO [1962], chap. 7. Provides a succinct and readable development of the more commonly used properties of characteristic functions. Cited at the end of our Chap. 3.

LOÈVE [1963], chap. 4. Provides a thorough treatment of characteristic functions. Cited at the end of our Chap. 4.

WIDDER [1941]: "The Laplace Transform." A thorough treatment of the Laplace-Stieltjes transform. Provides the basic theorems and properties of the moment-generating function for complex parameter s. See especially chap. 6, secs. 1 through 6.

chapter 6

Sequences and sums of random variables

In the brief discussion of random samples in Sec. 5-5, attention is directed to the importance of sums of random variables. The study of sums and averages of random variables, with particular attention to certain limiting properties of sequences and sums of variables, has played an important role in the development of probability theory. In this chapter, we discuss some of the classical theorems, such as the law of large numbers (in a variety of forms) and the central limit theorem. Because of the inherent character of the mathematical material, the presentation must appear recondite and somewhat tedious in many places. Some readers will want to scan the material quickly, picking up the gist of the development from comments and an occasional simply stated result; others will want to move into a more complete study of some points with the help of the references. In keeping with the introductory character of this work, the treatment of the topics varies greatly in completeness and rigor. It is the purpose of this exposition to give some comprehension of the nature of the existing mathematical theory and to provide some sampling of the kind of techniques and arguments upon which the central results rest.

In the first section, the so-called "weak" laws of large numbers are studied in some mathematical detail. For the most part, the proofs are simple, and they give important insight into the mathematical character of the problem.   Although these limit laws assert the convergence of certain sequences of partial sums, they do not ordinarily provide precise information on the rapidity of convergence.   Some special cases of bounds on probabilities for finite sums are studied.   The techniques are illustrative, and the results are interesting in themselves.   The limit theorems lead to a consideration of several types of convergence which are important in probability theory.   A brief discussion of the "strong" laws of large numbers is then carried out.   Proofs are limited to the simpler cases, to illustrate the mathematics without imposing a burden of excessive detail.   Finally, the central limit theorem is examined and discussed without proof.   This theorem asserts that sums of large numbers of independent random variables are approximately normally distributed in a certain important sense.

6-1   Law of large numbers (weak form)

In Sec. 5-5, the sum of n identically distributed random variables is studied.   These appear as a random sample of size n for a given random variable.   With the aid of the inequality of Chebyshev, it is shown that the average for large n has high probability of lying close to the sample mean (i.e., the common mean value for the random variables which are summed).   In this section, we study some generalizations of this fact, stated in a series of limit theorems.   These limit theorems provide a variety of expressions of the *law of large numbers*.   In two cases, the proofs are somewhat sophisticated.   They are included to provide illustrations of the kind of mathematics which are germane to this topic.   As in the investigation carried out in Sec. 5-5, the Chebyshev inequality plays a key role.

In order to simplify the statement of various theorems, we adopt some notation to be used consistently throughout this section.

Let $\{X_i(\cdot): 1 \le i < \infty\}$ be a sequence of random variables having mean $\mu[X_i] = \mu_i$ and standard deviation $\sigma[X_i] = \sigma_i$

$Y_n(\cdot) = \displaystyle\sum_{i=1}^{n} X_i(\cdot)$ with mean $\mu[Y_n] = m_n = \displaystyle\sum_{i=1}^{n} \mu_i$ and standard deviation $s_n = \sigma[Y_n]$

$A_n(\cdot) = \dfrac{1}{n} Y_n(\cdot) = \dfrac{1}{n} \displaystyle\sum_{i=1}^{n} X_i(\cdot)$ with mean $\mu[A_n] = m_n/n$ and standard deviation $\sigma[A_n] = s_n/n$

Particular conditions, such as independence or identity of the distributions, are noted specifically where they apply. It is convenient to refer to $Y_n(\cdot)$ as the nth *sum* and $A_n(\cdot)$ as the nth *average* of the sequence.

As a simple consequence of the Chebyshev inequality, we have the general limit theorem:

Theorem 6-1A

If $\lim\limits_{i \to \infty} \sigma_i^2 = 0$, then

$$\lim\limits_{i \to \infty} P(|X_i - \mu_i| < \epsilon) = 1 \qquad \text{for any } \epsilon > 0$$

or equivalently,

$$\lim\limits_{i \to \infty} P(|X_i - \mu_i| \geq \epsilon) = 0 \qquad \text{for any } \epsilon > 0$$

PROOF   By the Chebyshev inequality (Theorem 5-4B),

$$P(|X_i - \mu_i| \geq \epsilon) \leq \frac{\sigma_i^2}{\epsilon^2} \to 0 \text{ as } i \to \infty \quad \blacksquare$$

We may apply this theorem to obtain the following theorem on the nth average:

Theorem 6-1B

If $\lim\limits_{n \to \infty} s_n^2/n^2 = 0$,

$$\lim\limits_{n \to \infty} P\left(\left| A_n - \frac{m_n}{n} \right| < \epsilon \right) = 1 \qquad \text{for any } \epsilon > 0$$

PROOF $\sigma^2[A_n] = s_n^2/n^2$. $\mu[A_n] = m_n/n$.   Apply Theorem 6-1A. $\blacksquare$

Theorem 6-1B does not require independence of the $X_i(\cdot)$. The crucial condition $s_n^2/n^2 \to 0$, which we may call the *variance property*, makes the Chebyshev inequality applicable. If the random variables are uncorrelated, the variance property is met in two important special cases. These are presented in the following theorems, which may be considered corollaries to Theorem 6-1B.

Theorem 6-1C

If the $X_i(\cdot)$ are uncorrelated and if there exists a positive constant C such that, for all i, $\sigma_i^2 \leq C$, then

$$\lim\limits_{n \to \infty} P\left(\left| A_n - \frac{m_n}{n} \right| < \epsilon \right) = 1 \qquad \text{for any } \epsilon > 0$$

PROOF $s_n^2 = \sigma^2[Y_n] = \sum\limits_{i=1}^{n} \sigma_i^2 \leq nC$, so that $s_n^2/n^2 \to 0$ as $n \to \infty$. $\blacksquare$

Theorem 6-1D

If the $X_i(\cdot)$ are uncorrelated and all have the same mean $\mu_i = \mu$ and the same variance $\sigma_i^2 = \sigma^2$, then

$$\lim_{n \to \infty} P(|A_n - \mu| < \epsilon) = 1 \qquad \text{for any } \epsilon > 0$$

PROOF $m_n = n\mu$ and $s_n^2 = n\sigma^2$, so that $s_n^2/n^2 \to 0$ as $n \to \infty$. ∎

This last theorem is a slight generalization of the situation considered in the case of random sampling in Sec. 5-5.   The generalization consists in requiring only that the random variables be uncorrelated and have common mean and common variance.   Historically, it is this theorem which has been referred to as the law of large numbers.   The other theorems are generalizations of this law.

The following special case is the first known result of the type under consideration.   It is the celebrated theorem of James Bernoulli (1654–1705), who proved it near the end of his life, although it was not published until 1713, some eight years after his death.

Theorem 6-1E   Bernoulli's Theorem

If E_i is the event of a success on the ith trial of a sequence of Bernoulli trials and if $X_i(\cdot) = I_{E_i}(\cdot)$, then $A_n(\cdot)$ is the relative number of successes in n trials.   We may assert

$$\lim_{n \to \infty} P(|A_n - p| < \epsilon) = 1 \qquad \text{for any } \epsilon > 0$$

PROOF   According to the results of Examples 5-2-2 and 5-4-1, $\mu_i = p$, so that $m_n/n = p$, and $s_n^2 = npq$, so that the variance condition holds.   The result follows from Theorem 6-1B. ∎

Although, with the resources at hand, we find the proof almost trivial, Bernoulli did not have such mathematical resources.   His theorem is one of the landmarks of early probability theory, which had much to do with the further development of that theory.   The theorem states that if a long enough sequence is selected, the average number of successes will, with high probability, lie close to the probability of success on any given trial.   This idea paves the way for removing the restriction of "equally likely" in defining basic probabilities and provides mathematical support for the idea that probabilities may be determined from repeated trials.   It is easy to conclude too much from this theorem, however, and a stronger form known as the strong law of large numbers (Borel's theorem) is much to be desired in this respect.   The stronger form is considered in Sec. 6-4.

In the previous theorems, it is assumed that the mean and variance exist for each of the random variables $X_i(\cdot)$.   For independent random variables having the same distribution, it is not necessary that the variance exist.   The situation is as follows:

Theorem 6-1F

Suppose the class of random variables $\{X_i(\cdot): 1 \leq i < \infty\}$ is pairwise independent and each random variable in the class has the same distribution, with $\mu[X_i] = \mu$.   The variance may or may not exist. Then $\lim\limits_{n \to \infty} P(|A_n - \mu| < \epsilon) = 1$ for any $\epsilon > 0$.

The proof makes use of a classical device known as the *method of truncation*.   The somewhat delicate calculations involved are characteristic of much mathematical work in the area.   A proof is presented in Appendix D-3, to illustrate the type of analysis required.

The theorems of this section present a variety of sufficient conditions for the law of large numbers.   It is of mathematical interest to be able to characterize such a property by conditions which are both necessary and sufficient.   The following theorem presents such a condition (Gnedenko [1962]).

Theorem 6-1G

Let $B_n(\cdot) = A_n(\cdot) - m_n/n$.   Then

$$(A) \quad \lim_{n \to \infty} P(|B_n| < \epsilon) = 1 \qquad \text{for any } \epsilon > 0$$

iffi

$$(B) \quad E\left[\frac{B_n{}^2}{1 + B_n{}^2}\right] \to 0 \text{ as } n \to \infty$$

PROOF   Let $E = \{\xi: |B_n| \geq \epsilon\}$.   Since $a \geq b > 0$ implies

$$\frac{a}{1 + a} \cdot \frac{1 + b}{b} \geq 1$$

it follows that

$$P(E) = E[I_E] \leq E\left[I_E \frac{1 + \epsilon^2}{\epsilon^2} \cdot \frac{B_n{}^2}{1 + B_n{}^2}\right] \leq \frac{1 + \epsilon^2}{\epsilon^2} E\left[\frac{B_n{}^2}{1 + B_n{}^2}\right]$$

Hence (B) implies (A).   On the other hand,

$$P(E) \geq E\left[I_E \frac{B_n{}^2}{1 + B_n{}^2}\right] = E\left[\frac{B_n{}^2}{1 + B_n{}^2}\right] - E\left[I_{E^c} \frac{B_n{}^2}{1 + B_n{}^2}\right]$$

$$\geq E\left[\frac{B_n{}^2}{1 + B_n{}^2}\right] - \epsilon^2$$

since

$$I_{E^c}(\cdot) \frac{B_n{}^2(\cdot)}{1 + B_n{}^2(\cdot)} \leq \epsilon^2$$

Under condition (A), if we choose ϵ sufficiently small, we can make $E[B_n{}^2/(1 + B_n{}^2)]$ as small as desired for all n sufficiently large. ∎

The theorems of this section ensure that the probabilities converge as asserted under the appropriate conditions. They do not necessarily give the best bounds on the probabilities as n increases. Some special cases are studied in the next section.

6-2 Bounds on sums of independent random variables

In this section we consider certain problems of finding bounds on the probability that the average of a sum of n independent random variables having the same distribution differs from the mean by no more than a given amount. Specifically, we consider Bernoulli sequences, in which each random variable is the indicator function $I_{E_i}(\cdot)$ for the event of a success on the ith trial in a sequence. We compare the bound provided by Chebyshev's inequality with a bound derived by Chernoff [1952] and with direct calculations from the binomial distribution.

The particular Chernoff bound studied is a special case of an inequality developed, in the work cited above, for statistical tests of a hypothesis on the distribution of a random variable. Shannon [1957] has made extensive application of this bound to problems of coding in noisy channels. We develop a special form which is useful in the study of Bernoulli sequences. In particular, we use this form in studying the strong law of large numbers in Sec. 6-4.

As in previous sections, we let E_i be the event of a success on the ith trial in a sequence. We suppose the E_i form an independent class and assume that for each i we have $P(E_i) = p$. For notational convenience, in keeping with the convention of the previous section, we let

$$X_i(\cdot) = I_{E_i}(\cdot) \qquad \text{and} \qquad A_n(\cdot) = \frac{1}{n} \sum_{i=1}^{n} X_i(\cdot)$$

Then $A_n(\cdot)$ is binomially distributed with

$$P\left(A_n = \frac{r}{n}\right) = p_r = C_r{}^n p^r (1 - p)^{n-r} \qquad r = 0, 1, \ldots, n$$

$$\mu[A_n] = p \qquad \text{and} \qquad \sigma^2[A_n] = \frac{pq}{n} \qquad q = 1 - p$$

By Chebyshev's inequality

$$P(|A_n - p| \geq \epsilon) \leq \frac{pq}{n\epsilon^2}$$

If $p = k/n$ and $\epsilon = b/n$, where k and b are positive integers, the probability may be written in terms of the binomial expansion

$$P\left(\left|A_n - \frac{k}{n}\right| \leq \frac{b}{n}\right) = \sum_{k-b}^{k+b} C_r{}^n p^r q^{n-r} = \sum_{k-b}^{k+b} p_r$$

$$= \sum_{k-b}^{n} p_r - \sum_{k+b+1}^{n} p_r$$

Values may be found in tables of the summed binomial distribution.

Chernoff bounds

We shall develop a general formula, applicable to sequences of random variables having the same distribution, and then specialize this result to the case of Bernoulli sequences. Suppose $\{X_i(\cdot): 1 \leq i \leq n\}$ is an independent class of random variables, each having the same distribution. Put

$$Y_n(\cdot) = \sum_{i=1}^{n} X_i(\cdot)$$

Suppose $M(s) = E[e^{sX_i}]$ is the moment-generating function for each of the $X_i(\cdot)$ (this is the same function for each i). We restrict consideration to real values of s. Because of the independence of the $X_i(\cdot)$, the moment-generating function for $Y_n(\cdot)$ is the nth power of $M(\cdot)$.

Suppose $E = \{\xi: Y_n \geq c\}$ and $F = \{\xi: Y_n \leq c\}$. Then

$$e^{sY_n(\cdot)} \geq e^{sc} I_E(\cdot) \qquad \text{for } s \geq 0$$
$$e^{sY_n(\cdot)} \geq e^{sc} I_F(\cdot) \qquad \text{for } s \leq 0$$

Thus $M^n(s) \geq e^{sc} P(E)$ for $s \geq 0$ and $M^n(s) \geq e^{sc} P(F)$ for $s \leq 0$. We put $h(s) = \log_e M(s)$ and note that $h(0) = 0$. Then

$$P(Y_n \geq c) \leq e^{-sc+nh(s)} \qquad \text{for } s \geq 0$$
$$P(Y_n \leq c) \leq e^{-sc+nh(s)} \qquad \text{for } s \leq 0$$

Since s is arbitrary (but real), we seek a value which gives the best possible bound. Differentiating with respect to s and setting the derivative equal to zero gives

$$e^{-sc+nh(s)}[-c + nh'(s)] = 0$$

The exponential can never be zero. A minimum, if it exists, must be at one of the zeros of the expression in brackets; i.e., it must be a root of

$$h'(s) = \frac{c}{n}$$

Consider any root s_0.   For $s_0 \geq 0$

$$P[Y_n \geq nh'(s_0)] \leq e^{-n[s_0 h'(s_0) - h(s_0)]}$$

Exactly the same expression holds as a bound for $P[Y_n \leq nh'(s_0)]$ when the root $s_0 \leq 0$.

Chernoff bounds and Bernoulli trials

For the Bernoulli trials, $X_i(\cdot) = I_{E_i}(\cdot)$ with $P(E_i) = p$.   In this case

$$M(s) = pe^s + q \qquad \text{and} \qquad h(s) = \log_e (pe^s + q)$$

Differentiation with respect to s gives

$$h'(s) = \frac{pe^s}{pe^s + q}$$

It is convenient to put $c = na = nh'(s_0)$.   Then

$$a = \frac{pe^{s_0}}{pe^{s_0} + q}$$

Some algebraic manipulations give the result

$$e^{s_0} = \frac{qa}{pb} \qquad \text{where } b = 1 - a$$

or

$$s_0 = \log_e \frac{qa}{pb} \left. \begin{cases} \leq 0 & \text{for } a \leq p \\ \geq 0 & \text{for } a \geq p \end{cases} \right.$$

Then

$$h(s_0) = \log_e \left(p \, \frac{qa}{pb} + q \right) = \log_e q \left(\frac{a}{b} + 1 \right) = \log_e \frac{q}{b}$$

Since $h'(s_0) = a$, we have

$$n[s_0 h'(s_0) - h(s_0)] = n \left[a \log_e \frac{qa}{bp} - \log_e \frac{q}{b} \right] = ng(a, p)$$

Thus

$$\begin{aligned} P(Y_n \geq na) &= P(A_n \geq a) \leq e^{-ng(a,p)} & \text{for } a \geq p \\ P(Y_n \leq na) &= P(A_n \leq a) \leq e^{-ng(a,p)} & \text{for } a \leq p \end{aligned}$$

where $A_n(\cdot) = (1/n)Y_n(\cdot)$ is the average of the $X_i(\cdot)$ or the fraction of successes in n trials.

The logarithms used are to base e (i.e., the natural logarithms). Any other base may be used by virtue of the relation

$$e^{\log_e x} = 10^{\log_{10} x} = 2^{\log_2 x} = \cdots$$

In works on information theory, it is customary to use base 2. The expression for the exponent can be put into a form which exploits the binary entropy function $H(p) = -p \log p - q \log q$ for a binary scheme having two probabilities p and q. The expression $g(a, p)$ may be rearranged as follows:

$$
\begin{aligned}
g(a, p) &= a \log q + a \log a - a \log p - a \log b - \log q + \log b \\
&= a \log a + (1 - a) \log b - a \log p - (1 - a) \log q \\
&= [-a \log p - b \log q] - [-a \log a - b \log b]
\end{aligned}
$$

If we put $a = p + \epsilon$ and $b = 1 - a = q - \epsilon$, the last term in brackets is $H(p + \epsilon)$. Writing $g(p + \epsilon, p) = f(\epsilon, p)$, we have

$$
\begin{aligned}
f(\epsilon, p) &= [-(p + \epsilon) \log p - (q - \epsilon) \log q] - H(p + \epsilon) \\
&= H(p) + \epsilon \log \frac{q}{p} - H(p + \epsilon)
\end{aligned}
$$

Now it is easy to show that

$$
\log \frac{q}{p} = H'(p)
$$

so that

$$
f(\epsilon, p) = H(p) + \epsilon H'(p) - H(p + \epsilon)
$$

This is given a simple geometrical interpretation in Fig. 6-2-1 (cf. Wozencraft and Reiffen [1961, p. 14]). The curve of $H(p)$ has the general shape shown (logarithms are to base 2). The quantity $f(\epsilon, p)$ is the vertical distance between the $H(\cdot)$ curve at $p + \epsilon$ and the line which is tangent to the $H(\cdot)$ curve at p. An example is shown for positive ϵ. The result holds for negative ϵ as well. We suppose throughout that $0 < p + \epsilon < \frac{1}{2}$. Since the slope of the $H(\cdot)$ curve decreases with increasing p, it is apparent that, for positive ϵ, we must have $f(-\epsilon, p) > f(\epsilon, p)$ within the allowable range.

We now use the Chernoff bound to get a bound on the probability in the law of large numbers (Bernoulli's form). Using logarithms to base 2, we have

$$
\begin{aligned}
P(|A_n - p| \geq \epsilon) &= P(A_n \geq p + \epsilon) + P(A_n \leq p - \epsilon) \\
&\leq 2^{-nf(\epsilon, p)} + 2^{-nf(-\epsilon, p)} < 2 \cdot 2^{-nf(\epsilon, p)}
\end{aligned}
$$

A comparison of the Chernoff bound with the Chebyshev bound shows that, for fixed p and ϵ, the first bound decreases exponentially with n, while the latter decreases inversely with n. For large n the Chernoff bound can be expected to give much the superior estimate of the probability. This fact is exploited in the study of the strong law of large numbers in Sec. 6-4.

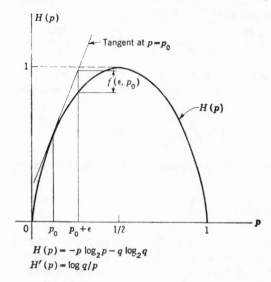

$$H(p) = -p \log_2 p - q \log_2 q$$
$$H'(p) = \log q/p$$

Fig. 6-2-1 The binary entropy function and the exponent for the Chernoff bound.

Since tables of $H(\cdot)$ and of logarithms to the base 2 are provided in many books on information theory (cf. Fano [1961, appendix B]), the expression for the exponent in terms of the entropy function may be used in calculations.   When it is desired that ϵ be smaller than p (we assume $p < q$ throughout), the form given has the disadvantage of requiring small differences of relatively large numbers.   In this case, it is more convenient to obtain a series expansion for $f(\epsilon, p)$ and to approximate it with a very few terms.   In order to obtain a series expansion, it is convenient to use logarithms to base e (i.e., to use natural logarithms) and to rewrite the expression as follows:

$$
\begin{aligned}
f(\epsilon, p) &= -(p + \epsilon) \log p - (q - \epsilon) \log q \\
&\qquad + (p + \epsilon) \log (p + \epsilon) + (q - \epsilon) \log (q - \epsilon) \\
&= p \left(1 + \frac{\epsilon}{p}\right) \log \left(1 + \frac{\epsilon}{p}\right) + q \left(1 - \frac{\epsilon}{q}\right) \log \left(1 - \frac{\epsilon}{q}\right)
\end{aligned}
$$

Use of well-known expressions for $\log (1 + x)$ and $\log (1 - x)$, which hold for $|x| < 1$, gives, after considerable manipulation,

$$
f(\epsilon, p) = \epsilon \left\{ \frac{1}{2} \left[\frac{\epsilon}{p} + \frac{\epsilon}{q} \right] - \frac{1}{6} \left[\left(\frac{\epsilon}{p}\right)^2 - \left(\frac{\epsilon}{q}\right)^2 \right] \right\} + \cdots
$$

The remaining terms may be paired:

$$\epsilon\left\{\frac{1}{(n+1)n}\left[\left(\frac{\epsilon}{p}\right)^n+\left(\frac{\epsilon}{q}\right)^n\right]-\frac{1}{(n+2)(n+1)}\left[\left(\frac{\epsilon}{p}\right)^{n+1}-\left(\frac{\epsilon}{q}\right)^{n+1}\right]\right\}$$

where n is odd. Since $1 > \epsilon/p > \epsilon/q > 0$, each of these pairs is positive. Hence $f(\epsilon, p)$ is greater than the first two terms. These may be rewritten to give

$$f(\epsilon, p) > \frac{\epsilon^2}{2pq} - \frac{\epsilon^2}{6pq}\left(\frac{\epsilon}{p} - \frac{\epsilon}{q}\right) > \frac{\epsilon^2}{3pq}$$

Recalling that base e is used in the development of the series expression, we have

$$P(|A_n - p| \geq \epsilon) \leq 2\exp\left(-\frac{n\epsilon^2}{3pq}\right)$$

This simple expression, which is a conservative one, may be compared with the Chebyshev bound. If we let $x = n\epsilon^2/pq$, the Chebyshev bound is $b_1 = 1/x$, while the Chernoff bound is $b_2 = 2e^{-x/3}$. Some illustrative values are

$x = n\epsilon^2/pq$	2	3	5	10	20	30
$b_1 = 1/x$	0.50	0.333	0.200	0.100	0.0500	0.03333
$b_2 = 2e^{-x/3}$	1.69	0.736	0.376	0.072	0.0036	0.00009

The two bounds are approximately equal at $x = 8.5$. The common value of the bounds for this value of x is $1/8.5 = 0.118$. For $x > 8.5$, the Chernoff bound b_2 should be used, since the ratio b_2/b_1 goes to zero rapidly with increasing x. A more extensive table of bounds can be obtained readily from tables of reciprocals and of values of e^{-x}, which are found in many handbooks. Such a table could be useful in estimating bounds or in determining what size sample (i.e., what value of n) is needed to ensure a given probability when p and ϵ are specified. It must be kept in mind that the approximate expression used requires that ϵ/p be less than unity. For larger values of ϵ/p, the form utilizing the entropy expressions should be used; numerical values, in this case, are obtained from tables of $H(p)$ and $\log_2 x$, where $x = q/p$.

6-3   Types of convergence

Several types of convergence have been identified and exploited in the theory of probability. In this section we shall define the more important of these and state, generally without proof, several basic theorems. For a careful treatment, see a standard work on measure theory such as Munroe [1953, chap. 6].

In order to simplify writing, we use a notational scheme adapted from that used by Munroe. Throughout this section we suppose $\{X_n(\cdot): 1 \leq n < \infty\}$ is a sequence of integrable random variables. Integrability implies that the $X_n(\cdot)$ are finite with probability 1 (that is, $|X_n(\cdot)| < \infty$ $[P]$). Convergence of the sequence to a random variable $X(\cdot)$ in one of the manners to be defined is indicated by writing

$$X_n(\cdot) \to X(\cdot) \ [\cdot\cdot\cdot] \qquad \text{or} \qquad \lim_{n \to \infty} X_n(\cdot) = X(\cdot) \ [\cdot\cdot\cdot]$$

where the appropriate entry is made in the brackets, to indicate the type of convergence.   If the assertions are limited to elements of some subset E of the basic space, the expression "on E" is added after the brackets.

Definition 6-3a

The sequence is said to *converge with probability* 1 to $X(\cdot)$, indicated

$$X_n(\cdot) \to X(\cdot) \ [P]$$

iffi there is a set E with $P(E^c) = 0$ such that $X_n(\xi) \to X(\xi)$ for each $\xi \in E$.   The term *almost surely* is also used.

Definition 6-3b

The sequence is said to *converge almost uniformly* to $X(\cdot)$, indicated

$$X_n(\cdot) \to X(\cdot) \qquad [\text{a. unif}]$$

iffi to each $\epsilon > 0$ there corresponds a set E with $P(E^c) < \epsilon$ such that $X_n(\xi)$ converges uniformly to $X(\xi)$ for all $\xi \in E$.

Definition 6-3c

The sequence is said to *converge in probability* to $X(\cdot)$, indicated

$$X_n(\cdot) \to X(\cdot) \qquad [\text{in prob}]$$

iffi $\lim_{n \to \infty} P(\{\xi: |X(\xi) - X_n(\xi)| \geq \epsilon\}) = 0 \qquad$ for each $\epsilon > 0$.

Definition 6-3d

The sequence is said to *converge in the mean of order* p $(p \geq 1)$, indicated

$$X_n(\cdot) \to X(\cdot) \qquad [\text{mean}^p]$$

iffi

$$\lim_{n \to \infty} \int |X - X_n|^p \, dP = 0$$

When $p = 1$, the terminology is usually simplified by deleting the p and referring to *convergence in the mean*. In this case, it is worth noting that

$$|\textstyle\int X\, dP - \int X_n\, dP| \le \int |X - X_n|\, dP$$

so that convergence in the mean implies convergence of the integrals.

It is fairly easy to show that the ordinary theorems for limits of sums, products, etc., hold for the first three kinds of convergence. Theorems relating the various kind of convergence are of considerable importance.

Set of convergence points

In order to gain some insight into the convergence process, we shall characterize it in terms of appropriate events and prove one basic theorem. Let D be the set on which the sequence fails to converge. We wish to characterize D (and hence its complement on which the series converges).

The sequence $X_i(\xi)$ fails to converge for any given ξ iff there is some k such that, if any n is chosen, there is at least one $i \ge n$ for which $|X(\xi) - X_i(\xi)| \ge 1/k$.

For any given k, let the set of all those ξ for which the above conditions hold be D_k. Then

$$D = \bigcup_{k=1}^{\infty} D_k$$

On the one hand, if $\xi \in D_k$ for some k, it must belong to D; on the other hand, if $\xi \in D$, it must belong to D_k for some k.

We next try to characterize the events D_k. In order to facilitate precise formulation, we introduce the notation

$$E_i(\epsilon) = \{\xi : |X(\xi) - X_i(\xi)| \ge \epsilon\} \qquad \text{for any } \epsilon > 0$$

For any fixed k, suppose an integer n is chosen. The set of those ξ for which $|X(\xi) - X_i(\xi)| \ge 1/k$ for at least one $i \ge n$ is the set

$$A_{kn} = \bigcup_{i=n}^{\infty} E_i\left(\frac{1}{k}\right)$$

Now the set D_k is the set for which this condition holds for every integer n. Thus D_k is given by

$$D_k = \bigcap_{n=1}^{\infty} A_{kn} = \bigcap_{n=1}^{\infty} \bigcup_{i=n}^{\infty} E_i\left(\frac{1}{k}\right)$$

We now have characterized D completely, for

$$D = \bigcup_{k=1}^{\infty} D_k = \bigcup_{k=1}^{\infty} \bigcap_{n=1}^{\infty} A_{kn} = \bigcup_{k=1}^{\infty} \bigcap_{n=1}^{\infty} \bigcup_{i=n}^{\infty} E_i\left(\frac{1}{k}\right)$$

It is worth noting that D is an event.

Conditions for convergence with probability 1

We now prove a fundamental theorem on convergence with probability 1.

Theorem 6-3A

(A) $X_n(\cdot) \to X(\cdot) \ [P]$

iffi

(B) $\lim\limits_{n \to \infty} P[\bigcup\limits_{i=n}^{\infty} E_i(\epsilon)] = 0$ for each $\epsilon > 0$

iffi

(C) $\lim P(A_{kn}) = 0$ for each positive integer k

PROOF   Assuming condition (A) is equivalent to assuming $P(D) = 0$. Now $P(D) \geq P[\bigcap\limits_{n=1}^{\infty} A_{kn}] = \lim\limits_{n \to \infty} P(A_{kn}) \geq 0$ for any k, so that the limit is zero. Given any $\epsilon > 0$, there is a k such that $1/k < \epsilon$. Since for such a k, $E_i(1/k) \supset E_i(\epsilon)$ for each i, we have $A_{kn} \supset \bigcup\limits_{i=n}^{\infty} E_i(\epsilon)$, so that the desired result holds for any $\epsilon > 0$. If condition (B) holds, we have $\lim\limits_{n \to \infty} P(A_{kn}) = 0$ for each k. Condition (A) follows from the fact that

$$0 \leq P(D) \leq \sum_{k=1}^{\infty} \lim_{n \to \infty} P(A_{kn}) = 0 \ \blacksquare$$

As a consequence of this theorem, we may prove two further important theorems.

Theorem 6-3B

If $\sum\limits_{i=1}^{\infty} P[E_i(1/k)]$ converges for each positive integer k, then

$$X_n(\cdot) \to X(\cdot) \ [P]$$

PROOF

$$P(A_{kn}) = P\left[\bigcup_{i=n}^{\infty} E_i\left(\frac{1}{k}\right)\right] \leq \sum_{i=n}^{\infty} P\left[E_i\left(\frac{1}{k}\right)\right]$$

Because of the assumed convergence of the series, we must have $\lim\limits_{n \to \infty} P(A_{kn}) = 0$ for each k. The argument in the proof of the previous theorem shows that this property implies convergence $[P]$ for the series. $\blacksquare$

The usefulness of this condition for convergence $[P]$ is illustrated in Sec. 6-4, in the proof of Borel's theorem on the strong law of large numbers.

The arguments in the preceding theorems show that:

1. Convergence $[P]$ is characterized by the property

$$\lim_{n \to \infty} P(A_{kn}) = 0 \qquad \text{for each positive integer } k$$

2. Convergence [in prob] is characterized by the property

$$\lim_{n \to \infty} P[E_n(1/k)] = 0 \qquad \text{for each positive integer } k$$

Since $A_{kn} \supset E_i(1/k)$ for each k and each $i \geq n$, the following theorem must hold:

Theorem 6-3C

Convergence $[P]$ implies convergence [in prob].

The converse of this theorem does not hold, however. The following example shows that convergence in probability does not imply convergence with probability 1.

We let the basic space S be the unit interval $0 \leq \xi \leq 1$. Probability is distributed uniformly on this interval, so that the probability of any subset E is the length (Lebesgue measure) of the set. We now define a sequence $\{X_i(\cdot): 1 \leq i < \infty\}$ of random variables, each of which is a step function on the unit interval S. These are formed in groups. The first group consists of 2 functions $X_1(\cdot)$ and $X_2(\cdot)$; the second group consists of $4 = 2^2$ functions; and in general the nth group consists of 2^n functions. The 2^n functions in the nth group are formed as follows (refer to Fig. 6-3-1):

1. Divide the unit interval into 2^n equal subintervals.

2. Let the rth function in the group have the value 1 over the rth subinterval and the value 0 elsewhere. Define the function at the jump points to be continuous from the right, so that the function is defined for all ξ.

Then, for any function in the nth group,

$$P(|X_i| > \epsilon) = (1/2)^n \qquad \text{for any } \epsilon < 1$$

For sufficiently large i the number n may be made as large as desired; hence $(1/2)^n$ may be made as small as desired. We may therefore assert that

$$X_i(\cdot) \to 0 \qquad \text{[in prob]}$$

On the other hand, consider any ξ in the unit interval. There is one and only one function in each group for which $X_i(\xi) = 1$. For all other functions in that group, $X_i(\xi) = 0$. This means that $\lim_{i \to \infty} X_i(\xi)$ does not exist for any ξ on the unit interval; hence it certainly is not true that the limit exists with probability 1.

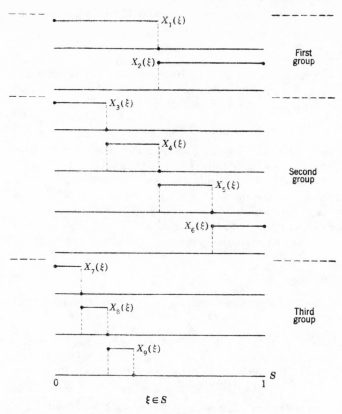

Fig. 6-3-1   A sequence of functions which converges in probability but
which does not converge with probability 1.

Relations between types of convergence

Other theorems relating the types of convergence to one another may be
proved.   We combine several of these into the two following theorems,
which we state without proof (cf. Munroe [1953, chap. 6]).

Theorem 6-3D

[a. unif] $\Rightarrow$ [P] $\Rightarrow$ [in prob] $\Leftarrow$ [mean^p]

Here we have simply indicated the types of convergence by the abbrevi-
ations.   Some of these relations depend upon the fact that $P(S)$ is finite.
Almost uniform convergence implies convergence [P], which in turn
implies convergence [in prob].   Convergence [mean^p], for any $p \geq 1$,
implies convergence [in prob].   Under a certain *condition of domination*,

the first and last implications in Theorem 6-3D hold both ways, as stated in the following

Theorem 6-3E

If $Y(\cdot)$ is a nonnegative random variable such that $Y^p(\cdot)$ is integrable $(p \geq 1)$ and $|X_n(\cdot)| \leq Y(\cdot)$ $[P]$, then for $p = 1$, $[P] \Rightarrow$ [a. unif], and for $p \geq 1$, [in prob] $\Rightarrow$ [mean^p], so that in this case

$$[\text{a. unif}] \Leftrightarrow [P] \Rightarrow [\text{in prob}] \Leftrightarrow [\text{mean}^p]$$

The composite relation may be stated as indicated, since the condition $Y^p(\cdot)$ is integrable implies the condition $Y(\cdot)$ is integrable. The assertion that, under the condition of domination, convergence in probability implies convergence in the mean of order p is often referred to as the *Lebesgue dominated-convergence theorem*. For $p = 1$, this implies convergence of the integrals of $X_n(\cdot)$ to the integral of $Y(\cdot)$.

One further theorem which is important in analysis may be stated as follows:

Theorem 6-3F

A sequence of random variables $\{X_n(\cdot)\colon 1 \leq n < \infty\}$ satisfies the condition

$\quad$ (A) $X_n(\cdot) \to X(\cdot)$ $\qquad$ [in prob]

iffi

$\quad$ (B) Each subsequence has a further subsequence which converges to $X(\cdot)$ with probability 1.

Examination of the weak law of large numbers, as developed in Sec. 6-1, shows that it can be stated in terms of convergence [in prob]. If we let $B_n(\cdot) = A_n(\cdot) - m_n/n$, this law may be stated as follows:

Weak law of large numbers: $B_n(\cdot) \to 0$ $\qquad$ [in prob]

For many purposes, the stronger convergence $[P]$ is desired. In this case it is customary to refer to the *strong law of large numbers*, stated as follows:

Strong law of large numbers: $B_n(\cdot) \to 0$ $[P]$

In view of Theorem 6-3C, it is apparent that the strong law implies the weak law, but not conversely. Because of the importance of the strong law, we give some attention to it in the next section.

6-4   The strong law of large numbers

The weak law of large numbers (especially Bernoulli's theorem) has played an important historical role in the development of the theory of probability.   It has pointed the way to applications, by indicating that probabilities may be determined by repeated sampling and averaging. It is easy to claim too much for this law, and the stronger form discussed in this section is highly desirable.   The weak law says that if a long sequence of independent Bernoulli trials is made, the probability is high that the average of this sequence is near the common probability of success on a single trial.   It does not ensure, however, that if the sequence is made longer, the average number of successes will remain near the common probability p.   The strong law asserts that for all sequences, except for a set of probability zero, the average number of successes will converge to the common probability.   It is this property that is really desired (and often tacitly assumed in discussing the law of large numbers).

About 1909, the French mathematician E. Borel formulated and proved the strong law in the case of Bernoulli trials.   A variety of proofs are given in the literature.   The proof given here is a very simple consequence of Theorem 6-3B and the special form of the Chernoff bound developed in Sec. 6-2.   In the development of the weak law of large numbers, the Chebyshev inequality suffices to show that

$$P[E_i(\epsilon)] \,=\, P(|X_i - p| \geq \epsilon)$$

goes to zero as i becomes infinite for each $\epsilon > 0$.   For the strong law, it is required to show that $P(A_{kn}) = P[\bigcup\limits_{i=n}^{\infty} E_i(1/k)]$ goes to zero as n becomes infinite, for each positive integer k.   A much stronger inequality than the Chebyshev inequality is needed.   Several have been developed and serve as the basis of proofs given in the literature.   We shall use the Chernoff inequality developed in Sec. 6-2.   The independence condition assumed in Bernoulli trials should be noted.

Theorem 6-4A   Borel's Theorem

If $X_i(\cdot)$ is the indicator function for the event of a success on the ith trial in a Bernoulli sequence and $A_n(\cdot)$ is the average number of successes in the first n trials, then

$$A_n(\cdot) \rightarrow p \;\; [P]$$

The number p is the probability of success on any trial.

PROOF   By Theorem 6-3B, it is sufficient to show that $\sum\limits_{i=1}^{\infty} P[E_i(1/k)]$ converges for each positive integer k.   By the Chernoff inequality in Sec. 6-2,

$$P\left[E_n\left(\frac{1}{k}\right) \right] \leq 2e^{-n/3pqk^2}$$

It is easy to show that a series of the form $2 \sum\limits_{n=1}^{\infty} e^{-na}$ converges for each $a > 0$.
Putting $a = \frac{1}{8} pqk^2$ completes the proof. ∎

The strong law may be applied to more general classes of random variables.

Definition 6-4a

If $\{X_i(\cdot) : 1 \leq i < \infty\}$ is a class of random variables with $\mu[X_i] = \mu_i$ and we put

$$B_n(\cdot) = \frac{1}{n} \sum_{i=1}^{n} X_i(\cdot) - \frac{1}{n} \sum_{i=1}^{n} \mu_i = A_n(\cdot) - m_n/n$$

we say the *strong law of large numbers holds* iffi

$$B_n(\cdot) \to 0 \quad [P]$$

We conclude the section by stating, without proof, several theorems giving sufficient conditions for the strong law.

Theorem 6-4B

If the $X_i(\cdot)$ form an independent class and the variances $\sigma^2[X_i] = \sigma_i{}^2$ satisfy the condition

$$\sum_{i=1}^{\infty} \frac{\sigma_i{}^2}{i^2} < \infty$$

the strong law holds.

Theorem 6-4C

If the $X_i(\cdot)$ form an independent class and the variances satisfy the condition $\sigma_i{}^2 < C$ for all i, the strong law holds.

Theorem 6-4D

If the $X_i(\cdot)$ form an independent class, if all have the same distribution, and if the mean value exists, the strong law holds.

It should be noted that, in each case, it is assumed that the $X_i(\cdot)$ form an independent class (and are not simply pairwise independent).

6-5   The central limit theorem

The normal, or gaussian, distribution plays a very important role in a wide variety of applications of probability theory to physical problems and to problems of statistics.   One of the reasons for this is found in a class of limit theorems known generically as the *central limit theorem*. A class of independent random variables may individually have distributions which are quite different from the normal distribution. But when these are summed and standardized in an appropriate manner, the resulting random variable has a distribution which is approximately normal.

Consider a class of *independent* random variables $\{X_i(\cdot): 1 \leq i < \infty\}$ having respective means μ_i and variances σ_i^2. Let

$$Y_n(\cdot) = \sum_{i=1}^{n} X_i(\cdot) \text{ with mean } m_n = \sum_{i=1}^{n} \mu_i \text{ and variance } s_n^2 = \sum_{i=1}^{n} \sigma_i^2$$

Put $Z_n(\cdot) = [Y_n(\cdot) - m_n]/s_n$ so that $\mu[Z_n] = 0$ and $\sigma^2[Z_n] = 1$. Let $F_n(\cdot)$ be the distribution function for $Z_n(\cdot)$.

Central limit theorem

Under very general conditions

$$\lim_{n \to \infty} F_n(t) = \Phi(t) \qquad \text{uniformly in } t$$

where $\Phi(\cdot)$ is the distribution function for the random variable which is normal $(0, 1)$.

Generally, the mathematics required to establish the various theorems is detailed and sophisticated. We state some results without proof.

Theorem 6-5A

Let $\{X_i(\cdot): 1 \leq i < \infty\}$ be a sequence of *independent* random variables. The central limit theorem holds if the $X_i(\cdot)$ satisfy any of the following sets of conditions:

1. *Liapounov.* There exists a number $\delta > 0$ such that

$$\lim_{n \to \infty} \frac{1}{s_n^{2+\delta}} \sum_{i=1}^{n} E[|X_i - \mu_i|^{2+\delta}] = 0$$

2. There exist positive numbers c and C such that

$$\sigma_i^2 \geq c > 0 \qquad \text{and} \qquad E[|X_i - \mu_i|^3] < C \qquad \text{for all } i$$

3. All $X_i(\cdot)$ have the same distribution, and $0 < \sigma^2[X_i] < \infty$.

It should be noted that conditions 2 and 3 require that $s_n^2 \to \infty$ as $n \to \infty$. It is known that this condition is necessary for the central limit theorem to hold.

Condition 3 is of particular interest for the theory of large sample statistics. If the $X_i(\cdot)$ are the random variables defined in studying samples from a given random variable $X(\cdot)$ with mean μ and variance σ^2, condition 3 applies.   In this case

$$Z_n(\cdot) = \frac{Y_n(\cdot) - n\mu}{\sqrt{n}\,\sigma}$$

Upon dividing numerator and denominator by n and setting

$$A_n(\cdot) = \frac{1}{n} Y_n(\cdot)$$

as before, we have

$$Z_n(\cdot) = \frac{A_n(\cdot) - \mu}{\sigma/\sqrt{n}}$$

Since $Z_n(\cdot)$ is approximately normal $(0, 1)$ for sufficiently large n, the *sample mean* $A_n(\cdot)$ *is approximately normal* $(\mu, \sigma/\sqrt{n})$.

The normal distribution plays a central role in the theory of observational errors and in the theory of electrical noise (i.e., disturbances). In each case, the perturbance at any moment of observation can usually be attributed to the combined influence of a large number of unrelated physical causes.   If this combination has the character of the sum of a large number of random phenomena, the central limit theorem leads one to expect that the resulting random phenomenon will have a distribution which is approximately normal.   Experience shows that this is true in a great many physical situations.

Problems

6-1. Suppose $\{X_i(\cdot): 1 \leq i < \infty\}$ is a class of pairwise independent random variables, each with zero mean value; and suppose there exists a random variable $Y(\cdot)$ with $E[Y^2] < \infty$ and such that $|X_i(\cdot)| \leq Y(\cdot)$ $[P]$ for each i.   Show that the weak law of large numbers holds for the sequence.

6-2. Consider a sequence $\{X_n(\cdot): 1 \leq n < \infty\}$ of pairwise independent random variables with the properties that $\mu[X_n] = 0$ and $|X_n(\cdot)| \leq n^\alpha$ $[P]$ for each $n = 1$, $2, \ldots$.   Show that the weak law of large numbers holds if $\alpha < \frac{1}{2}$.   (*Suggestion:*

Use the fact that if $\displaystyle\sum_{n=1}^{\infty} (1/n^2)a_n$ converges, then $\displaystyle\lim_{n\to\infty} (1/n^2) \sum_{i=1}^{n} a_i = 0$.)

6-3. The condition of pairwise independence in Theorem 6-1C may be replaced by a more general condition.   Consider the correlation coefficients defined as follows:

$$\rho_{ij} = \frac{E[(X_i - \mu_i)(X_j - \mu_j)]}{\sigma_i \sigma_j}$$

Use of the Schwarz inequality shows that $0 \leq |\rho_{ij}| \leq 1$.   Also, it is apparent that $\rho_{ii} = 1$ for each i.   Show that the $X_i(\cdot)$ obey the weak law of large numbers if (1) there exists a positive constant C such that, for every i, $\sigma_i{}^2 \leq C$, and (2) $\rho_{ij} \to 0$ as $|i - j| \to \infty$. [*Suggestion:* Show the variance property holds. Note that in the double sum $\sum_{i=1}^{n} \sum_{j=1}^{n} \rho_{ij}$, there are n^2 terms; for $n_0 < n$, there are $(n - n_0)^2$ terms for which $|i - j| > n_0$.]

6-4. It is known that the probability p of a success on a given trial in a Bernoulli sequence lies between 0.5 and 0.6. What size sample must be taken in order to have probability 0.95 or better that the mean value of the sample will lie within 5 percent of the true value p?

(a) Make an estimate of the sample size based on the Chebyshev inequality.

ANSWER: 8,000

(b) Make an estimate of the sample size based on the Chernoff estimate developed in the text.

ANSWER: 4,440

6-5. Suppose each random variable in the independent class $\{X_n(\cdot): 1 \leq n < \infty\}$ is normal (μ, σ). Let $A_n(\cdot) = (1/n) \sum_{i=1}^{n} X_i(\cdot)$. Obtain a Chernoff bound for $P(|A_n(\cdot) - \mu| \geq \epsilon)$. {*Suggestion:* Note that $X_i(\cdot) - \mu$ is normal $(0, \sigma)$ and that $|A_n(\cdot) - \mu| \geq \epsilon$ iffi either $\sum_{i=1}^{n} [X_i(\cdot) - \mu] \geq n\epsilon$ or $\sum_{i=1}^{n} [X_i(\cdot) - \mu] \leq -n\epsilon$.}

ANSWER: $P \leq 2 \exp\left(\dfrac{-n\epsilon^2}{2\sigma^2}\right)$

6-6. It is known that a random variable $X(\cdot)$ is approximately normal (μ, σ) and that $1 \leq \mu \leq 2$ and $2 \leq \sigma^2 \leq 10$. How large a random sample must be taken to have probability 0.98 or better that the sample average differs from the mean value by less than 5 percent?

(a) Estimate n with the aid of the Chebyshev inequality. ANSWER: 200,000

(b) Estimate n by the Chernoff inequality derived in Prob. 6-5. ANSWER: 37,000

6-7. Consider the sequence described in Prob. 6-2, but suppose the class is an independent class (and not simply pairwise independent). Show that the strong law of large numbers holds if $\alpha < \frac{1}{2}$.

6-8. Prove the *Borel-Cantelli lemma*, which may be stated as follows. Let $\{E_i: 1 \leq i < \infty\}$ be a sequence of events, and let

$$E = \limsup E_i = \bigcap_{n=1}^{\infty} \bigcup_{i=n}^{\infty} E_i$$

(a) If $\sum_{i=1}^{\infty} P(E_i) < \infty$, then $P(E) = 0$.

(b) If the sequence forms an independent class and if $\sum_{i=1}^{\infty} P(E_i) = \infty$, then $P(E) = 1$.

{*Suggestions:* (a) Let $A_n = \bigcup_{i=n}^{\infty} E_i$; note that the A_n form a decreasing sequence; use properties (P9) and (P10) for probabilities. (b) Show $P(E^c) = 0$. Put $B_n = \bigcap_{i=n}^{\infty} E_i{}^c$;

note that the B_n form an increasing sequence so that (P9) and (P10) may be applied;

use the fact that $\displaystyle\prod_{i=n}^{m} [1 - P(E_i)] \leq \exp \left[- \sum_{i=n}^{m} P(E_i) \right]$ for each n and m.$\}$

6-9. Use the Borel-Cantelli lemma [Prob. 6-8] to show the following:

(a) If $\{E_i : 1 \leq i < \infty\}$ is an independent class such that $\lim E_i = E$ exists,. then either $P(E) = 0$ or $P(E) = 1$.

(b) If $\{X_n(\cdot) : 1 \leq n < \infty\}$ is an independent class of random variables and if $X_n(\cdot) \to 0$ [P], then, for any positive constant c, $\displaystyle\sum_{n=1}^{\infty} P(|X_n| \geq c) < \infty$.

6-10. The following is a model for *random digits*, which play an important role in certain types of computational schemes. Let $X(\cdot)$ be a simple random variable whose range is the set of integers $0, 1, 2, \ldots, r - 1$. In the decimal system $r = 10$, and in the binary system $r = 2$. We suppose $P[X(\xi) = k] = 1/r$ for each $k = 0, 1, 2, \ldots,$ $r - 1$. Now consider an independent sequence $\{X_i(\cdot) : 1 \leq i < \infty\}$, each member of which has the same distribution as $X(\cdot)$. A realization of this sequence is called a sequence of random digits. Define a random variable $R_{kn}(\cdot)$ whose value for any ξ is the fraction of the first n integers $X_1(\xi), X_2(\xi), \ldots, X_n(\xi)$ whose values are k $(0 \leq k \leq r - 1)$. Show that $\lim_{n \to \infty} R_{kn}(\cdot) = 1/r$ [P], for each k. [*Suggestion:* Let A_{ki} be the event $X_i(\xi) = k$. Express $R_{kn}(\cdot)$ in terms of the indicator functions for the A_{ki}.]

6-11. Suppose $Y(\cdot)$ is uniformly distributed on the interval $[0, 1]$. Let $Y(\cdot) = \displaystyle\sum_{k=1}^{\infty} X_k(\cdot)10^{-k}$. Thus $X_k(\cdot)$ is the kth digit in the decimal expansion of the number $Y(\cdot)$. Note that $1 = 0.9999 \ldots$, an unending sequence of nines.

(a) Show that the $X_k(\cdot)$ form an independent class and that each $X_k(\cdot)$ is uniformly distributed over the integers $0, 1, 2, \ldots, 9$. (*Suggestion:* The event $\{\xi : X_i(\xi) = k\}$ is the event that $Y(\cdot)$ takes on one of those values whose decimal expansion has the integer k in the ith place. Show that the 10^n sets of numbers obtained by fixing the first n digits in the decimal expansion determine a partition, each of whose members has probability 10^{-n}. From this fact, show that events determined by fixing any finite subset of the integers have the probabilities required.)

(b) Show that with probability 1 (i.e., that for all choices except possibly a set whose probability is 0), each of the integers $k = 0, 1, 2, \ldots, 9$ will appear in $\frac{1}{10}$ of the places in the decimal expansion of $Y(\xi)$. Use the results of Prob. 6-10. Note that special numbers like $0.000, \ldots, 0.11111, \ldots, 0.123000, \ldots$, etc., must all belong to a set assigned zero probability mass by the random variable $Y(\cdot)$.

Selected references

Law of large numbers

FISZ [1963], chap. 6.   Cited at the end of our Chap. 2.

GNEDENKO [1962], chap. 6.   Cited at the end of our Chap. 3.

LOÈVE [1963], chap. 6.   Provides a thorough and rigorous treatment of the law of large numbers and of the central limit theorem.

McCORD AND MORONEY [1964], chaps. 10 through 12. Provides a readable treatment, tracing the central ideas but omitting the more difficult proofs. Recommended collateral reading.

Chernoff bounds

CHERNOFF [1952]: A Measure of Asymptotic Efficiency for Tests of a Hypothesis Based on Sums of Observations, *Annals of Mathematical Statistics*.

SHANNON [1957]: Certain Results in Coding Theory for Noisy Channels, *Information and Control*. Applications to problems in information theory.

WOZENCRAFT AND REIFFEN [1961]: "Sequential Decoding," chap. 2, sec. 2.

Types of convergence

HALMOS [1950]: "Measure Theory," secs. 21, 22. A standard work in measure theory. Somewhat more abstract in its treatment than Munroe.

MUNROE [1953], chap. 6. Cited at the end of our Chaps. 2 and 4.

Central limit theorem

BRUNK [1964], sec. 57. Some brief but important comments. Cited at the end of our Chap. 2.

CRAMÉR [1946], secs. 17.4, 17.5. Cited at the end of our Chap. 5.

GNEDENKO [1962], chap. 6.

LOÈVE [1963]. See above.

chapter 7

Random processes

The concept of a random variable has been introduced as a mathematical device for representing a physical quantity whose value is characterized by a quality of "randomness," in the sense that there is uncertainty as to which of the possible values will be observed. The concept is essentially static in nature, for it does not take account of the fact that, in real-life situations, phenomena are usually dependent upon the passage of time. Most of the problems to which probability concepts are applied involve a discrete sequence of observations in time or a continuous process of observing the behavior of a dynamic system. At any one instant of time, the outcome has the quality of "randomness." Knowledge of the outcomes at previous times may or may not reduce the uncertainty. The quantity under observation may be the fluctuating magnitude of a signal from a communication system, the quotations on the stock market, the incidence of particles from a radiation source, a critical parameter of a manufactured item, or any one of a large number of other items that one can readily call to mind. There are, of course, situations in which the outcome to be observed may depend upon a parameter other than (or in addition to) time. The random quantity to be observed may depend in some manner upon such parameters as position, temperature, age of a patient, etc. Mathematically, these dependences are not essentially different from the dependence upon time, and it is helpful to think in terms of a time variable and to develop our notation as if the parameter indicated time.

The problem is to find a generalization of the notion of random variable which allows the extension of probability models to dynamic systems. We have, in fact, already moved in this direction in considering vector-valued random variables or, equivalently, finite families of random variables considered jointly. The components of the vector or the members of the family can be conceived to represent observed values of a physical quantity at successive instants of time. The full generalization of these ideas leads naturally to the concept of a random, or stochastic, process.

The treatment of random processes in this chapter must of necessity be introductory and incomplete. Full-sized treatises on the subject usually have an introductory or prefatory remark to the effect that only selected topics can be treated therein. We introduce the general concept and discuss its meaning as a mathematical model of actual processes. Then we consider briefly some special processes and classes of processes which are often encountered in practice and which illustrate concepts, relationships, and properties that play an important role in the theory.

7-1 The general concept of a random process

The general concept of a random process is a simple extension of the idea of a random variable. We suppose there is defined a suitable probability space, consisting of a basic space S, a class of events $\mathcal{E}$, and a probability measure $P(\cdot)$ defined for each event $E \in \mathcal{E}$.

Definition 7-1a

Let T be an infinite set of real numbers (countable or uncountable). A *random process* X is a family of random variables $\{X(\cdot, t): t \in T\}$. T is called the *index set*, or the *parameter set*, for the process.

The terms *stochastic process* and *random function* are used by various authors instead of the term random process. Some authors include finite families of random variables in the category of random processes, but they then point out that the term is usually reserved for the infinite case. For most of the processes discussed in this chapter, T is either the set of positive integers, the set of positive integers and zero, the set of all integers, or an interval on the real line (possibly the whole real line). In the case of a countable index set, the term *random sequence* is sometimes employed. We shall use the term random process to designate either the countable or uncountable case.

A random process X may be considered from *three points of view,* each of which is natural for certain considerations.

1. As a *function* $X(\cdot, \cdot)$ on $S \times T$.

$$x = X(\xi, t)$$

To know the value of a random process as defined above, it is necessary to determine the parameter t and the choice variable ξ. Thus, from a mathematical point of view, it is entirely natural to view the process as a function. Mathematically, the manner in which ξ and t are determined is of no consequence.

2. As a *family of random variables* $\{X(\cdot, t): t \in T\}$. It is in these terms that the definition is framed. For each choice of $t \in T$, a random variable is chosen. It is usual to think of the choice of t as somehow deterministic. If t represents the time of observation, the choice of this time is taken to be deliberate and premeditated. The choice of ξ is, however, in some sense "random." This means that there is some uncertainty before the choice as to which of the possible ξ will be chosen. It is this choice which is the mathematical counterpart of selection from a jar, sampling from a population, or otherwise selecting by a trial or experiment. For this reason, it is helpful to think of ξ as the *random choice variable,* or simply the *choice variable.* Selection of subsets (finite or infinite) of the index set T amounts to selection of subfamilies of the process. One important way to study random processes is to study the joint distributions of finite subfamilies of the process.

3. As a *family of functions on* T, $\{X(\xi, \cdot): \xi \in S\}$. For each choice of ξ, a function $X(\xi, \cdot)$ is determined. The domain of this function is T. Should T be countable, the function so defined is a sequence. Each such function is called a *sample function,* or a *realization of the process,* or a *member of the process.*

It is the sample function which is observed in practice. The random process can serve as a model of a physical process only if, in observing the behavior of the physical process, one can take the point of view that *the function observed is one of a family of possible realizations of the process.* The concept of randomness is associated precisely with the uncertainty of the choice of one of the sample functions in the family of such functions. The mechanism of this choice is represented in the choice of ξ from the basic space. The concept of randomness has nothing to do with the character of the function chosen. This sample function may be quite smooth and regular, or it may have the wild fluctuations popularly and naïvely associated with "random" phenomena. One can conceive of a simple generator of a specific curve which has this "random" character; each time the generator is set in motion, the same curve is produced.

There is nothing random about the function represented by the curve. The outcome of the operation is completely determined.

As an example of a simple random process whose sample functions have quite regular characteristics, consider the following

Example 7-1-1   Random Sinusoid

Let $A(\cdot)$, $\omega(\cdot)$, and $\theta(\cdot)$ be random variables. Define the process X by the relation

$$X(\xi, t) = A(\xi) \cos [\omega(\xi)t + \theta(\xi)]$$

For each choice of ξ a choice of amplitude $A(\xi)$, angular frequency $\omega(\xi)$, and phase angle $\theta(\xi)$ is made. Once this choice is made, a perfectly regular sinusoidal function of time is determined. The uncertainty preceding the choice is the uncertainty as to which of the possible functions in the family is to be chosen. On the other hand, if a value of t is determined, each choice of ξ determines a value of $X(\cdot, t)$. As a Borel function of three random variables, $X(\cdot, t)$ is itself a random variable. ∎

Before continuing to further examples, we take note of some common variations of notation used in the literature. Most authors suppress the choice variable, since it plays no role in practical computations. They write X_t or $X(t)$ where we have used $X(\cdot, t)$, above. Since there are distinct conceptual advantages in thinking of ξ as the choice variable, we shall commonly use the notation $X(\cdot, t)$ to emphasize that, for any given determination of the parameter t, we have a function of ξ (i.e., a random variable). In some situations, however, particularly in the writing of expectations, we find it notationally convenient to suppress the choice variable. In designating mathematical expectations we shall usually write $E[X(t)]$.

The following examples indicate some ways in which random processes may arise in practice and how certain analytical or stochastic assumptions give character to the process.

Example 7-1-2   A Coin-flipping Sequence

A coin is flipped at regular time intervals $t = k, k = 0, 1, 2, \ldots$. If a head appears at $t = k$, the sample function has the value 1 for $k \leq t < k + 1$. The value during this interval is 0 if a tail is thrown at $t = k$.

SOLUTION   Each sample function is a step function having values 0 or 1, with steps occurring only at integral values of time. If the sequence of 0's or 1's resulting from a sequence of trials is written down in succession after the binary point, the result is a number in binary form in the interval [0, 1]. To each element ξ there corresponds a point on the unit interval and a sample function $X(\xi, \cdot)$. We may conveniently take the unit interval to be the basic space S and let ξ be the number whose binary equivalent describes the sample function $X(\xi, \cdot)$. If we let $A_n = \{\xi: X(\xi, t) = 1, n \leq t < n + 1\}$, we have, under the usual assumptions, $P(A_n) = \frac{1}{2}$ for any n. It may be shown that the probability mass is distributed uniformly on the unit interval, so that the probability measure coincides with the Lebesgue measure, which assigns to each interval a mass equal to the length of that interval. ∎

Example 7-1-3   Shot Effect in a Vacuum Tube

Suppose a vacuum tube is connected to a linear circuit. Let the response of the system to a single electron striking the plate at $t = 0$ be given at any time by the value $g(t)$, where $g(\cdot)$ is a suitable response function. Then the resulting current in the tube is

$$I(\xi, t) = \sum_k g[t - \tau_k(\xi)]$$

where $\{\tau_k(\cdot) : -\infty < k < \infty\}$ is a random sequence which describes the arrival times of the electrons. A choice of ξ amounts to a choice of the sequence of arrival times. Once these are determined, the value of the sample function (representing the current in the tube) is determined for each t. Ways of describing in probability terms the arrival times in terms of *counting processes* are considered in Sec. 7-3. ■

Example 7-1-4   Signal from a Communication System

An important branch of modern communication theory applies the theory of random processes to the analysis and design of communication systems. To use this approach, one must take the point of view that the signal under observation is one of the set of signals that are conceptually possible from such a system; that is, the signal is a sample signal from the process. Signals from communication systems can usually be characterized in terms of a variety of analytical properties. The transmitted signal may be taken from a known class of signals, as in the case of certain types of pulse-code modulation schemes. The dynamics of the transmission system place certain restrictions on the character of the signals. Certainly they must be bounded. The "frequency spectrum" of the signals is limited in specified ways. The mechanism producing unwanted "noise" can often be described in probabilistic terms. Utilization of these analytical properties enables the communication theorist to develop and study various processes as models of communication systems. ■

This simple list of examples does not begin to indicate the extent of the applications of random processes to physical and statistical problems. For a more adequate picture of the scope of the theory and its applications, one should examine such works as those by Parzen [1962], Rosenblatt [1962], Bharucha-Reid [1960], Wax [1954], and Middleton [1960]. Examples included in these works, as well as the voluminous literature cited therein, should serve to convey an impression of the extent and importance of the field.

7-2   Constant Markov chains

The simplest kind of a random process is a sequence of discrete random variables. The simplest sequences to deal with are those in which the random variables form an independent family. Much of the early work on sequences of trials has dealt with the case of independent random variables. There are many systems, however, for which the independence assumption is not satisfactory. In a sequence of trials, the outcome of any trial may be influenced by the outcome for one or more previous trials in the sequence. Early in this century, A. A. Markov (1856–1922) considered an important case of such dependence in which the outcome of any trial in a sequence is conditioned by the outcome of the trial immediately preceding, but by no earlier ones. Such dependence has come to be known as *chain* dependence.

Such a chain of dependences is common in many important practical situations. The result of one choice affects the next. This is true in many games of chance, when a sequence of moves is considered rather than a single, isolated play. Many physical, psychological, biological, and economic phenomena exhibit such a chain dependence. Even when the dependence extends further back than one step, a reasonable approximation may result from considering only one-step dependence. Some simple examples are given later in this section.

The systems to be studied are described in terms of the results of a sequence of trials. Each trial in the sequence is performed at a given *transition time*. During the *period* between two transition times, the system is in one of a discrete set of *states*, corresponding to one of the possible outcomes of any trial in the sequence. This set of states is called the state space $\mathcal{S}$. We thus have a random process $\{X(\cdot, n) : 0 \leq n < \infty\}$, in which the value of the random variable $X(\cdot, n)$ is the state of the system (or a number corresponding to that state) in the period following the nth trial in the sequence. We take the zeroth period to be the *initial period*, and the value of $X(\cdot, 0)$ is the *initial state* of the system. The process may be characterized by giving the probabilities of the various states in the initial period [i.e., the distribution of $X(\cdot, 0)$] and the appropriate conditional probabilities of the various states in each succeeding period, given the states of the previous periods. The precise manner in which this is done is formulated below.

Suppose the *state space* $\mathcal{S}$ is the set $\{s^i : i \in J\}$, where J is a finite or countably infinite index set. The elements s^i of the state space represent possible states of the system and appear in the model as possible values of the various $X(\cdot, n)$; that is, $\mathcal{S}$ is the range set for each of the random variables in the process. We may take the s^i to be real numbers without loss of generality. The random variables may be written

$$X(\cdot, n) = \sum_{i \in J} s^i I_{E_n{}^i}(\cdot)$$

where $E_n{}^i = \{\xi : X(\xi, n) = s^i\}$. In words, $E_n{}^i$ is the event that the system is in the ith state during the nth period or interval. The probability distributions for the $X(\cdot, n)$ are determined by specifying (1) the initial probabilities $P(E_0{}^i)$ for each $i \in J$ and (2) the conditional probabilities $P(E_n{}^k | E_{n-1}{}^j \cdots E_0{}^i)$ for all the possible combinations of the indices.

A wide variety of dependency arrangements is possible, according to

the nature of the conditional probabilities.   We shall limit consideration
to processes in which the probabilities for the state after a transition are
conditioned only by the state immediately before the transition.   This
may be stated precisely as follows:

(MC1)   *Markov property.* $P(E_n{}^k|E_{n-1}^j \cdots E_{n-r}^i) = P(E_n{}^k|E_{n-1}^j)$ for each per-
missible n, r, k, and j.

Definition 7-2a

The conditional probabilities in property (MC1) are called the *transition probabilities* for the process.

In most of the processes studied, it is assumed further that the transition probabilities do not depend upon n; that is, we have the

(MC2)   *Constancy property.* $P(E_n{}^k|E_{n-1}^j) = P(E_r{}^k|E_{r-1}^j)$ for each permissible n,
r, k, and j.

It is convenient, in the case of constant transition probabilities, to designate them by the shorter notation

$$\pi(j, k) = P(E_r{}^k|E_{r-1}^j)$$

In a similar manner, the total probability for any state i in the rth
period is designated by

$$\pi_r(i) = P(E_r{}^i)$$

These probabilities define the probability distribution for the random
variable $X(\cdot, r)$.

Definition 7-2b

A random process $\{X(\cdot, n): 0 \leq n < \infty\}$ with finite or countably
infinite *state space* $\mathcal{S} = \{s^i: i \in J\}$ and which has the Markov
property (MC1) and the constancy property (MC2) is called a
Markov chain with constant transition probabilities (or more briefly,
a *constant Markov chain*).   If the state space $\mathcal{S}$ is finite, the process
is called a *finite Markov chain*.   The set of probabilities $\{\pi_0(i): i \in J\}$
is called the set of *initial state probabilities*.   The matrix $\mathcal{P} = [\pi(j, k)]$
is called the *transition matrix*.

Constant Markov chains are also called *homogeneous* chains, or *stationary*
chains.   Most often the reference is simply to Markov chains, the constancy property being tacitly assumed.

A very extensive literature has developed on the subject of constant
Markov chains.   We can touch upon only a few basic ideas and results
in a manner intended to introduce the subject and to facilitate further
reference to the pertinent literature.

Since a conditional probability measure, given a fixed event, is itself a probability measure, and since the events $\{E_r{}^j: j \in J\}$ form a partition for each r, we must have, for each i,

$$0 \leq \pi(i, j) \leq 1 \qquad \text{and} \qquad \sum_{j \in J} \pi(i, j) = 1$$

The last equation means that the elements on each row of the transition matrix $\mathcal{P}$ sum to 1, and hence form a probability vector. Such matrices are of sufficient importance to be given a name as follows:

Definition 7-2c

A square matrix (finite or infinite) whose elements are nonnegative and whose sum along any row is unity is called a *stochastic matrix*. If, in addition, the sum of the elements along any column is unity, the matrix is called *doubly stochastic*.

The transition matrix for a constant Markov chain is a stochastic matrix. In special cases, it may be doubly stochastic.

We now note that a constant Markov process is characterized by the initial probabilities and the transition matrix. As a direct consequence of properties (CP1), (MC1), and (MC2), we have

(MC3) $P(E_n{}^k E_{n-1}^i \cdot \cdot \cdot E_{r+1}^i E_r{}^h) = \pi_r(h)\pi(h, i) \cdot \cdot \cdot \pi(j, k)$

If we put $r = 0$, we see that specification of the initial state probabilities and of the matrix of transition probabilities serves to specify the joint probability mass distribution for any finite subfamily of random variables in the process. In particular, we may use property (P6) to derive the following expression for the *state probabilities* in the nth period:

(MC4)   *State probabilities*

$$\pi_n(k) = \sum_{h, i, \ldots, j} \pi_0(h)\pi(h, i) \cdot \cdot \cdot \pi(j, k)$$

PROOF   This follows from (MC3) and the fundamental relation

$$P(E_n{}^k) = \sum_{h, i, \ldots, j} P(E_n{}^k E_{n-1}^i \cdot \cdot \cdot E_1{}^i E_0{}^h) \quad \blacksquare$$

Before developing further the theoretical apparatus, we consider some simple examples of Markov processes and their characteristic probabilities. To simplify writing, we shall consider finite chains.

Example 7-2-1   Urn Model

Feller [1957] has shown that a constant Markov chain is equivalent to the following urn model. There is an urn for each state of the system. Each urn has balls marked with numbers corresponding to the states. The probability of drawing a ball marked k from urn j is $\pi(j, k)$. Sampling is with replacement, so that the distribution in any given urn is not disturbed by the sampling. The process consists in making a sequence of samplings according to the following procedure. The first sampling, to start the process, consists of making a choice of an urn according to some prescribed set of probabilities. Call this the zeroth choice, and let $E_0{}^j$ be the event that the jth urn is chosen on this initial step. The choice of urn j is made with probability $\pi_0(j)$ so that $P(E_0{}^j) = \pi_0(j)$. From this urn make the first choice of a ball. A ball numbered k is chosen, on a random basis, from urn j with probability $\pi(j, k)$. The next choice is made from urn k. A ball numbered h is chosen with probability $\pi(k, h)$. The process, once started, continues indefinitely. If we let $E_n{}^k$ be the event that a ball numbered k is chosen at the nth trial, then because of the independence of the successive samples, we have

$$P(E_0{}^j E_1{}^k E_2{}^h \cdot \cdot \cdot) = \pi_0(j)\pi(j, k)\pi(k, h) \cdot \cdot \cdot$$

This is exactly the Markov property with constant transition probabilities. ∎

Example 7-2-2

Any square stochastic matrix plus a set of initial probabilities defines a constant Markov chain. Consider a finite chain with three states. Let

$$\mathcal{P} = \begin{bmatrix} 0.5 & 0.5 & 0 \\ 0.8 & 0.2 & 0 \\ 0.3 & 0.3 & 0.4 \end{bmatrix} \qquad \pi_0(1) = \pi_0(2) = 0, \ \pi_0(3) = 1$$

SOLUTION The set of initial probabilities is chosen so that, with probability 1, the system starts in state 3. Thus, in the first period, it is sure that the system is in state 3. At the first transition time, the system remains in state 3 with probability 0.4, moves into state 1 with probability 0.3, or into state 2 with probability 0.3. Once in state 1 or state 2, the probability of returning to state 3 is zero. Thus, once the system leaves the state 3, into which it is forced initially, it oscillates between states 1 and 2. If it is in state 1, it remains there with probability 0.5 and changes to state 2 with probability 0.5. When in state 2, it remains there with probability 0.2 or changes to state 1 with probability 0.8. The probability of being in states 3, 3, 1, 2, 2, 1 in that order is $\pi_0(3)\pi(3, 3)\pi(3, 1)\pi(1, 2)\pi(2, 2)\pi(2, 1) = 1 \times 0.4 \times 0.3 \times 0.5 \times 0.2 \times 0.8$. ∎

For the next example, we consider a simple problem of the "random-walk" type, which is important in the analysis of many physical problems. The example is quite simple and schematic, but it illustrates a number of important ideas and suggests possibilities for processes.

Example 7-2-3

A particle is positioned on the grid shown in Fig. 7-2-1 in a "random manner" as follows. If it is in position 1 or position 9, it remains there with probability 1. If it is in any other position, at the next transition time it moves with equal probability to one of its adjacent neighbors along one of the grid lines (but not diagonally). In position 5 there are four possible moves, each with probability $a = \frac{1}{4}$. In positions

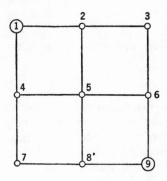

Fig. 7-2-1   Grid for the random-walk problem of Example 7-2-3.

2, 4, 6, or 8 there are three possible moves, each with probability $b = \frac{1}{3}$. In positions 3 and 7 there are two possible moves, each with probability $c = \frac{1}{2}$.

SOLUTION

$$
\begin{array}{c}
\\1\\2\\3\\4\\5\\6\\7\\8\\9
\end{array}
\begin{array}{ccccccccc}
1 & 2 & 3 & 4 & 5 & 6 & 7 & 8 & 9 \\
\left[\begin{array}{ccccccccc}
1 & 0 & 0 & 0 & 0 & 0 & 0 & 0 & 0 \\
b & 0 & b & 0 & b & 0 & 0 & 0 & 0 \\
0 & c & 0 & 0 & 0 & c & 0 & 0 & 0 \\
b & 0 & 0 & 0 & b & 0 & b & 0 & 0 \\
0 & a & 0 & a & 0 & a & 0 & a & 0 \\
0 & 0 & b & 0 & b & 0 & 0 & 0 & b \\
0 & 0 & 0 & c & 0 & 0 & 0 & c & 0 \\
0 & 0 & 0 & 0 & b & 0 & b & 0 & b \\
0 & 0 & 0 & 0 & 0 & 0 & 0 & 0 & 1
\end{array}\right]
\end{array}
$$

The numbers outside the matrix are simply an aid to identifying the rows and columns. To formulate (or to read) such a matrix, the following procedure is helpful.   Go to a position on the main diagonal, say, in the jth row and column.   Enter (or read) $\pi(j, j)$, the conditional probability of remaining in that position once in it.   For the present example, this is zero except in positions 1, 1 and 9, 9 on the matrix, corresponding to positions 1 and 9 on the grid.   Then consider various possible positions to be reached from position j.   Suppose position k is such a position; this corresponds to the matrix element in the kth column but on the same (i.e., the jth) row.   Enter $\pi(j, k)$, the conditional probability of moving to position k, given that the present position is j.   For example, in the fifth row of the transition matrix above (corresponding to present position 5), $\pi(5, 5)$ is zero; but $\pi(5, 2)$, $\pi(5, 4)$, $\pi(5, 6)$, and $\pi(5, 8)$ each has the same value $a = \frac{1}{4}$.   Thus the number a is entered into the positions in columns 2, 4, 6, and 8 on the fifth row.   In the third row there are only two nonzero entries, each of value $c = \frac{1}{2}$, corresponding to the fact that from position 3 only two moves are possible.   The matrix should be examined to see that the other entries correspond to the physical system assumed.

Once the initial state probabilities are specified, the Markov chain is determined. ■

Higher-order transition probabilities

By the use of elementary probability relations, we may develop formulas for conditional probabilities which span several periods.

Definition 7-2d

The *mth-order transition probability from state* s^i *to state* s^q is the quantity

$$\pi_m(i, q) = P(E_{r+m}^q | E_r^i)$$

It is apparent that, for constant chains, the higher-order transition probabilities are independent of the beginning period and depend only upon m and the two states involved. For $m = 1$, the first-order transition probability is just the ordinary transition probability; that is, $\pi_1(i, q) = \pi(i, q)$. A number of basic relations are easily derived.

(**MC5**) $\quad \pi_m(h, s) = \sum_{i, \ldots, r} \pi(h, i) \cdot \cdot \cdot \pi(r, s) \qquad\qquad \pi_0(i, q) = \delta_{iq}$

This property can be derived easily by applying properties (CP1), (MC1), and (P6). In a similar way, the following result can be developed.

(**MC6**) $\quad \pi_{m+n}(h, s) = \sum_k \pi_m(h, k)\pi_n(k, s) \qquad\qquad$ [m factors]

This equation is a special case of the *Chapman-Kolmogorov* equation. The state probabilities in the $(n + m)$th period can be derived from those for the mth period by the following formula:

(**MC7**) $\quad \pi_{m+n}(s) = \sum_h \pi_m(h)\pi_n(h, s)$

Use of the definitions and application of fundamental probability theorems suffices to demonstrate the validity of this expression.

In the case of finite Markov processes, in which the state space index set $J = \{1, 2, \ldots, N\}$, several of the properties for Markov processes can be expressed compactly in terms of the matrix of transition probabilities.

Let $\mathcal{Q}_r$ be the *row* matrix of state probabilities in the rth period, and let $\mathcal{O}$ be the matrix of transition probabilities. Since the latter matrix is square, the nth power $\mathcal{O}^n = \mathcal{O} \times \mathcal{O} \times \cdots \times \mathcal{O}$ (n factors) is defined. We may then write in matrix form

(**MC4**) $\quad \mathcal{Q}_n = \mathcal{Q}_0 \mathcal{O}^n$
(**MC5**) $\quad [\pi_m(h, s)] = \mathcal{O}^m$
(**MC6**) $\quad [\pi_{m+n}(h, s)] = \mathcal{O}^m \mathcal{O}^n = \mathcal{O}^{m+n}$
(**MC7**) $\quad \mathcal{Q}_{m+n} = \mathcal{Q}_m \mathcal{O}^n$

Because of this matrix formulation, it has been possible to exploit properties of matrix algebra to develop a very general theory of finite chains. Also, a major tool for analysis is a type of generating function which is identical with the z *transform*, much used in the theory of sampled-data control systems. We shall not attempt to describe or exploit these special techniques, although we shall assume familiarity with the simple rules of matrix multiplication. For theoretical developments and applications employing these techniques, see Feller [1957], Kemeny and Snell [1960], and Howard [1960].

Example 7-2-4   *The Success-Failure Model*

The model is characterized by two states: state 1 may correspond to "success," and state 2 to "failure." This may be used as a somewhat crude model of marketing success. A manufacturer features a specific product in each marketing period. If he is successful in sales during one period, he has a reasonable probability of being successful in the succeeding period; if he is unsuccessful, he may have a high probability of remaining unsuccessful. For example, if he is successful, he may have a 50-50 chance of remaining successful; if he is unsuccessful in a given period, his probability of being successful in the next period may only be 1 in 4. Under these assumptions, suppose he is successful in a given sales period. What are the probabilities of success and lack of it in the first, second, and third subsequent sales periods?

SOLUTION   Under the assumptions stated above, $\pi(1, 1) = \pi(1, 2) = \frac{1}{2}$, $\pi(2, 1) = \frac{1}{4}$, and $\pi(2, 2) = \frac{3}{4}$. The matrix of transition probabilities is thus

$$\mathcal{O} = \begin{bmatrix} \frac{1}{2} & \frac{1}{2} \\ \frac{1}{4} & \frac{3}{4} \end{bmatrix} = \frac{1}{4} \begin{bmatrix} 2 & 2 \\ 1 & 3 \end{bmatrix}$$

Direct matrix multiplication shows that

$$\mathcal{O}^2 = \frac{1}{16} \begin{bmatrix} 6 & 10 \\ 5 & 11 \end{bmatrix} \qquad \mathcal{O}^3 = \frac{1}{64} \begin{bmatrix} 22 & 42 \\ 21 & 43 \end{bmatrix}$$

As a check, we note that $\mathcal{O}^2$ and $\mathcal{O}^3$ are stochastic matrices. The assumption that the system is initially in state 1 is equivalent to the assumption $\pi_0(1) = 1$ and $\pi_0(2) = 0$. Thus $Q_0 = [1 \quad 0]$, so that

$$Q_1 = Q_0\mathcal{O} = \frac{1}{4}[1 \quad 0]\begin{bmatrix} 2 & 2 \\ 1 & 3 \end{bmatrix} = [\frac{1}{2} \quad \frac{1}{2}]$$

$$Q_2 = Q_0\mathcal{O}^2 = \frac{1}{16}[1 \quad 0]\begin{bmatrix} 6 & 10 \\ 5 & 11 \end{bmatrix} = [\frac{3}{8} \quad \frac{5}{8}]$$

$$= Q_1\mathcal{O}$$

$$Q_3 = Q_0\mathcal{O}^3 = \frac{1}{64}[1 \quad 0]\begin{bmatrix} 22 & 42 \\ 21 & 43 \end{bmatrix} = [\frac{11}{32} \quad \frac{21}{32}]$$

$$= Q_1\mathcal{O}^2 = Q_2\mathcal{O}$$

In the first period (after the initial one) the system is equally likely to be in either state. In the second and third periods, state 2 becomes the more likely. The probabilities for subsequent periods may be calculated similarly. ■

Example 7-2-5   *Alternative Routing in Communication Networks*

A communication system has trunk lines connecting five cities, which we designate by numbers. There are central-office switching facilities in each city. If it is desired to establish a line from city 1 to city 5, the switching facilities make a search for a line. Three possibilities exist: (1) a direct line is available, (2) a line to an intermediate city is found, or (3) no line is available. In the first case, the line is established. In the

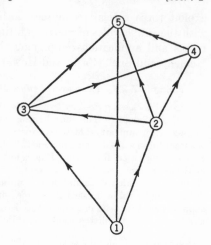

Fig. 7-2-2   Possible trunk-line connections for establishing communication
from city 1 to city 5.

second alternative, a new search procedure is set up in the intermediate city to find
a line to city 5 or to another intermediate city.   In the third alternative, the system
begins a new search after a prescribed delay (which may be quite short in a modern,
high-capacity system).   A fundamental problem is to design the system so that the
waiting time (in terms of numbers of searching operations) is below some prescribed
level.   Searching operations mean not only delay in establishing communication, but
also use of extensive switching equipment in the central offices.   Traffic conditions
and the number of trunk lines available determine probabilities that various connections will be made on any search operation.   We may set up a Markov-chain
model by assuming that the state of the system corresponds to the connection established.   Consider the network shown in Fig. 7-2-2.   The system is in state 1 if no line
has been found at the originating city.   The system is in state 5 if a connecting link—
either direct or through intermediate cities—is established with city 5.   The system
is in state 4 if a connecting link is established to city 4, etc.   Suppose the probabilities
of establishing connections are those indicated in the transition matrix

$$\mathcal{O} = \begin{array}{c} \\ \\ \\ \\ \\ \\ \end{array} \begin{array}{ccccc} 1 & 2 & 3 & 4 & 5 \\ \begin{bmatrix} 0.1 & 0.3 & 0.3 & 0 & 0.3 \\ 0 & 0.1 & 0.1 & 0.2 & 0.6 \\ 0 & 0 & 0.1 & 0.1 & 0.8 \\ 0 & 0 & 0 & 0.1 & 0.9 \\ 0 & 0 & 0 & 0 & 1.0 \end{bmatrix} \end{array}$$

Since it is determined that the call originates in city 1, the initial probability matrix is
[1   0   0   0   0].   The transition matrix reflects the fact that in city 1 and in each
intermediate city the probability is 0.1 of not finding a line on any search operation.
Cities 3 and 4 are assumed to be closer to city 5 and to have more connecting lines,
so that the probabilities of establishing connection to that city are higher than for
cities 1 and 2.   The connection network is such as to provide connections to higher-
numbered cities.   The probability of establishing connection in n or fewer search
operations is $\pi_n(5)$.   Since $\pi_0(1) = 1$, we must have $\pi_n(5) = \pi_n(1, 5)$.   Using the
fundamental relations, we may calculate the following values:

n	$\pi_n(1, 5)$
1	0.3
2	0.75
3	0.942
4	0.989

Thus only about 1 call in 19 takes more than three search operations; and only about 1 percent of the calls take more than 4 search operations. About 1 call in 5 (0.192 = 0.942 − 0.750) takes three search operations. Other probabilities may be determined similarly. ∎

Closed state sets and irreducible chains

In the analysis of constant Markov chains, it is convenient to classify the various states of the system in a variety of ways. We introduce some terminology to facilitate discussion and then consider the concept of closure.

Definition 7-2e

We say that state s^k *is accessible from state* s^i (indicated $s^i \to s^k$) iffi there is a positive integer n such that $\pi_n(i, k) > 0$. States s^i and s^k are said to *communicate* iffi both $s^i \to s^k$ and $s^k \to s^i$. We designate this condition by writing $s^i \leftrightarrow s^k$.

It should be noted that the case $i = k$ is included in the definition above. From the basic Markov relation (MC6) it is easy to derive the following facts:

Theorem 7-2A

1. $s^i \to s^j$ and $s^j \to s^k$ implies $s^i \to s^k$.
2. $s^i \leftrightarrow s^j$ and $s^j \leftrightarrow s^k$ implies $s^i \leftrightarrow s^k$.
3. $s^i \leftrightarrow s^j$ for some j implies $s^i \leftrightarrow s^i$.

In the case of finite chains (or even some special infinite chains), these relations can often be visualized with the aid of a *transition diagram*.

Definition 7-2f

A *transition diagram* for a finite chain is a linear graph with one *node* for each state and one *directed branch* for each one-step transition with positive probability. It is customary to indicate the transition probability along the branch.

Figure 7-2-3 shows a transition diagram corresponding to the three-state process described in Example 7-2-2. The nodes are numbered to correspond to the number of the state represented. The probability $\pi(i, j)$ is

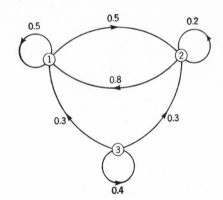

Fig. 7-2-3   The transition diagram for
the process of Example 7-2-2.

placed along the branch directed from node i to node j.   If for any i and j
the transition probability $\pi(i, j) = 0$, the branch directed from node i to
node j is omitted.

With such a diagram, the relation $s^i \rightarrow s^k$ is equivalent to the existence of at least one path through the diagram (traced in the direction of
the arrows) which leads from node i to node k.   It is apparent in Fig.
7-2-3 that nodes 1 and 2 communicate and that each node can be reached
from node 3; but node 3 cannot be reached from node 1 or from node 2.

Definition 7-2g

A set C of states is said to be *closed* if no state outside C may be
reached from any state in C.   If a closed set C consists of only one
member state s^i, that state is called an *absorbing state*, or a *trapping
state*.   A chain is called *irreducible* (or *indecomposable*) iffi there are
no closed sets of states other than the entire state space.

In the chain represented in Fig. 7-2-3, the set of states $\{s^1, s^2\}$ is a closed
set, since the only other state in the state space cannot be reached from
either member.   The chain is thus reducible.   It has no absorbing states,
however.   In order to be absorbing, a state s^k must be characterized by
$\pi(k, k) = 1$.   In terms of the transition diagram, this means that the
only branch leaving node k must be a self-loop branch beginning and
ending on node k.   The transition probability associated with this branch
is unity.

We note that a set of states C is closed iffi $s^i \in C$ and $s^k \notin C$ implies
$\pi_n(i, k) = 0$ for all positive integers n.   In this case we must have
$\pi(i, k) = 0$.   This means that the nonzero entries in row i of a transition matrix of any order must lie in columns corresponding to states
in C.   This statement is true for every row corresponding to a state in C.
If from each $\mathcal{P}^n$ we form the submatrix $\mathcal{P}_C{}^n$ containing only rows and

columns corresponding to states in C, this submatrix is a stochastic matrix. In fact, from the basic properties it follows that:

Theorem 7-2B

If C is a closed set, a stochastic matrix $\mathcal{P}_{C^n}$ may be derived from each matrix $\mathcal{P}^n$ by taking the rows and columns corresponding to the states in C. The elements of $\mathcal{P}_C$ and $\mathcal{P}_{C^n}$ satisfy the fundamental relations (MC1) through (MC7).

This theorem implies that a Markov chain is defined on any closed state set. Once the closed set is entered, it is never left. The behavior of the system, once this closed set of states is entered, can be studied in isolation, as a separate problem.

It is easy to show that the set R_i of all states which can be reached from a given state s^i is closed. From this fact, the following theorem is an immediate consequence.

Theorem 7-2C

A constant Markov chain is irreducible iffi all states communicate.

For finite chains (i.e., finite state spaces) a systematic check for irreducibility can be made to discover if every state can be reached from each given state. This may be done either with the transition matrix or with the aid of the transition diagram for the chain. If the chain is reducible, it is usually desirable to identify the closed subsets of the state space.

Waiting-time distributions

Suppose the system starts in state s^i at the rth period and reaches state s^j for the first time in the $(r + n)$th period. For each n, i, and j we can designate a conditional probability which we define as follows:

Definition 7-2h

The *waiting-time distribution function* is given by

$$f_n(i, j) = P[E_{n+r}^j (E_{n+r-1}^j)^c \cdots (E_{r+1}^j)^c | E_r^i]$$

In words, $f_n(i, j)$ is the conditional probability that, starting with the system in state s^i, it will reach state s^j for the first time n steps later. It is apparent that for constant Markov processes these conditional probabilities are independent of the choice of r. The waiting-time distribution is determined by the following recursion relations:

$$f_1(i, j) = \pi(i, j)$$

$$f_n(i, j) = \pi_n(i, j) - \sum_{r=1}^{n-1} f_{n-r}(i, j)\pi_r(j, j)$$

The first expression is obvious. The second expression can be derived by recognizing that the system moves from state s^i to state s^j in n steps by reaching s^j for the first time in n steps; or by reaching s^j for the first time in $n - 1$ steps and returning to s^j in one more step; or by reaching s^j for the first time in $n - 2$ steps and then returning in exactly two more steps; etc. The product relation in each of the terms of the final sum can be derived by considering carefully the events involved and using the fundamental Markov properties.

Let $f(i, j)$ be the conditional probability that, starting from state s^i, the system ever reaches state s^j. It is apparent that

$$f(i, j) = \sum_{n=1}^{\infty} f_n(i, j)$$

If the initial state and final state are the same, it is customary to call the waiting times *recurrence times*. We shall define some quantities which serve as the basis of useful classifications of states.

Definition 7-2i

The quantity $f(i, i)$ is referred to as the *recurrence probability* for s^i. The *mean recurrence time* $\tau(i)$ for a state s^i is given by

$$\tau(i) = \begin{cases} \sum_{n=1}^{\infty} n f_n(i, i) & \text{for } f(i, i) = 1 \\ \infty & \text{for } f(i, i) < 1 \end{cases}$$

If state s^i communicates with itself, the *periodicity parameter* $\omega(i)$ is the greatest integer such that $\pi_n(i, i)$ can differ from zero only for $n = k\omega(i)$, k an integer; if s^i does not communicate with itself, $\omega(i) = 0$.

In terms of the quantities just derived, we may classify the states of a Markov process as follows:

Definition 7-2j

A state is said to be

1. *Transient* iffi $f(i, i) < 1$.
2. *Persistent* iffi $f(i, i) = 1$.
3. *Periodic* with period $\omega(i)$ iffi $\omega(i) > 1$.
4. *Aperiodic* iffi $\omega(i) \leq 1$.
5. *Null* iffi $f(i, i) = 1$, $\tau(i) = \infty$.
6. *Positive* iffi $f(i, i) = 1$, $\tau(i) < \infty$.
7. *Ergodic* iffi $f(i, i) = 1$, $\tau(i) < \infty$, $\omega(i) = 1$.

Example 7-2-6

Classify the states in the process described in Example 7-2-2 and Fig. 7-2-3.

SOLUTION   We may use the transition diagram as an aid to classification and in calculating waiting-time distributions.

State 3.   This state is obviously transient, for once the system leaves this state, it does not return.   We thus have

$$f_1(3, 3) = \pi(3, 3) = 0.4 \qquad f_n(3, 3) = 0 \qquad \text{for } n > 1$$

From this it follows that $f(3, 3) = \sum\limits_{n=1}^{\infty} f_n(3, 3) = 0.4 < 1$. It is obvious that $\omega(3) = 1$, so that the state is aperiodic.   By definition, the mean recurrence time $\tau(3) = \infty$.

State 1.   It would appear that this state is persistent and aperiodic.   For one thing, it is obvious that $\omega(1) = 1$.   We also note that the possibility of a return for the first time to state 1 after starting in that state must follow a simple pattern. There is either (1) a return in one step, (2) a transition to state 2, then back on the second step, or (3) a transition to state 2, one or more returns to state 2, and an eventual return to state 1.   We thus have

$$f_1(1, 1) = \pi(1, 1) = 0.5 \qquad f_2(1, 1) = \pi(1, 2)\pi(2, 1) = 0.4$$
$$f_n(1, 1) = \pi(1, 2)[\pi(2, 2)]^{n-2}\pi(2, 1) = 0.4(0.2)^{n-2}$$

Now $f(1, 1) = 0.5 + 0.4 \sum\limits_{k=0}^{\infty} (0.2)^k = 1$, since $\sum\limits_{k=0}^{\infty} (0.2)^k = 1/(1 - 0.2) = 5/4$.   The mean recurrence time $\tau(1)$ is given by $\tau(1) = 0.5 + 0.4 \sum\limits_{k=0}^{\infty} (0.2)^k(k + 2) = 1.625$. In obtaining this result, we put $k + 2 = (k + 1) + 1 = n + 1$, and write

$$\sum_{k=0}^{\infty} (0.2)^k(k + 2) = \sum_{n=1}^{\infty} n(0.2)^{n-1} + \sum_{k=0}^{\infty} (0.2)^k$$
$$= \frac{1}{(1 - 0.2)^2} + \frac{1}{1 - 0.2}$$

The other state has similar properties, and the numerical results for that state may be obtained by similar computations.   The return probability $f(2, 2) = 1$, and the mean recurrence time $\tau(2) = 2.6$. ∎

Two fundamental theorems

Next we state two fundamental theorems which give important characterizations of constant Markov chains.   These are stated without proof; for proofs based on certain analytical, nonprobabilistic properties of the so-called *renewal equation*, see Feller [1957].

Theorem 7-2D

In an irreducible, constant Markov chain, all states belong to the same class: they are either all transient, all null, or all positive.   If any state is periodic, all states are periodic with the same period.

In every constant Markov chain, the persistent states can be partitioned uniquely into closed sets $C_1, C_2, \ldots$, such that, from any state in a given set C_k, all other states in that set can be reached. All states belonging to a given closed set C_k belong to the same class.

In addition to the states in the closed sets, the chain may have transient states from which the states of one or more of the closed sets may be reached.

According to Theorem 7-2B, each of the closed sets C_k described in the theorem just stated may be studied as a subchain. The behavior of the entire chain may be studied by examining the behavior of each of the subchains, once its set of states is entered, and by studying the ways in which the subchains may be entered from the transient states.

Before stating the next theorem, we make the following

Definition 7-2k

A set of probabilities $\{\pi_i : i \in J\}$, where J is the index set for the state space $\mathcal{S}$, is called a *stationary distribution* for the process iffi

$$\pi_j = \sum_{i \in J} \pi_i \pi(i, j) \qquad \text{for each } j \in J$$

The significance of this set of probabilities and the justification for the terminology may be seen by noting that if $\pi_r(j) = \pi_j$ for some r and all $j \in J$, then

1. $\pi_{r+n}(j) = \pi_j$ for all $n \geq 1, j \in J$
2. $\pi_j = \sum_{i \in J} \pi_i \pi_n(i, j)$ for all $n \geq 1, j \in J$

Theorem 7-2E

An irreducible, aperiodic, constant Markov chain is characterized by one of two conditions: either

1. All states are transient or null, in which case $\lim\limits_{n \to \infty} \pi_n(i, j) = 0$ for any pair i, j and there is no stationary distribution for the process, or

2. All states are ergodic, in which case

$$\lim_{n \to \infty} \pi_n(i, j) = \pi_j = \frac{1}{\tau(j)} > 0$$

for all pairs i, j, and $\{\pi_j : j \in J\}$ is a stationary distribution for the chain.

In the ergodic case, we have $\lim\limits_{n \to \infty} \pi_n(j) = \pi_j$ regardless of the choice of initial probabilities $\pi_0(i)$, $i \in J$.

The validity of the last statement, assuming condition 2, may be seen from the following argument:

$$\pi_n(j) = \sum_{i \in J} \pi_0(i)\pi_n(i, j)$$

If, for some n, all $|\pi_n(i, j) - \pi_j| < \epsilon$, then

$$|\pi_n(j) - \pi_j| \leq \sum_{i \in J} \pi_0(i)|\pi_n(i, j) - \pi_j| < \epsilon$$

since the $\pi_0(i)$ sum to unity.

To determine whether or not an irreducible, aperiodic chain is ergodic, the problem is to determine whether there is a set of π_j to satisfy the equation in Definition 7-2*k*. In the case of finite chains, this may be approached algebraically.

Example 7-2-7

Consider once more the two-state process described in Example 7-2-4. This process is characterized by the transition matrix

$$\mathcal{P} = \begin{bmatrix} \frac{1}{2} & \frac{1}{2} \\ \frac{1}{4} & \frac{3}{4} \end{bmatrix}$$

The equation in the definition for stationary distributions can be written

$$[\pi_1 \, \pi_2](\mathcal{P} - \mathcal{U}) = 0$$

where $\mathcal{U}$ is the unit matrix. In addition, we have the condition $\pi_1 + \pi_2 = 1$. Writing out the matrix equation, we get

$$-2\pi_1 + \pi_2 = 0$$
$$2\pi_1 - \pi_2 = 0$$

From this we get $\pi_1 = \pi_2/2$, so that $\pi_1 = \frac{1}{3}$ and $\pi_2 = \frac{2}{3}$. A check shows that this set is in fact a stationary distribution. It is interesting to compare these stationary probabilities with those in $\mathcal{Q}_2$, $\mathcal{Q}_3$, $\mathcal{P}^2$, and $\mathcal{P}^3$ in Example 7-2-4. It appears that the convergence to the limiting values is quite rapid. After a relatively few periods, the probability of being in state 2 (the failure state) during any given period approaches $\frac{2}{3}$. This condition prevails regardless of the initial probabilities. ∎

As a final example, we consider a classical application which has been widely studied (cf. Feller [1957, pp. 343 and 358f] or Kemeny and Snell [1960, chap. 7, sec. 3]).

Example 7-2-8   *The Ehrenfest Diffusion Model*

This model assumes $N + 1$ states, $s^0, s^1, \ldots, s^N$. The transition probabilities are $\pi(i, i) = 0$ for all i, $\pi(i, j) = 0$ for $|i - j| > 1$, $\pi(i, i - 1) = i/N$, and $\pi(i, i + 1) = 1 - i/N$. This model was presented in a paper on statistical mechanics (1907) as a representation of a conceptual experiment in which N molecules are distributed in two containers labeled A and B. At any given transition time, a molecule is chosen at random from the total set and moved from its container to the other. The state of the system is taken to be the number of molecules in container A. If the system is in state i, the molecule chosen is taken from A and put into B, with probability i/N, or is taken from B and put into A, with probability $1 - i/N$. The first case changes the state from s^i to s^{i-1}; the second case results in moving to state s^{i+1}.

SOLUTION   The defining equations for the stationary distributions become

$$\pi_j = \sum_{i=0}^{N} \pi_i \pi(i, j) = \pi_{j-1}\pi(j-1, j) + \pi_{j+1}\pi(j+1, j)$$

$$= \pi_{j-1}\left(1 - \frac{j-1}{N}\right) + \pi_{j+1}\left(\frac{j+1}{N}\right) \qquad 1 \le j \le N - 1$$

$$\pi_0 = \frac{\pi_1}{N} \qquad \pi_N = \frac{\pi_{N-1}}{N}$$

Solving recursively, we can obtain expressions for each of the π_j in terms of π_0. We have immediately

$$\pi_1 = N\pi_0$$

The basic equation may be rewritten in the form

$$\pi_{j+1} = \frac{N\pi_j - (N - j + 1)\pi_{j-1}}{j + 1}$$

so that

$$\pi_2 = \frac{N\pi_1 - N\pi_0}{2} = \frac{N(N - 1)}{2}\pi_0$$

It may be noted that this is of the form $C_2{}^N\pi_0$. If $\pi_r = C_r{}^N\pi_0$ for $0 \le r \le j$, the same formula must hold for $r = j + 1$. We have

$$\pi_{j+1} = \frac{NC_j{}^N - (N - j + 1)C_{j-1}^N}{j + 1}\pi_0$$

Use of the definitions and some easy algebraic manipulations show that the expression on the right reduces to the desired form. Hence, by induction, $\pi_j = C_j{}^N\pi_0$ for all $j = 1, 2, \ldots, N$. Now

$$\sum_{j=0}^{N} C_j{}^N\pi_0 = 1 \qquad \text{implies} \qquad \pi_0 = 2^{-N}$$

so that the stationary distribution is the binomial distribution with $p = \frac{1}{2}$. The mean value of a random variable with this distribution is thus $N/2$ (Example 5-2-2). ∎

Feller [1957] has given an alternative physical interpretation of this random process. For a detailed study of the characteristics of the probabilities, waiting times, etc., the work by Kemeny and Snell [1960], cited above, may be referred to.

A wide variety of physical and economic problems have been studied with Markov-chain models. The present treatment has sketched some of the fundamental ideas and results. For more details of both application and theory, the references cited in this section and at the end of the chapter may be consulted. Extensive bibliographies are included in several of these works, notably that by Bharucha-Reid [1960].

7-3   Increments of processes; the Poisson process

In some applications it is natural to consider the differences between values taken on by a sample function at specific instants of time.   For such purposes, it is natural to make the following

Definition 7-3a

If X is a random process and t_1 and t_2 are any two distinct values of the parameter set T, then the random variable $X(\cdot, t_2) - X(\cdot, t_1)$ is called an *increment* of the process.

Increments of random processes are frequently assumed to have one or both of two properties, which we consider briefly.

Definition 7-3b

A random process X is said to have *independent increments* iffi, for any n and any $t_1 < t_2 < \cdots < t_n$, with each $t_i \in T$, it is true that the increments $X(\cdot, t_2) - X(\cdot, t_1)$, $X(\cdot, t_3) - X(\cdot, t_2)$, . . . , $X(\cdot, t_n) - X(\cdot, t_{n-1})$ form an independent class of random variables.

One may construct a simple example of a process with independent increments as follows.   Begin with any sequence of independent random variables $\{Y_i(\cdot) : 1 \leq i < \infty\}$ and an arbitrary random variable $X_0(\cdot)$.   Put

$$X(\cdot, n) = X_0(\cdot) + \sum_{i=1}^{n} Y_i(\cdot) \qquad n = 1, 2, \ldots$$

The resulting process $\{X(\cdot, n) : 1 \leq n < \infty\}$ has independent increments.

Definition 7-3c

A random process X is said to have *stationary increments* iffi, for each $t_1 < t_2$ and $h > 0$ such that $t_1, t_2, t_1 + h, t_2 + h$ all belong to T, the increments $X(\cdot, t_2) - X(\cdot, t_1)$ and $X(\cdot, t_2 + h) - X(\cdot, t_1 + h)$ have the same distribution.

The Poisson process considered below has increments that are both independent and stationary.

Counting processes

A counting process N is one in which the value $N(\xi, t)$ of a sample function "counts" the number of occurrences of a specified phenomenon in an interval $[0, t)$.   For each choice of ξ, a particular sequence of occurrences of this phenomenon results.   The sample function $N(\xi, \cdot)$ for the process is the counting function corresponding to this particular sequence.   The number of occurrences in the interval $[t_1, t_2)$ is given by the increment $N(\xi, t_2) - N(\xi, t_1)$ of that particular sample function.   Obviously, a counting process has values which are integers.   Counting processes have been derived which count a wide variety of phenomena, e.g., the number of telephone calls in a given time interval, the number of errors in decoding a message from a communication channel in a given time period, the number of particles from a radioactive source striking a target in a given time, etc.

The Poisson process

One of the most common of the counting processes is the Poisson process, so named because of the nature of the distributions for its random variables.   Since it is a counting process, it has integer values.   Its parameter or index set T is usually taken to be the nonnegative real numbers $[0, \infty)$.   The process appears in situations for which the following assumptions are valid.

1. $N(\cdot, 0) = 0 \ [P]$; that is, $P[N(\cdot, 0) = 0] = 1$.
2. The increments of the process are independent.
3. The increments of the process are stationary.
4. $P[N(\cdot, t) > 0] > 0 \qquad$ for all $t > 0$
5. $\lim\limits_{h \to 0} \dfrac{P[|N(\cdot, t + h) - N(\cdot, t)| > 1]}{P[|N(\cdot, t + h) - N(\cdot, t)| = 1]} = 0$

We have thus described a counting process with stationary and independent increments which has unit probability of zero count in an interval of length zero.   The probability of a positive count in an interval of positive length is positive, and the probability of more than one count in a very short interval is of smaller order than the probability of a single count.

With the aid of a standard theorem from analysis, it may be shown that assumptions 4 and 5 may be replaced with the following:

4'. $\lim\limits_{t \to 0} \dfrac{P[N(\cdot, t) = 1]}{t} = \lambda \qquad$ a constant

5'. $\lim\limits_{t \to 0} \dfrac{P[N(\cdot, t) > 1]}{t} = 0$

In order to see this, consider

$$p_0(t) = P[N(\cdot, t) = 0]$$

Then

$$
\begin{aligned}
p_0(t_1 + t_2) &= P[N(\cdot, t_1 + t_2) = 0] \\
&= P[N(\cdot, t_1 + t_2) - N(\cdot, t_1) = 0 \qquad \text{and} \qquad N(\cdot, t_1) = 0] \\
&= P[N(\cdot, t_1 + t_2) - N(\cdot, t_1) = 0]P[N(\cdot, t_1) = 0] \\
&= P[N(\cdot, t_2) = 0]P[N(\cdot, t_1) = 0] \\
&= p_0(t_2)p_0(t_1)
\end{aligned}
$$

If we consider $g(t) = -\log_e p_0(t)$, for $t \geq 0$, we must have

$$g(t_1 + t_2) = g(t_1) + g(t_2) \qquad \text{with } g(t) \geq 0$$

This linear functional equation is known to have the solution

$$g(t) = g(1) \cdot t \qquad \text{for } t \geq 0$$

Putting $g(1) = \lambda$, a positive constant, we have

$$p_0(t) = e^{-\lambda t} \qquad \text{for } t \geq 0$$

This being the case,

$$P[N(\cdot, t) > 0] = 1 - e^{-\lambda t}$$

which implies

$$\lim_{t \to 0} \frac{P[N(\cdot, t) > 0]}{t} = \lambda$$

Assumption 5 then requires

$$\lim_{t \to 0} \frac{P[N(\cdot, t) = 1]}{t} = \lambda \qquad \text{and} \qquad \lim_{t \to 0} \frac{P[N(\cdot, t) > 1]}{t} = 0$$

It is also apparent that conditions 1, 2, 3, 4', and 5' imply the original set. We now wish to determine the probabilities

$$p(i, t) = P[N(\cdot, t) = i]$$

We begin by noting that we may have i occurrences in an interval of length $t + \Delta t$ by having

i occurrences in the first interval of length t and
0 occurrences in the second interval of length Δt,
or $i - 1$ in the first interval and 1 in the second,
or $i - 2$ in the first interval and 2 in the second,
etc., to the case
0 in the first interval and i in the second.

These events are mutually exclusive, so that their probabilities add. Thus

$$
\begin{aligned}
p(i, t + \Delta t) \\
= \sum_{k=0}^{i} P[N(\cdot, t + \Delta t) - N(\cdot, t) = k \quad &\text{and} \quad N(\cdot, t) = i - k] \\
= \sum_{k=0}^{i} p(k, \Delta t) p(i - k, t) \qquad &\text{for } \Delta t > 0
\end{aligned}
$$

We have used conditions 2 and 3 to obtain the last expression.   Conditions 4' and 5' imply

$$p(1, \Delta t) = \lambda \, \Delta t + o(\Delta t)$$

$$\sum_{k=2}^{i} p(k, \Delta t) = o(\Delta t)$$

$$p(0, \Delta t) = 1 - \lambda \, \Delta t + o(\Delta t)$$

where $o(\Delta t)$ indicates a quantity which goes to zero faster than Δt as the latter goes to zero.   Substituting these expressions into the sum for $p(i, t + \Delta t)$ and rearranging, we obtain

$$p(i, t + \Delta t) - p(i, t) = -\lambda \, \Delta t \, p(i, t) + \lambda \, \Delta t \, p(i - 1, t) + o(\Delta t)$$

or

$$\frac{p(i, t + \Delta t) - p(i, t)}{\Delta t} = \lambda[p(i - 1, t) - p(i, t)] + \frac{o(\Delta t)}{\Delta t}$$

Taking the limit as Δt goes to zero, we obtain the right-hand derivatives. A slightly more complicated argument holds for $\Delta t < 0$, so that we have the differential equations

$$\frac{\partial p(i, t)}{\partial t} = \lambda[p(i - 1, t) - p(i, t)] \qquad i = 0, 1, 2, \ldots$$

Condition 1 implies the boundary conditions

$$p(0, 0) = 1 \qquad p(i, 0) = 0 \qquad i = 1, 2, \ldots$$

The solution of the set of differential equations under the boundary conditions imposed is readily found to be

$$p(k, t) = e^{-\lambda t} \frac{(\lambda t)^k}{k!} \qquad \text{for } t > 0$$

which means that, for each positive t, the random variable $N(\cdot, t)$ has the Poisson distribution with $\mu = \lambda t$ (Example 3-4-7).   Since the mean value for the Poisson distribution is μ, we have

$$E[N(\cdot, t)] = \lambda t \qquad \text{or} \qquad E\left[\frac{1}{t} N(\cdot, t)\right] = \lambda \qquad \text{for each } t > 0$$

so that λ is the *expected frequency*, or *mean rate*, of occurrence of the phenomenon counted by the process.

The Poisson process has served as a model for a surprising number of physical and other phenomena.   Some of these which are commonly found in the literature are

Radiation phenomena: the number of particles emitted in a given time t

Accidents, traffic counts, misprints

Demands for service, maintenance, sales of units of merchandise, admissions, etc.

Counts of particles suspended in liquids

Shot noise in vacuum tubes

A fortunate feature of the Poisson process is the fact that it is governed by a single parameter which has a simple interpretation, as noted above.

A number of generalizations and modifications of the Poisson process have been studied and applied in a variety of situations. For a discussion of some of these, one may refer to Parzen [1962, chap. 4] or Bharucha-Reid [1960, sec. 2.4] and to references cited in these works.

7-4   Distribution functions for random processes

In Sec. 7-1 it was pointed out that one way of describing and studying random processes is by using joint distributions for finite subfamilies of random variables in the process. Analytically, this description is made in terms of distribution functions, just as in the case of several random variables considered jointly.

Definition 7-4a

The *first-order distribution function* $F(\cdot, \cdot)$ for a random process X is the function defined by

$$F(x, t) = P[X(\cdot, t) \leq x] \qquad \text{for every real } x \text{ and every } t \in T$$

The *second-order distribution function* for the process is the function defined by

$$F(x, t; y, s) = P[X(\cdot, t) \leq x, X(\cdot, s) \leq y] \qquad \text{for every pair of real}$$

x and y and every pair of numbers t and s from the index set T.

Distribution functions of any order n are defined similarly.

The first-order distribution function for any fixed t is the ordinary distribution function for the random variable $X(\cdot, t)$ selected from the family that makes up the process. The second-order distribution is the ordinary joint distribution function for the pair of random variables $X(\cdot, t)$ and $X(\cdot, s)$ selected from the process. These distribution functions have the properties generally possessed by joint distribution functions, as discussed in Secs. 3-5 and 3-6.

The following schematic example may be of some help in grasping the significance of the distribution functions for the process.

Example 7-4-1

A process has four sample functions (hence the basic space may be considered to have four elementary events). The functions are shown in Fig. 7-4-1. We let

$$P(\{\xi_k\}) = p_k \qquad \text{with} \sum_{k=1}^{4} p_k = 1$$

SOLUTION   Examination of the figure shows that none of the sample functions lies below the value a at $t = t_1$, so that $F(a, t_1) = 0$. Only the function $X(\xi_1, \cdot)$ lies below b at $t = t_1$ so that $F(b, t_1) = p_1$. Examination of the other cases indicated shows that

$$\begin{array}{lll}
F(a, t_1) = 0 & F(a, t_2) = 0 & F(a, t_3) = 0 \\
F(b, t_1) = p_1 & F(b, t_2) = p_1 + p_2 & F(b, t_3) = p_2 + p_4 \\
F(c, t_1) = 1 & F(c, t_2) = p_1 + p_2 + p_3 & F(c, t_3) = 1
\end{array}$$

For the second-order distribution, some values are

$$F(a, t_1; a, t_2) = 0 \qquad F(b, t_1; c, t_2) = p_1 \qquad F(c, t_2; b, t_3) = p_2$$

In the first case, none of the sample functions lies below a at both t_1 and t_2; in the second case, only $X(\xi_1, \cdot)$ lies below b at t_1 and c at t_2; in the third case, only $X(\xi_2, \cdot)$ lies below c at t_2 and below b at t_3. Other cases may be evaluated similarly. ■

That this process is not realistic is obvious; almost any process in which the sample functions are continuous will have an infinity of such sample functions.

It is apparent that if a process is defined, the distribution functions of all finite orders are defined. Kolmogorov [1933, 1956] has shown that if a family of such distribution functions is defined, there exists a proc-

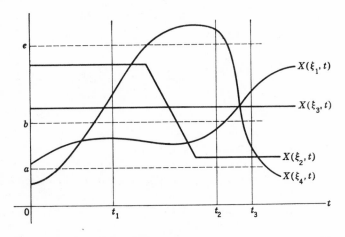

Fig. 7-4-1   A simple schematic random process with four sample functions.

ess for which they are in fact the distribution functions. The events
involved in the definition are of the form

$$\bigcap_{t \in T_0} \{\xi : X(\xi, t) \leq a_t\} \qquad T_0 \subset T$$

The distribution functions of finite order assign probabilities to these
events for any finite subset T_0 of the parameter set. Fundamental
measure theory shows that the probability measure so defined ensures
the unique definition of probability for all such sets when T_0 is countably infinite. Serious technical mathematical problems arise when T_0 is
uncountably infinite. It usually occurs in practice, however, that analytic conditions upon the sample functions are such that they are determined by values at a countable number of points t in the index set T.
A discussion of these conditions is well beyond the scope of the present
treatment. For a careful development of these mathematical conditions,
the treatise by Doob [1953, chap. 2] may be consulted. Fortunately, in
practice it is rarely necessary to be concerned with these problems, so that
we can forgo a discussion of them in an introductory treatment.

7-5 Processes consisting of step functions

In this section we consider a random process (actually a class of processes) whose sample functions have a step-function character; i.e., the
graph of a member function has jumps in value at discrete instants of
time and remains constant between these transition times. The distribution of amplitudes in the various time intervals and the distribution
of transition, or jump, times at which the sample function changes value
is described in rather general terms, so that a class of processes is thus
included.

Such a step process is commonly assumed in practice, e.g., in the
theory of control systems with "randomly varying" inputs. The derivation of the first and second distribution functions illustrates a number
of interesting points of theory. These processes are used in later sections to illustrate a number of concepts and ideas developed there.

Description of the process

Consider the step-function process

$$X(\cdot, t) = \sum_{n = -\infty}^{\infty} a_n(\cdot)[u_+(t - \tau_n(\cdot)) - u_+(t - \tau_{n+1}(\cdot))]$$

where $u_+(t) = 0$ for $t < 0$ and $= 1$ for $t \geq 0$. For any choice of ξ, a step
function is defined. The steps occur at times $t = \tau_n(\xi)$. The value of
the function in the nth interval $[\tau_n(\xi), \tau_{n+1}(\xi))$ is $a_n(\xi)$. We make the

following assumptions about the random sequences which characterize the process.

1. $\{a_n(\cdot): -\infty < n < \infty\}$ is a random process whose members form an independent class; each of these random variables has the same distribution, described by the distribution function $F_A(\cdot)$.

2. $\{\tau_n(\cdot): -\infty < n < \infty\}$ is a random process satisfying the following conditions:

    a. For every ξ and every n, $\tau_n(\xi) < \tau_{n+1}(\xi)$.

    b. The distribution of the jump points $\tau_n(\xi)$ is described by a counting process $N(\cdot, t)$. For any pair of real numbers t_1, t_2 we put $N(\xi, t_1, t_2) = |N(\xi, t_2) - N(\xi, t_1)|$ and suppose

$$P[N(\cdot, t_1, t_2) = 0] = \pi(t_1, t_2)$$

    is a known function of t_1, t_2. For example, if the counting process is Poisson, $\pi(t_1, t_2) = e^{-\lambda|t_2 - t_1|}$. We must also have

$$P[N(\cdot, t_1, t_2) > 0] = 1 - \pi(t_1, t_2)$$

3. The $a_n(\cdot)$ process and the $\tau_n(\cdot)$ process must be independent in an appropriate sense. We may state the independence conditions precisely in terms of certain events defined as follows. Let t_1, t_2 be any two arbitrarily chosen real numbers $(t_1 \neq t_2)$. Let

$$\begin{aligned}
I_{kn} &= \{\xi: t_k \in [\tau_n(\xi), \tau_{n+1}(\xi))\} \\
&= \{\xi: t_k \text{ is in } n\text{th interval}\} \qquad k = 1, 2
\end{aligned}$$

If x_1, x_2 are any two real numbers, let

$$A_{jm} = \{\xi: a_m(\xi) \leq x_j\} \qquad j = 1, 2$$

Then $\{A_{1m}, A_{2n}, I_{1p}I_{2q}\}$ is an independent class for each choice of the indices m, n, p, q, provided $m \neq n$. The significance of the independence assumption in this particular form will appear in the subsequent analysis.

Determination of the distribution functions

We wish to determine the first and second distribution functions for the process X. In order to state the arguments with precision, it is expedient to define some events. Let

$$\begin{aligned}
I &= \{\xi: N(\xi, t_1, t_2) = 0\} \\
&= \{\xi: t_1 \text{ and } t_2 \text{ belong to same interval}\} \\
B_k &= \{\xi: X(\xi, t_k) \leq x_k\} \qquad k = 1, 2
\end{aligned}$$

Then

(A) $F(x_1, t_1) = P(B_1)$
(B) $F(x_1, t_1; x_2, t_2) = P(B_1 B_2)$

We shall use the relation

(C) $P(B_1 B_2) = P(B_1 B_2 | I) P(I) + P(B_1 B_2 | I^c) P(I^c)$

Under the assumed independence conditions, it seems intuitively likely that $P[X(\cdot, t_1) \leq x_1] = F_A(x_1)$. A problem arises, however, from the fact that, for each ξ, the point t_1 may be in a different interval. To validate the expected result, we establish a series of propositions.

1. For fixed k, $\{ I_{kn} : -\infty < n < \infty \}$ is a partition, since, for each ξ, t_k belongs to one and only one interval. We may then assert that for each k, j, n, fixed, $\{ I_{kn} I_{jm} : -\infty < m < \infty \}$ is a partition of I_{kn}.

2. $I = \underset{n}{\uplus} I_{1n} I_{2n}$ $I^c = \underset{n \neq m}{\uplus} I_{1n} I_{2m}$

This follows from the fact that t_1 and t_2 belong to the same interval iffi, for some n, I_{1n} and I_{2n} both occur. The disjointedness follows from property 1.

3. Use of Theorem 2-6G shows that for each permissible j, k, m, n, each class $\{ A_{jm}, I_{kn} \}$, $\{ A_{jm}, I \}$, and $\{ A_{1m} A_{2n}, I \}$ is an independent class.

4. $B_k = \underset{n}{\uplus} A_{kn} I_{kn}$

5. $F(x_k, t_k) = F_A(x_k)$. We may argue as follows:

$$F(x_k, t_k) = P(B_k) = \sum_n P(A_{kn} I_{kn}) = \sum_n P(A_{kn}) P(I_{kn})$$

Now $P(A_{kn}) = F_A(x_k)$ for any n; hence

$$F(x_k, t_k) = F_A(x_k) \sum_n P(I_{kn}) = F_A(x_k)$$

We have thus established the first assertion to be proved. A somewhat more complicated argument, along the same lines, is used to develop an expression for the second distribution. We shall argue in terms of the conditional probabilities in expression (C), above.

6. B_k and I are independent for $k = 1$ or 2.

$$B_1 I = \biguplus_{n,m} A_{1n} I_{1n} I_{1m} I_{2m} = \biguplus_{n} A_{1n} I_{1n} I_{2n}$$

from which the result follows as in the previous arguments. An exactly similar argument follows for $B_2 I$.

7. $P(B_1 B_2 | I^c) = P(B_1)P(B_2) = P(B_1 | I^c)P(B_2 | I^c)$

$$B_1 B_2 I^c = [\biguplus_{n} A_{1n} I_{1n}][\biguplus_{m} A_{2m} I_{2m}] I^c$$

We note that $I_{1n} I_{2n} I^c = \emptyset$ and $I_{1n} I_{2m} \subset I^c$ for $n \neq m$; hence

$$B_1 B_2 I^c = \biguplus_{n \neq m} A_{1n} A_{2m} I_{1n} I_{2m} I^c = \biguplus_{n \neq m} A_{1n} A_{2m} I_{1n} I_{2m}$$

By the independence assumptions

$$
\begin{aligned}
P(B_1 B_2 I^c) &= \sum_{n \neq m} P(A_{1n})P(A_{2m})P(I_{1n} I_{2m}) \\
&= F_A(x_1)F_A(x_2) \sum_{n \neq m} P(I_{1n} I_{2m}) \\
&= P(B_1)P(B_2)P(I^c)
\end{aligned}
$$

Thus

$$P(B_1 B_2 | I^c) = P(B_1)P(B_2) = P(B_1 | I^c)P(B_2 | I^c)$$

For the last assertion we have used the independence of B_k and I^c, which follows from the independence of B_k and I.

8. $P(B_1 B_2 | I) = F_A(\min \{x_1, x_2\})$

If I occurs, $X(\xi, t_1) = X(\xi, t_2)$.

For $x_1 \leq x_2$, $B_1 I \subset B_2 I$ and $B_1 B_2 I = B_1 I$; similarly, for $x_2 \leq x_1$, $B_2 I \subset B_1 I$ and $B_1 B_2 I = B_2 I$. Because of the independence shown earlier, if $x_1 \leq x_2$, $P(B_1 B_2 I) = P(B_1 I) = P(B_1)P(I) = F_A(x_1)P(I)$. For $x_2 \leq x_1$, interchange subscripts 1 and 2 in the previous statements. Thus

$$P(B_1 B_2 I) = F_A(\min \{x_1, x_2\})P(I)$$

from which the asserted relation follows immediately.

We thus may assert

$$
\begin{aligned}
F(x_1, t_1; x_2, t_2) &= P(B_1 B_2) \\
&= F_A(\min \{x_1, x_2\})P(I) + F_A(x_1)F_A(x_2)P(I^c)
\end{aligned}
$$

When the counting process describing the distribution of jump points is known, the probabilities $P(I) = \pi(t_1, t_2)$ and $P(I^c) = 1 - \pi(t_1, t_2)$ are known. If the distribution of jump points is Poisson, we have

$$
\begin{aligned}
F(x_1, t_1; x_2, t_2) = F_A(\min \{x_1, x_2\})e^{-\lambda |t_1 - t_2|} \\
+ F_A(x_1)F_A(x_2)(1 - e^{-\lambda |t_1 - t_2|})
\end{aligned}
$$

It is of interest to note that the first distribution function is independent of t and the second depends upon the difference between the two values t_1 and t_2 of the parameter. The significance of this fact is discussed in Sec. 7-8.

The following special case represents a situation found in many pulse-code transmission systems and computer systems. We deal with it as a special case, although it could be dealt with directly by simple arguments based on fundamentals.

Example 7-5-1

Suppose each member of the random process consists of a voltage wave whose value in any interval is either $\pm a$ volts. Switching may occur at any of the prescribed instants $\tau_n = nT$, where T is a positive constant. The values $\pm a$ are taken on with probability $\frac{1}{2}$, each. Values in different intervals are independent. This process may be considered a special case of the general step process. The τ process consists of the constants $\tau_n(\cdot) = nT$. The process $\{a_n(\cdot): -\infty < n < \infty\}$ is an independent class with

$$P[a_n(\xi) = a] = P[a_n(\xi) = -a] = \frac{1}{2}$$

Since constant random variables are independent of any combination of other random variables, the required independence conditions follow. For any t_1, t_2, $P[N(\cdot, t_1, t_2) = 0] = \pi(t_1, t_2)$ is either 0 or 1 and is a known function of t_1, t_2. Thus

$$F(x, t) = F_A(x)$$
$$F(x, t; y, u) = F_A(\min\{x, y\})\pi(t, u) + F_A(x)F_A(y)[1 - \pi(t, u)]$$

$\pi(t, u) = 1$, whenever t, u are in the same interval, and has the value 0 whenever t, u are in different intervals. Note that in this case we do not have $\pi(t, u)$ a function of the difference $t - u$ alone, as in the case of a Poisson distribution of the transition times. ■

In many control situations, say, an on-off control, the value of the control signal may be limited to two values, as in the pulse-code example above, but the transition times are not known in advance. In this case, it is necessary to assume (usually on the basis of experience) a counting-function distribution. When the expected number of occurrences in any interval is a constant, the assumptions leading to the Poisson counting function are often valid.

Example 7-5-2

The first stage of an electronic binary counter produces a voltage wave which takes on one of two values. Each time a "count" occurs, the stage switches from one of its two states to the other. The output is a constant voltage corresponding to the state. The state, and hence the voltage, is maintained until the next count. If the counter should be counting cars passing a given point, the output wave would have the character of a two-value step function with a Poisson distribution of the transition times. ■

7-6   Expectations; correlation and covariance functions

It should not be surprising that the important role of mathematical expectations in the study of random variables should extend to random processes, since the latter are infinite families of random variables.   In this section, we shall consider briefly some special expectations which are important in the analytical treatment of many random processes. In particular, we shall be concerned with the covariance function and with the closely related correlation functions.   These functions play a central role in the celebrated theory of extrapolation and smoothing of "time series," developed somewhat independently in Russia by A. N. Kolmogorov and in the United States by Norbert Wiener, about 1940. The first applications of this theory were to wartime problems of "fire control" of radar and antiaircraft guns and to related problems of communicating in the presence of noise.   Through his teaching and writing on the subject, Y. W. Lee pioneered in the task of making this abstruse theory available to engineers.   In this country the work is often referred to as the Wiener-Lee statistical theory of communication.   The original theoretical development laid the groundwork for further intensive development of the theory, as well as for applications of the theory to a wide range of real-world problems.   Unfortunately, to develop this theory to the point that significant applications could be made would require the development of topics outside the scope of this book.   Technique becomes difficult, even for simple problems.   The complete theory requires the use of spectral methods based on the Fourier, Laplace, and other related transforms.   We shall limit our discussion to an introduction of the probability concepts which underlie the theory.   Examples must therefore be limited essentially to simple illustrations of the concept.   For applications, one may consult any of a number of books in the field of statistical theory of communication and control (e.g., Laning and Battin [1956] or Davenport and Root [1958]).

We proceed to define and discuss several concepts.

Definition 7-6a

A process X is said to be of *order p* if, for each $t \in T$, $E[|X(t)|^p] < \infty$ (p is a positive integer).

From an elementary property of random variables it follows that if the process X is of order p, it is of order k for all positive integers $k < p$.

Definition 7-6b

The *mean-value function* for a process is the first moment

$$\mu_X(t) = E[X(t)]$$

The mean-value function for a process X exists if the process is of first order (or higher).

In the consideration of any two real-valued random variables of order 2, a convenient measure of their relatedness is provided by the covariance factor, defined as follows:

Definition 7-6c

The *covariance factor* cov $[X, Y]$ for two random variables is given by

$$\text{cov } [X, Y] = E[(X - \mu_X)(Y - \mu_Y)]$$

Simple manipulations show cov $[X, Y] = E[XY] - \mu_X\mu_Y$. If the random variables are independent, cov $[X, Y] = 0$; if one of the random variables is a linear function of the other—say, $Y(\cdot) = aX(\cdot) + b$—then

$$\text{cov } [X, Y] = a\sigma_X{}^2$$

The covariance factor may be normalized in a useful way to give the correlation coefficient, defined as follows:

Definition 7-6d

The *correlation coefficient* for two random variables is given by

$$\rho[X, Y] = \frac{\text{cov } [X, Y]}{\sigma_X\sigma_Y}$$

Use of the Schwarz inequality (E5) shows that

$$|\text{cov } [X, Y]| \leq \sigma_X\sigma_Y$$

so that $-1 \leq \rho[X, Y] \leq 1$. When $\rho[X, Y] = 0$, we say the random variables are *uncorrelated*. Independent random variables are always uncorrelated; however, if two random variables are uncorrelated, it does not follow that they are independent, as may be shown by simple examples. By the use of properties (V2) and (V4), one may show that

$$\sigma^2\left[\frac{X(\cdot)}{\sigma_X} \pm \frac{Y(\cdot)}{\sigma_Y}\right] = 2 \pm 2\rho[X, Y] \geq 0$$

Equality can occur iffi $X(\cdot)/\sigma_X \pm Y(\cdot)/\sigma_Y = c$ $[P]$, where c is an appropriate constant. This means that

$$Y(\cdot) = \pm \frac{\sigma_Y}{\sigma_X} X(\cdot) + \alpha \ [P]$$

The plus sign corresponds to $\rho = \rho[X, Y] = 1$, and the minus sign corresponds to $\rho = -1$.

Let us look further at the significance of the correlation coefficient by considering the joint probability mass distribution. If $\rho = +1$, the

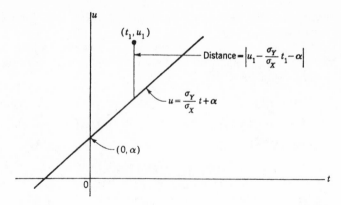

Fig. 7-6-1 Geometric relations for interpreting the correlation coefficient ρ.

probability mass must all lie on the straight line

$$u = \frac{\sigma_Y}{\sigma_X} t + \alpha$$

This line is shown in Fig. 7-6-1. Now suppose $0 < \rho < 1$. Image points for the mapping, and hence probability mass, may lie at points off the straight line. Consider the image point t_1, u_1 shown in Fig. 7-6-1. The distance of the point from the line is proportional to the vertical distance $|v_1|$, where

$$v_1 = u_1 - \left[\frac{\sigma_Y}{\sigma_X} t_1 + \alpha \right]$$

It is convenient, in fact, to consider the quantity w_1, proportional to v_1, determined by the relation

$$w_1 = \frac{v_1}{\sigma_Y} = \frac{u_1}{\sigma_Y} - \frac{t_1}{\sigma_X} - c$$

An examination of the argument above shows that we must have

$$c = \frac{\mu_X}{\sigma_X} - \frac{\mu_Y}{\sigma_Y}$$

so that

$$w_1 = \frac{u_1 - \mu_Y}{\sigma_Y} - \frac{t_1 - \mu_X}{\sigma_X}$$

Since t_1 and u_1 are supposed to be simultaneous observations of the random variables $X(\cdot)$ and $Y(\cdot)$, respectively, w_1 must be a corresponding observation of the random variable

$$W(\cdot) = \frac{Y(\cdot) - \mu_Y}{\sigma_Y} - \frac{X(\cdot) - \mu_X}{\sigma_X} = Y'(\cdot) - X'(\cdot)$$

This is the difference between the standardized random variables $X'(\cdot)$ and $Y'(\cdot)$. The mean value $\mu_W = 0$, and the variance $\sigma_W{}^2 = 2(1 - \rho)$. Thus ρ serves to measure the tendency of the standardized random variables $X'(\cdot)$ and $Y'(\cdot)$ to take on different values. The expression for $\sigma_W{}^2$ holds for negative as well as positive values of ρ.

As an illustration of these ideas, consider two random variables distributed uniformly on the interval $[-a, a]$. Several joint distributions are shown in Fig. 7-6-2. For $\rho = 1$, the mass is concentrated along the line $u = t$. For $\rho = -1$, the mass is concentrated along the line $u = -t$. For uniform distribution of the mass over the square, the random variables are independent and $\rho = 0$. For the case shown in Fig. 7-6-2d, the probability mass is distributed uniformly over the portions of the square in the first and third quadrants. For this distribution, $\rho = \frac{3}{4}$.

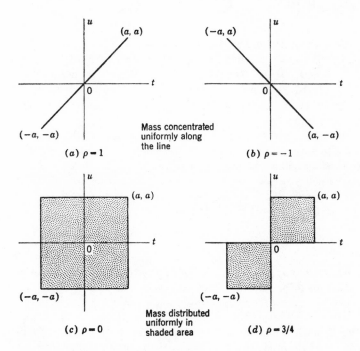

Fig. 7-6-2   Joint mass distributions for two uniformly distributed random variables $X(\cdot)$, $Y(\cdot)$.

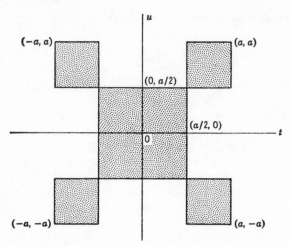

Probability mass distributed uniformly over shaded area

Fig. 7-6-3   Probability mass distribution for uncorrelated
random variables which are not independent.

The joint distribution in Fig. 7-6-3 shows that $\rho = 0$ does not imply
independence.   Both $X(\cdot)$ and $Y(\cdot)$ are uniformly distributed on the
interval $[-a, a]$ and $\rho = 0$.   But the test for independence discussed in
Sec. 3-7 (Example 3-7-3) shows that the variables cannot be independent.
Note the symmetry of the mass distribution which gives rise to the
uncorrelated condition.

These concepts may be extended to pairs of random variables $X(\cdot, s)$
and $X(\cdot, t)$, which are members of a random process.   In this case, the
covariance is a function of the pair of parameters s, t.   In order to allow
for complex-valued processes, we modify the definition slightly by using
the complex conjugate of the second factor.   Also, we modify the use of
the term correlation, somewhat, in deference to widespread practice in
the literature.

Definition 7-6e

The *covariance function* $K_X(\cdot, \cdot)$ of a process X, if it exists, is the
function defined by

$$K_X(s, t) = E[(X(s) - \mu_X(s))(\overline{X(t) - \mu_X(t)})] = \text{cov}\,[X(s), \overline{X(t)}]$$

The bar denotes the complex conjugate.   The *autocorrelation func-
tion* $\varphi_{XX}(\cdot, \cdot)$ of a process, if it exists, is the function defined by

$$\varphi_{XX}(s, t) = E[X(s)\overline{X(t)}]$$

It is apparent from the definition that the covariance function and the autocorrelation function coincide if the mean-value function is identically zero. Use of the linearity property for expectations shows that these two functions are related by

$$K_X(s, t) = \varphi_{XX}(s, t) - \mu_X(s)\overline{\mu_X(t)}$$

If the autocorrelation function exists, the process must be of second order, since on setting $s = t$ we have

$$\varphi_{XX}(s, s) = E[|X(s)|^2]$$

On the other hand, if the process is of second order, the mean-value function, the autocorrelation function, and the covariance function, all exist. The existence of the mean-value function has already been noted. By the Schwarz inequality

$$|E[X(s)\overline{X(t)}]|^2 \leq E[|X(s)|^2]E[|X(t)|^2]$$

so that the finiteness of the expectations on the right-hand side, ensured by the second-order condition, implies the finiteness of the autocorrelation function. The existence of the covariance function follows immediately.

The above discussion of the correlation coefficient as a measure of the similarity of two random variables may be extended to pairs of random variables from a random process. In order to simplify discussion, we restrict our attention to real-valued processes. It is convenient to introduce the notation

$$\rho_X(s, t) = \rho[X(s), X(t)]$$

and

$$\sigma_X{}^2(t) = \sigma^2[X(t)] = K_X(t, t)$$

We may therefore write

$$K_X(s, t) = \rho_X(s, t)\sigma_X(s)\sigma_X(t)$$

We wish to relate the behavior of the covariance function to the character of the sample functions $X(\xi, \cdot)$ from the process. This can be done in a qualitative way, which gives some valuable insight. First, suppose all the sample functions are constant over a given interval. This requires $X(\cdot, s) = X(\cdot, t)$ for each s and t in the interval. In this case $\rho_X(s, t) = 1$ and $K_X(s, t) = K_X(s, s)$. Now suppose, for some pair of values s and t, the correlation coefficient is near zero. The two random variables $X(\cdot, s)$ and $X(\cdot, t)$ would show little tendency to take on the same values. If this situation occurs for s and t quite close together, the sample functions of the process might well be expected to fluctuate quite rapidly in value from point to point, even though the mean-value function $\mu_X(\cdot)$ and the variance $\sigma_X{}^2(\cdot)$ may not change much. If $\rho_X(s, t)$ should be negative, there would be a tendency for the sample functions to take on values of opposite sign.

In many processes, one would expect the random variables $X(\cdot, s)$ and $X(\cdot, t)$ to become uncorrelated for s and t at great distances from one another. If the process represents an actual physical process, there are many situations in which the value taken on by the process at one time s would have little influence on the value taken on at some very much later time t. Suppose we start with a process whose random variables are $X(\cdot, t)$ and whose covariance function has the property that $K_X(s, t) \to 0$ as $|s - t| \to \infty$. Now consider a new process whose random variables are $Y(\cdot, t) = X(\cdot, \lambda t)$, with $\lambda > 1$. Then $K_Y(s, t) = K_X(\lambda s, \lambda t)$. In the Y process, any sample function $Y(\xi, \cdot)$ will fluctuate more rapidly with time than does the corresponding function $X(\xi, \cdot)$ in the X process. Associated with this is the fact that $K_Y(s, t) \to 0$ more rapidly than does $K_X(s, t)$ as $|s - t| \to \infty$. This suggests that processes whose member functions fluctuate rapidly are likely to have a rapid drop-off of correlation between random variables $X(\cdot, s)$ and $X(\cdot, t)$ as s and t become separated. Experience—supported by some mathematics outside the scope of this treatment—shows that this is frequently the case.

Example 7-6-1

A step process X has jumps at integral values of t. In each interval $k \leq t < k + 1$, the process X has the value zero or one, each with probability $\frac{1}{2}$. Values in different intervals are independent.

DISCUSSION   The mean-value function $\mu_X(t) = E[X(t)] = \frac{1}{2}$ for all t.

For s, t in the same interval: $\varphi_{XX}(s, t) = E[X^2(s)] = \frac{1}{2}$
For s, t in different intervals: $\varphi_{XX}(s, t) = E[X(s)]E[X(t)] = \frac{1}{4}$

Fig. 7-6-4 shows values of $\varphi_{XX}(\cdot, \cdot)$ in the appropriate regions in the plane. The covariance function is given by

$$K_X(s, t) = \varphi_{XX}(s, t) - \mu_X(s)\mu_X(t) = \varphi_{XX}(s, t) - \frac{1}{4}$$

Note that $K_X(s, t) = 0$ for $|s - t| > 1$. This condition ensures that s and t are in different intervals, so that, by assumption, $X(\cdot, s)$ and $X(\cdot, t)$ are independent. ∎

Example 7-6-2

Consider the general step process of Sec. 7-5. The autocorrelation function may be determined by using conditional expectations.

$$\varphi_{XX}(s, t) = E[X(s)X(t)|I]P(I) + E[X(s)X(t)|I^c]P(I^c)$$

Now

$$E[X(s)X(t)|I^c] = \iint xy \, dF(x, s; y, t|I^c)$$
$$= \iint xy \, dF_A(x)F_A(y) = E^2[A]$$

where $A(\cdot)$ is a random variable having the common distribution of the $a_n(\cdot)$. To evaluate the first conditional expectation, we must approach the problem in a some-

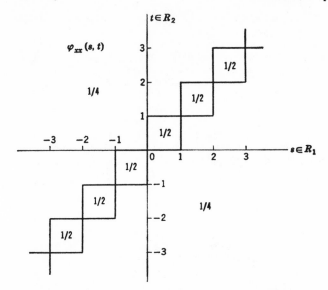

Fig. 7-6-4   Values of $\varphi_{XX}(s, t)$ for the process in Example 7-6-1 in various regions of the plane.

what more fundamental manner.   We note that, for any Borel function $g(\cdot, \cdot)$,

$$E[g(X, Y)|I] = \int g(X, Y)\, dP\, (\cdot|I)$$

It is a property of the conditional probability that $P(C|I) = 0$ for any event C contained in I^c or for any event C such that $P(CI) = 0$.   This implies that the values of $g[X(\cdot), Y(\cdot)]$ for any $\xi \in I^c$ do not affect the value of the integral.   Hence we may replace $g(X, Y)$ by any convenient function which has the same values for $\xi \in I$; the values for $\xi \in I^c$ may be assigned in any desired fashion.   Now for $\xi \in I$, $X(\xi, s)X(\xi, t) = X^2(\xi, s)$.   We may thus assert that

$$E[X(s)X(t)|I] = E[X^2(s)|I] = \int x^2\, dF_A(x) = E[A^2]$$

We may therefore write

$$\varphi_{XX}(s, t) = E[A^2]P(I) + E^2[A]P(I^c)$$

Since $\mu_X(t) = E[A]$ for all t, we have

$$\begin{aligned}
K_X(s, t) &= \varphi_{XX}(s, t) - E^2[A] \\
&= E[A^2]P(I) + E^2[A][1 - P(I)] - E^2[A] \\
&= (E[A^2] - E^2[A])P(I) \\
&= \sigma^2[A]P(I)
\end{aligned}$$

For a Poisson distribution of the jump points, we have

$$K_X(s, t) = \sigma^2[A]e^{-\lambda|t-s|} \qquad \lambda > 0$$

Several things may be noted about this process.   For one thing, $K_X(s, t) \to 0$ as $|s - t| \to \infty$.   The rapidity with which this occurs increases with increasing values of the parameter λ.   Because of the nature of the Poisson counting function, the average number of jumps in any interval of time (hence the rapidity of fluctuation of the sample functions) increases proportionally with λ.   The process is also characterized by the fact that the mean-value function does not vary with t, and the covariance (or the autocorrelation) function depends only upon the difference $s - t$ of the parameters s and t.   The importance of the last two conditions is discussed in the next section. ∎

Example 7-6-3

Suppose the outputs of several generators of sinusoidal signals of the same frequency are added together.   Although the frequencies are the same, the phases are selected at random.   We may represent the resulting signal by the random process whose value is given by

$$X(\xi, t) = \sum_{k=1}^{m} a_k \cos [\omega t + \theta_k(\xi)]$$

where ω is 2π times the frequency in cycles per second, and the class $\{\theta_k(\cdot) : 1 \leq k \leq m\}$ is an independent class of random variables, each of which is uniformly distributed on the interval $[0, 2\pi]$.   The mean-value function is given by

$$\mu_X(t) = E\left[\sum_{k=1}^{m} a_k \cos (\omega t + \theta_k) \right]$$

$$= \sum_{k=1}^{m} a_k E[\cos (\omega t + \theta_k)]$$

Applying the results of Prob. 5-11 to each term, we find that $\mu_X(t) = 0$ for all t. To obtain the autocorrelation function, we utilize the linearity of the expectation function to write

$$\varphi_{XX}(s, t) = \sum_{k,j} a_k a_j E[\cos (\omega s + \theta_k) \cos (\omega t + \theta_j)]$$

By the results of Prob. 5-11 and the independence of the random variables $\theta_k (\cdot)$ and $\theta_j(\cdot)$ for $k \neq j$, the only terms which survive are those for $k = j$.   For this case, we use the trigonometric identity $\cos x \cos y = \frac{1}{2}[\cos (x + y) + \cos (x - y)]$.   The kth term becomes

$$\tfrac{1}{2} a_k^2 E[\cos (\omega s + \omega t + 2\theta_k) + \cos \omega(s - t)]$$

Hence, using the result of Prob. 5-11 once more, we have

$$K_X(s, t) = \varphi_{XX}(s, t) = \tfrac{1}{2} \sum_{k=1}^{m} a_k^2 \cos \omega(s - t)$$

For $s = t$

$$K_X(t, t) = E[X^2(t)] = \tfrac{1}{2} \sum_{k=1}^{m} a_k^2 = K_X(0, 0) \qquad \text{a constant}$$

Thus

$$K_X(s, t) = K_X(0, 0) \cos \omega(s - t)$$

Again we have a constant mean-value function (with value zero) and a covariance or autocorrelation function which depends only on the difference $s - t$. In this case, however, we do not have the condition $K_X(s, t) \to 0$ as $|s - t| \to \infty$. For fixed t, $K_X(s, t)$ is periodic in s. This is a reflection of the fact that each sample function has period $2\pi/\omega$ (Prob. 7-20). The average power, in time, in the kth-component sine wave is proportional to $a_k{}^2$. The probability average of the instantaneous power in the members of the process at any time t is proportional to $E[X^2(t)]$; we have seen that this is the sum of the $a_k{}^2$. This suggests a certain interchangeability of time averages on sample functions and probability averages among the various sample functions. We shall discuss these matters more fully in Sec. 7-8. ■

Joint processes

The ideas introduced above for a single process may be extended to a joint consideration of two random processes. Each process is a family of random variables. The two processes, considered jointly, constitute a new family of random variables. Joint distribution functions of various orders may be defined. Mixed moments of various kinds may be described. One of the most natural is described in the following

Definition 7-6f

The *cross-correlation functions* for two random processes X and Y are defined by

$$\varphi_{XY}(s, t) = E[X(s)\overline{Y(t)}]$$
$$\varphi_{YX}(s, t) = E[Y(s)\overline{X(t)}]$$

Note the order in which the parameters and the conjugated random variables are listed. Other orders could be used; in fact, when reading an article, one must check to see which of the possible orders is used.

The following simple argument based on the Schwarz inequality shows that if the processes X and Y are of second order, the correlation functions exist.

$$|E[X(s)\overline{Y(t)}]|^2 \leq E[|X(s)|^2]E[|Y(t)|^2] < \infty$$

A parallel argument holds for the other case.

Properties of correlation functions

We may list a number of properties of correlation functions which have been found useful in the study of random processes. As a direct consequence of the definition, we may assert

$(\varphi 1)$ $\varphi_{XY}(s, t) = \overline{\varphi_{YX}(t, s)}$

The argument above, based on the Schwarz inequality, shows that

$(\varphi 2)$ $|\varphi_{XY}(s, t)|^2 \leq \varphi_{XX}(s, s)\, \varphi_{YY}(t, t)$

These two properties imply (among other things) that $\varphi_{XX}(s, s)$ is real-valued and that, if X is real, $\varphi_{XX}(s, t) = \varphi_{XX}(t, s)$.

Before introducing the next property, we define a class of functions as follows:

Definition 7-6g

A function $g(\cdot, \cdot)$ defined on $T \times T$ is *positive semidefinite* (or non-negative definite) iffi, for every finite subset T_n contained in T and every function $h(\cdot)$ defined on T_n, it follows that

$$\sum_{s,t \in T_n} g(s, t)h(s)\overline{h(t)} \geq 0$$

(φ3)   The autocorrelation function $\varphi_{XX}(\cdot, \cdot)$ is positive semidefinite.

PROOF   Suppose $T_n = \{t_1, t_2, \ldots, t_n\}$. Then

$$\sum_{s,t \in T_n} \varphi_{XX}(s, t)h(s)\overline{h(t)} = \sum_{j,k} E[X(t_j)h(t_j)\overline{X(t_k)h(t_k)}]$$
$$= E\left[\sum_{j,k} X(t_j)h(t_j)\overline{X(t_k)h(t_k)}\right]$$

The sum in the last expression is of the form

$$\sum_{j,k} a_j \bar{a}_k = \left[\sum_j a_j\right]\left[\sum_k \bar{a}_k\right] = \left[\sum_j a_j\right]\left[\overline{\sum_k a_k}\right]$$
$$= \left|\sum_j a_j\right|^2$$

Hence the last expression in the previous development becomes

$$E\left[\left|\sum_j X(t_j)h(t_j)\right|^2\right] \geq 0 \quad \blacksquare$$

The concept of continuity in analysis is tied to the concept of a limit.   In Sec. 6-3, several types of convergence to limits are considered. So long as the ordinary properties of limits hold for these types of convergence, it is possible to define corresponding concepts of continuity. One of the most useful in the theory of random processes is given in the following

Definition 7-6h

A random process X is said to be *continuous* [*mean²*] *at* t iffi $t \in T$ and

$$\lim_{h \to 0} X(\cdot, t + h) = X(\cdot, t) \qquad [\text{mean}^2]$$

The assertion concerning the limit is equivalent to the assertion

$$\lim_{h \to 0} E[|X(t + h) - X(t)|^2] = 0$$

The following property gives an important characterization of processes which are continuous [mean²].

($\varphi 4$)   The random process X is continuous [mean²] at t iffi the autocorrelation function $\varphi_{XX}(\cdot, \cdot)$ is continuous at the point t, t.

PROOF   Suppose X is continuous [mean²] at t.   Then

$$
\begin{aligned}
|\varphi_{XX}(t + h, t + k) - \varphi_{XX}(t, t)| &= |E[X(t + h)\overline{X(t + k)} - X(t)\overline{X(t)}]| \\
&\leq |E[(X(t + h) - X(t))(\overline{X(t + k)} - \overline{X(t)})]| \\
&\quad + |E[(X(t + h) - X(t))\overline{X(t)}]| \\
&\quad + |E[X(t)(\overline{X(t + k)} - \overline{X(t)})]|
\end{aligned}
$$

We may apply the Schwarz inequality to the first term in the last expression to assert

$$
\begin{aligned}
|E[(X(t + h) - X(t))(\overline{X(t + k)} &- \overline{X(t)})]|^2 \\
&\leq E[|X(t + h) - X(t)|^2]E[|X(t + k) - X(t)|^2]
\end{aligned}
$$

Because of the continuity, the factors in the last expression go to zero as h goes to zero and k goes to zero.   A similar argument holds for the other two terms in the previous inequality, so that the continuity of the autocorrelation function is established.   If the autocorrelation is continuous, we may use the following identity to show the continuity [mean²] of the process.

$$
\begin{aligned}
E[|X(t + h) - X(t)|^2] &= E[(X(t + h) - X(t))(\overline{X(t + h)} - \overline{X(t)})] \\
&= E[|X(t + h)|^2] - E[X(t + h)\overline{X(t)}] \\
&\quad - E[X(t)\overline{X(t + h)}] + E[|X(t)|^2]
\end{aligned}
$$

In the limit as h goes to zero, each term approaches (or remains at) the value $\varphi_{XX}(t, t)$ so that the sum goes to zero.   ■

An argument very similar to that used for the first part of the proof above, using $X(s + h)\overline{X(t + k)}$ rather than $X(t + h)\overline{X(t + k)}$, etc., leads to the following property.

($\varphi 5$)   If $\varphi_{XX}(s, t)$ is continuous at all points t, t, it is continuous for all s, t.

In establishing this theorem, one uses the fact that continuity at t, t guarantees continuity [mean²] of the process X.

Mean-square calculus

For second-order processes, it is possible to develop a calculus in which the limits are taken in the mean-square sense.   For example, we may make the following

Definition 7-6i

A second-order process X has a *derivative* [mean²] at t, denoted by $X'(\cdot, t)$ iffi

$$
\lim_{h \to 0} \frac{X(\cdot, t + h) - X(\cdot, t)}{h} = X'(\cdot, t) \qquad \text{[mean²]}
$$

One may use continuity arguments similar in character to those carried out above (although more complicated in detail) to derive the following properties of the autocorrelation function.

($\varphi 6$) If $\dfrac{\partial^2}{\partial s\, \partial t}\, \varphi_{XX}(s, t)$ exists for all points t, t, then $X'(\cdot, t)$ exists for all t.

($\varphi 7$) If $X'(\cdot, s)$ exists for all s and $Y'(\cdot, t)$ exists for all t, then the following correlation functions and partial derivatives exist and the equalities indicated hold.

$$\varphi_{X'Y}(s, t) = \frac{\partial}{\partial s}\, \varphi_{XY}(s, t)$$

$$\varphi_{XY'}(s, t) = \frac{\partial}{\partial t}\, \varphi_{XY}(s, t)$$

$$\varphi_{X'Y'}(s, t) = \frac{\partial^2}{\partial s\, \partial t}\, \varphi_{XY}(s, t)$$

Properties ($\varphi 6$) and ($\varphi 7$) may be specialized to the case $X = Y$ and combined to give a number of corollary statements.

Riemann and Riemann-Stieltjes integrals of random processes may also be defined, using mean-square limits. Approximating sums are defined as in the case of ordinary functions, except that, in dealing with random processes, the choice of a parameter determines a random variable rather than a value of the function. The approximating sum for a definite integral is thus a random variable. The integral [mean²] is the limit [mean²] of the approximating sums. Thus the "value" of the integral is a random variable. For a detailed development of this calculus, see Loève [1963, secs. 34.2 through 34.4].

The importance of the autocorrelation function rests in part on its close connection with gaussian processes, which are encountered in many physical applications. A brief discussion of these processes and of the role of the autocorrelation function is presented in the last section of this chapter.

7-7 Stationary random processes

In many applications it is possible to assume that the process under study has the property that its statistics do not change with time. Equivalently, these processes have the property that a shift in time origin does not affect the behavior to be expected in observing the process. The following definition gives a precise formulation of the manner in which this invariance with change of time origin is assumed to hold.

Definition 7-7a

A random process X is said to be *stationary* iffi, for every choice of any finite number n of elements $t_1, t_2, \ldots, t_n$ from the parameter set T and of any h such that $t_1 + h, t_2 + h, \ldots, t_n + h$ all belong to T, we have the *shift property*

$$F(\cdot, t_1; \cdot, t_2; \cdots ; \cdot, t_n) = F(\cdot, t_1 + h; \cdot, t_2 + h; \cdots ; \cdot, t_n + h)$$

In order to simplify discussion, we shall suppose that T is the entire real line, so that we do not have to be concerned about which values of the parameters are actually in the index set T. We note from the definition that if we take $h = -t_1$, we have

$$F(\cdot, t_1; \cdot, t_2; \cdots ; \cdot, t_n) = F(\cdot, 0; \cdot, t_2 - t_1; \cdots ; \cdot, t_n - t_1)$$

That is, the distribution really depends upon the differences of the parameters from any one taken as reference (or origin on the time scale).

For a stationary process of second order, we have a simplification of the mean-value function and the autocorrelation function. For all $t \in T$,

$$\mu_X(t) = \mu_X(0) = \mu_X \qquad \text{a constant}$$

In a similar fashion, for the autocorrelation function we have, for all permissible s, t,

$$\varphi_{XX}(s, t) = \varphi_{XX}(0, t - s) = \varphi_{XX}(\tau) \qquad \text{where } \tau = t - s$$

We have used here the same functional notation $\varphi_{XX}(\cdot)$ for two different functions, one a function of two variables and the other a function of a single variable. If the fact is kept in mind, no confusion need result. It should be noted that a second notational possibility exists as follows:

$$\varphi_{XX}(s, t) = \varphi_{XX}(s - t, 0) = \varphi_{XX}(u) \qquad \text{where } u = s - t$$

This form is employed by some writers, and the particular choice should be checked when reading any given article.

The results may be extended to a pair of stationary processes X and Y. In the stationary case

$$\mu_X(s) = \mu_X \qquad \text{and} \qquad \mu_Y(t) = \mu_Y$$
$$\varphi_{XY}(s, t) = \varphi_{XY}(0, t - s) = \varphi_{XY}(\tau) \qquad \text{where } \tau = t - s$$

The same alternative exists with respect to $\varphi_{XY}(s - t, 0)$ as for the autocorrelation function. The properties of the correlation functions $[(\varphi 1)$ through $(\varphi 7)]$, derived in the previous section, may be specialized to the stationary case. In reformulating these properties one should note that the point t, t corresponds to $\tau = 0$ and if the point s, t corresponds to τ, the point t, s corresponds to $-\tau$.

In investigations involving only the mean-value function and the correlation function, the definition of stationarity is much more restrictive than necessary. The definition given above requires the shift property for distribution functions of all orders. Only the first and second distributions enter into an analysis based on the mean-value function and the correlation function.   Hence it is natural to make the following

Definition 7-7b

A random process X is said to be *second-order stationary* if it is of second order and if its first and second distribution functions have the shift property

$$F(x, t) = F(x) \qquad F(x, s; y, t) = F(x, 0; y, t - s)$$

If the process is of second order and $\varphi_{XX}(t, t + \tau) = \varphi_{XX}(0, \tau)$ for all t, τ, the process is said to be *stationary in the wide sense* (Doob [1953]).

It is easy to show that if a process is second-order stationary it is also wide-sense stationary.   The converse is not necessarily true.

Example 7-7-1

Consider once more the step process derived in Sec. 7-5.   In the case that the jump points have a Poisson counting process, we have

$$F(x, s) = F_A(x)$$
$$F(x, s; y, t) = F_A(\min \{x, y\})e^{-\lambda|\tau|} + F_A(x)F_A(y)(1 - e^{-\lambda|\tau|}) \qquad \text{where } \tau = t - s$$

The process is thus second-order stationary (stationary in the wide sense).   As noted in Example 7-6-2, the mean-value function is constant and the autocorrelation function is a function of $\tau = t - s$. ∎

Example 7-7-2

Consider a process of the form

$$X(\xi, t) = A(\xi)f(t) \qquad \text{with } E[X(t)] \equiv 0 \text{ for all real } t$$

We wish to consider the possibility that $\varphi_{XX}(t, t + \tau) = \varphi_{XX}(\tau)$.

DISCUSSION   First we note that $E[X(t)] = f(t)E[A]$, so that $E[A] = 0$ except for the trivial case $f(t) \equiv 0$.   In order for the process to be of second order, we must have $E[|A|^2] < \infty$.   Now $\varphi_{XX}(t, t + \tau) = f(t)\overline{f(t + \tau)}E[|A|^2]$.   To make the autocorrelation function independent of t, we must have $f(t)\overline{f(t + \tau)}$ invariant with t.   Let $f(t) = r(t)e^{i\varphi(t)}$ with $r(t) \geq 0$.   Then we must have

$$\frac{\partial}{\partial t} r(t)r(t + \tau) = r'(t)r(t + \tau) + r(t)r'(t + \tau) = 0 \qquad \text{for all } t, \tau$$

For $\tau = 0$, $r'(t)r(t) \equiv 0$, which implies $r'(t) \equiv 0$, which in turn implies $r(t) \equiv r$, a positive constant.   Since

$$f(t)\overline{f(t + \tau)} = r^2 e^{i[\varphi(t) - \varphi(t+\tau)]}$$

we must have

$$\frac{\partial}{\partial t} [\varphi(t) - \varphi(t + \tau)] = 0 \qquad \text{or} \qquad \varphi'(t) = \varphi'(t + \tau)$$

This is satisfied only by $\varphi'(t) = \lambda$, a constant, so that $\varphi(t) = \lambda t + \theta$, with θ an arbitrary constant.

If we put $B(\xi) = A(\xi)re^{i\theta}$, we have $X(\xi, t) = B(\xi)e^{i\lambda t}$. In this case we have

$$\varphi_{XX}(\tau) = E[|B|^2]e^{-i\lambda\tau} = be^{-i\lambda\tau} \quad \blacksquare$$

Example 7-7-3

A more general process for which $\varphi_{XX}(t, t + \tau) = \varphi_{XX}(\tau)$ may be formed by taking a linear combination of processes of the type in the previous example. Consider

$$X(\xi, t) = \sum_{k \in J} B_k(\xi)e^{i\lambda_k t} \quad \text{with } E[B_k] = 0 \text{ and } E[|B_k|^2] < \infty \text{ for each } k \text{ in } J$$

DISCUSSION

$$\begin{aligned}\varphi_{XX}(t, t + \tau) &= E[X(t)\overline{X(t + \tau)}] \\ &= \sum_{j,k \in J} E[B_k\bar{B}_j]e^{i[(\lambda_k-\lambda_j)t - \lambda_j\tau]}\end{aligned}$$

In order for the autocorrelation function to be independent of t, we must have $E[B_k\bar{B}_j] = 0$ for $k \neq j$. In this case

$$\varphi_{XX}(\tau) = \sum_{k \in J} b_k e^{-i\lambda_k\tau} \quad \text{where } b_k = E[|B_k|^2]$$

Now $|\varphi_{XX}(\tau)| \leq \varphi_{XX}(0) = \sum_{k \in J} b_k$. If the index set J is infinite, it is necessary that the sum of the b_k converge.

If the process is real-valued, so also is $\varphi_{XX}(\cdot)$. In this case, the complex terms must appear in conjugate pairs. Suppose $J = \{\ldots, -2, -1, 0, 1, 2, \ldots\}$, $\lambda_{-k} = -\lambda_k$, and $B_{-k}(\cdot) = \overline{B_k(\cdot)}$. If we let $B_k(\cdot) = C_k(\cdot) - iD_k(\cdot)$, we may write

$$\begin{aligned}X(\xi, t) &= B_0(\xi) + \sum_{k=1}^{\infty} [B_k(\xi)e^{i\lambda_k t} + \overline{B_k(\xi)}e^{-i\lambda_k t}] \\ &= B_0(\xi) + 2 \sum_{k=1}^{\infty} [C_k(\xi) \cos \lambda_k t + D_k(\xi) \sin \lambda_k t] \\ &= B_0(\xi) + 2 \sum_{k=1}^{\infty} |B_k(\xi)| \cos [\lambda_k t + \theta_k(\xi)]\end{aligned}$$

$B_0(\cdot)$ must be real-valued, and $\theta_k(\cdot)$ is the phase angle for the sinusoid with angular frequency λ_k.

The autocorrelation function must be given by

$$\begin{aligned}\varphi_{XX}(\tau) &= b_0 + \sum_{k=1}^{\infty} b_k[e^{i\lambda_k\tau} + e^{-i\lambda_k\tau}] \\ &= b_0 + 2 \sum_{k=1}^{\infty} b_k \cos \lambda_k\tau\end{aligned}$$

with $b_k = E[|B_k|^2] = E[C_k^2 + D_k^2]$.

It is of interest that the autocorrelation function does not depend in any way on the phases of the sinusoidal components of the process; it depends only upon the amplitudes and frequencies of these components.   This fact is important in certain problems of statistical communication theory. ∎

Processes of the type considered in the last example are important in the spectral analysis of random processes (cf. Doob [1953] or Yaglom [1962]). Further investigation of this important topic would require a knowledge of harmonic analysis and would take us far beyond the. scope of the present work.

7-8   Expectations and time averages; typical functions

In the introduction and the first section of this chapter it is pointed out that the "randomness" of a random process is a matter of the uncertainty about which sample function has been selected rather than the "random" character of the shape of the graph of the function.   Yet in both popular thought and in actual experimental practice the notion of randomness is associated in an important way with the fluctuating characteristic of the graph of the sample function.

The situation that usually exists is that one obtains experimentally a record of successive values of the physical quantity represented by the values of the random process.   The record so obtained is a graph over some finite interval of one of the sample functions.   It is sometimes possible, because several devices or systems are available, to obtain partial records of several sample functions.   When several recordings are made from one device or system, these must be viewed as graphs of the same sample function over the various finite intervals.   With partial information on one or a few sample functions, how is one to characterize a random process which has an infinity of sample functions?

What one assumes in practice is that the sample function available is *typical* of the process in some appropriate sense.   One seeks to determine from the record of this function information on the statistics of the process—probability distributions, expectations, etc.   In order to do this, one has to suppose that the time sequence of observed values corresponds in some way to the distribution of the set of values of the whole ensemble of sample functions.

The practical approach is to take whatever sample functions are available and determine various time averages.   These time averages are then assumed to give at least approximate values of corresponding probability averages (expectations).   For one thing, this assumes that the probability distributions are *stationary*—at least over the period in which the time averaging is carried out.   *In this section we shall assume the processes are stationary* to whatever degree is needed for the problem at hand.

To facilitate discussion of the problem, we need to make precise the notion of time average.

Definition 7-8a

The *time average* (or *time mean*) of a function $f(\cdot)$ over the interval (a, b) is given by

$$A[f; a, b] = \frac{1}{b - a} \int_a^b f(t)\, dt$$

If either limit of the interval is infinite, we take the limit of the average over finite intervals as a, or b, or both, become infinite.

If $f(\cdot)$ is defined only for discrete values of t, we replace the integral by the sum and divide by the number of terms.

To simplify discussion, we shall suppose the interval of averaging is the whole real line and shall shorten the notation to $A[f]$.

Time averages have a number of properties which we may list for convenience. Since time averaging involves an integration process, it should be apparent that the operation of time averaging has properties in common with the property of taking mathematical expectations (probability averaging).

The following properties hold for complex-valued functions, except in the cases in which inequalities demand real-valued functions. We state these properties without proof, since most of them are readily derivable from properties of integrals and may be visualized in terms of the intuitive idea of averaging the values of a function. These properties are stated and numbered so that the first five properties correspond to the first five properties listed for mathematical expectation. The first five properties hold for averaging over any interval.

   (A1)   If c is any constant, $A[c] = c$.
   (A2)   *Linearity.*   If a and b are real or complex constants,

$$A[af + bg] = aA[f] + bA[g]$$

   (A3)   *Positivity.*   If $f(\cdot) \geq 0$, then $A[f] \geq 0$. If $f(\cdot) \geq 0$ and the interval of averaging is finite, $A[f] = 0$ iff $f(t) = 0$ for almost all t in the interval. If $f(\cdot) \geq g(\cdot)$, then $A[f] \geq A[g]$.
   (A4) $A[f]$ exists iff $A[|f|]$ does, and $|A[f]| \leq A[|f|]$.
   (A5)   *Schwarz inequality.* $|A[fg]|^2 \leq A[|f|^2]A[|g|^2]$. In the real case, equality holds for finite intervals of averaging iff there is a real constant λ such that $\lambda f(t) + g(t) = 0$ for almost all t in the interval.

The following properties hold for infinite intervals of averaging.

(A$_\infty$6)   If $\int_{-\infty}^{\infty} f(t)\, dt$ exists, then $A[f] = 0$.

(A$_\infty$7)   If $A[|f|]$ exists, then $A[f(t + a)] = A[f(t)]$, where a is any real constant.

(A$_\infty$8)   If $f(\cdot)$ is periodic with period T,

$$A[f] = \frac{1}{T} \int_a^{a+T} f(t)\, dt \qquad \text{for any real number } a$$

It should be noted that the time average is defined for any suitable function of time, whether the function is a member of a random process or not.   If the time function is a sample function from a random process, the average will, in general, depend upon which choice variable ξ is selected.   In most cases of interest, we are safe in assuming that this average, as a function of ξ, is a random variable.

In many investigations, one is interested in the relationship between time averages for a given sample function from a random process and the mathematical expectation for the process.   One of the first things that may be noted is that under very general conditions we may interchange the order of taking time averages and taking the mathematical expectation.   Thus

$$E\{A[X(\xi, t)]\} = A\{E[X(\xi, t)]\}$$

This follows from the fact that both operations are integration processes. The expression above represents the following integral expression:

$$\int_S \left[\frac{1}{b-a} \int_a^b X(\xi, t)\, dt \right] dP(\xi) = \frac{1}{b-a} \int_a^b \left[\int_S X(\xi, t)\, dP(\xi) \right] dt$$

The very general theorem of Fubini allows the interchange of order of integration under most practical conditions.   This result may be extended to averages and expectations of functions of the members of the process and to functions of members of more than one process.

Among the most useful time averages in the study of stationary random processes are the time correlation functions defined as follows:

Definition 7-8b

Suppose $f(\cdot)$ and $g(\cdot)$ are real or complex-valued functions defined on the whole real line.   The *time autocorrelation function* $R_{ff}(\cdot)$ for the function $f(\cdot)$ is given by

$$R_{ff}(\tau) = A[f(t)\overline{f(t + \tau)}]$$

The *time cross-correlation function* for $f(\cdot)$ and $g(\cdot)$ is given by

$$R_{fg}(\tau) = A[f(t)\overline{g(t + \tau)}]$$

As we shall see, these are closely associated, in the theory of stationary random processes, with the corresponding correlation functions defined in Sec. 7-6 as mathematical expectations.   To distinguish the two kinds of correlation, we shall speak of *time correlation* and *stochastic correlation*.

The time autocorrelation is a very simple operation, which may be viewed graphically in the case of real-valued functions.   The graph of $f(t + \tau)$ is obtained from the graph of $f(t)$ by shifting the latter τ units to the left for $\tau > 0$ or $|\tau|$ units to the right for $\tau < 0$.   The product of the function $f(\cdot)$ and the shifted function $f(\cdot + \tau)$ is a new function of

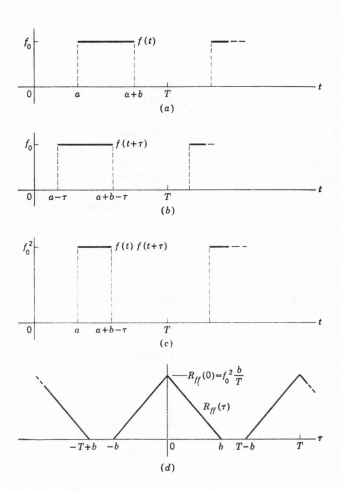

Fig. 7-8-1   Time autocorrelation function $R_{ff}(\cdot)$ for a rectangular wave function.

time, which is then averaged. The process may be illustrated by the
following simple example.

Example 7-8-1   *Time Autocorrelation of a Rectangular Wave*

We consider the real-valued function whose graph is a periodic succession of rectangular waves, as shown in Fig. 7-8-1. In dealing with a periodic wave, we need
deal with only one period and extend the result periodically. Figure 7-8-1a shows
one period of the graph of $f(\cdot)$. Figure 7-8-1b shows one period of the graph of
$f(\cdot + \tau)$, in the case $0 < \tau$. Figure 7-8-1c shows one period of the product function
$f(\cdot)f(\cdot + \tau)$. The time average, which is the time average over one period by property
$(A_\infty 8)$ is given by

$$A[f(\cdot)f(\cdot + \tau)] = f_0{}^2 \frac{b - \tau}{T} \qquad \text{for } 0 < \tau < b$$

Other cases may be argued similarly, to give the function $R_{ff}(\cdot)$ shown in Fig. 7-8-1d. ∎

If the value $f(t)$ of the function fluctuates rapidly in a "random"
manner as t varies, in such a way that $A[f] = 0$, it is to be expected that,
after a suitable shift τ, the product function will be positive as much as it
is negative, so that the average $R_{ff}(\tau)$ becomes small for large shift τ.
The more rapidly the function fluctuates, the more rapidly $R_{ff}(\tau)$ might
be expected to drop off from its maximum value of $R_{ff}(0) = A[f^2]$ at
$\tau = 0$.   If, however, the function contains a periodic component hidden
in the fluctuations, this periodic component will line up when the shift is
a multiple of the period.   This value will show up in the time correlation
function as a periodic term.   It is properties such as these, plus some
important theorems in harmonic analysis, which direct attention to the
time correlation functions.   We cannot deal in any complete way with
this topic.   For a careful and extensive treatment, one may consult the
important book by Lee [1960, chap. 2].

Suppose $X(\cdot, \cdot)$ is a stationary second-order process.   We have considered two operations which are designated by the term correlation.
Stochastic correlation:

$$\varphi_{XX}(\tau) = E[X(\cdot, t)X(\overline{\cdot, t + \tau})] \qquad \text{(a probability average)}$$

Time correlation:

$$R_{XX}(\xi, \tau) = A[X(\xi, t)\overline{X(\xi, t + \tau)}] \qquad \text{(a time average, depending} \\ \text{upon choice of sample function)}$$

In most cases, $R_{XX}(\cdot, \cdot)$ will be a random process; that is, for each τ,
we have a random variable.   Intuition and experience indicate the possibility that $\varphi_{XX}(\tau)$ and $R_{XX}(\xi, \tau)$ should be the same, at least for some
functions which are *typical* of the process, in the sense that they reflect in
their time variations the statistics of the process.

Let us examine the concept of a *typical function* more closely.   Suppose we have such a function $X(\xi, \cdot)$ from the process, or at least a graph of the function as observed over a finite period of time.   In what sense can the analytical character of this sample function give meaningful information about the statistical or probabilistic character of the process?   For one thing, we should suppose that the time average of the sample function should agree with the mean-value function (which is constant for stationary processes).   That is,

$$A[X(\xi, \cdot)] = \mu_X$$

We should suppose further that

$$\varphi_{XX}(\tau) = R_{XX}(\xi, \tau)$$

That is, the stochastic autocorrelation function for the process should coincide with the time autocorrelation function for the typical function.   In order to emphasize the importance of the idea, we formalize it in the following

Definition 7-8c

A sample function $X(\xi, \cdot)$ from a stationary random process is said to be *typical* (in the wide sense) iffi

$$A[X(\xi, \cdot)] = \mu_X$$
$$R_{XX}(\xi, \tau) = \varphi_{XX}(\tau) \qquad \text{for all } \tau$$

When other expectations are under consideration, the concept of a typical function may be modified in an appropriate way.   Also, it may be that a sample function is typical over certain finite intervals, in which case we say it is *locally typical*.   This is the sort of thing that occurs in experimental work.   A record is taken of the behavior of some physical system.   At times, "something is wrong"—perhaps the equipment is not working in the proper manner.   At these times, the record is not typical.   When the record is locally typical over a sufficiently long interval, the time averages over that interval may provide reasonable approximations to the time average of a typical function over the whole real line.

The key to empirical work with random processes is in finding a function which is typical over an interval of sufficient length so that the time averages provide suitable approximations to the corresponding probability averages.   When one has only a few short records—either of various intervals on the same sample function or of intervals for a few different sample functions—he has to assume the records are typical.   From these he must draw conclusions about the entire process which serve as the basis for analysis of the system under study.   In order to test for the condition of being locally typical, one may perform the time averaging over various subintervals of the records available and compare the results with the average over the entire record or set of records. If sufficient record is available, one may be able to check the tendency of the various averages to approximate the same function. If they do approximate the same function, one must then proceed on the assumption that this function is the desired function for the process.

There are processes for which no sample function is typical. As a trivial example, consider‸the case in which every sample function is a constant. This means that every random variable in the process is identical (as a function of the choice variable ξ) with every other. There are no variations in any sample function to provide information about the ensemble variations in the process.

On the other hand, there are important processes for which the probability of selecting a typical function is quite high. Processes which have this property have been singled out and given a special name.

Definition 7-8d

A random process X is said to be *ergodic* (in the wide sense) iffi it is stationary in the wide sense and

$$(A) \quad \lim_{T \to \infty} \frac{1}{2T} \int_{-T}^{T} X(\xi, t)\, dt = \mu_X \; [\cdot \; \cdot \; \cdot]$$

$$(B) \quad \lim_{T \to \infty} \frac{1}{2T} \int_{-T}^{T} X(\xi, t)\overline{X(\xi, t + \tau)}\, dt = \varphi_{XX}(\tau) \; [\cdot \; \cdot \; \cdot]$$

The symbol $[\cdot \; \cdot \; \cdot]$ is intended to indicate that the limit may be taken in any one of the senses which are common in probability theory (Sec. 6-3). Thus we may have convergence [mean²], which implies convergence [in prob]; or we may have convergence $[P]$, which is ordinary convergence for all but possibly a set of sample functions with zero probability of occurrence.

In the case of random sequences, the integral in expression (A) becomes a sum. If the convergence is [in prob], we then have a form of the weak law of large numbers discussed in Sec. 6-1. If the convergence is $[P]$, we have a form of the strong law of large numbers, examined briefly in Sec. 6-4. The ergodic property (B) may thus be viewed as an extension of the law of large numbers—weak or strong, depending upon the kind of convergence assumed.

The strong form (convergence $[P]$) is highly desirable. With this condition, if we choose a sample function "at random," the probability is 1 that the time averages for the sample function chosen will give the mathematical expectations corresponding to expressions (A) and (B). We are quite free to operate, in this case, with whatever sample function is available. The weak forms of convergence are not quite so desirable.

In the case of convergence [in prob], we are assured that for sufficiently large T we have a high probability of choosing a sample function for which the integral differs from the limiting value by an arbitrarily small number. The weak convergence does not, however, assure us that if we are successful in choosing such a desirable function for a given T, the value of the integral for this sample function will continue to approach the desired limit as T is increased.

A variety of conditions have been studied which ensure ergodicity, either in the strong or the weak sense. The difficulty with these conditions is that they are necessarily expressed in terms of probability distributions or mathematical expectations (such as the autocorrelation function). It is precisely this probability information which is missing when one has only a single sample function or a small number of sample functions available for study. The mathematical problem is to express the conditions in such a manner that one can tell from physical considerations or past experience whether or not the assumptions are likely to lead to a useful probability model. For a readable discussion of the conditions which ensure convergence [mean^2], see Yaglom [1962, Sec. 4].

In spite of the theoretical limitations, the practice of utilizing various time averages to obtain estimates of mathematical expectations has proved successful in many practical situations. *The success of this approach depends upon success in obtaining a function which is typical of the process in the sense that its time variations exhibit the properties of the "ensemble variations" characteristic of the process.* Here, as in all applications of theoretical models to physical situations, the final appeal must be to experience, to determine the adequacy of the model.

These remarks should not be interpreted as detracting from the importance of theoretical developments. Very often, assumptions based on physical experience can be used to derive important characteristics of the underlying probability model. This is the case for the Poisson counting process discussed in Sec. 7-3. A careful theoretical study of the process so derived often leads to the identification of parameters or characteristics, which may then be checked experimentally. On the basis of experience, the model is postulated. Conclusions are drawn which are checked experimentally, to test the hypothesis. If the "fit" is satisfactory, one uses the model as an aid to analysis and further experimentation.

7-9   Gaussian random processes

The class of random processes known as *gaussian processes* (or *normal processes*) plays a strategic role in both mathematical theory and in the application of the theory to real-world problems. This is notably true in the study of electrical "noise" and related phenomena in other physical

systems.   Because of the central limit theorem, one might expect to
encounter the gaussian distribution in random processes whose member
functions are the result of the superposition of many component functions.   The "shot noise" in thermionic vacuum tubes, which places limitations on their use as amplifiers of electrical signals, is the result of the
superposition of a large number of very small current pulses.   Each pulse
is due to the emission from the cathode of single electrons and their passage across the interelectrode space to the anode.   Since these electrons
are emitted "randomly"—in many situations the emission times are
described by a Poisson counting function—and the current pulses produced in the associated circuit overlap and add up, the instantaneous
current has a gaussian distribution at any time t.   Under certain natural
assumptions, the random variables of the process are jointly gaussian,
so that the resulting current is a sample function from a gaussian process.
In a similar manner, the "thermal noise," which consists of the voltage
fluctuations produced by thermal agitation of ions in a resistive material,
is represented by a gaussian process.

Gaussian processes are characteristic of electrical noise and other
noise phenomena resulting from the superposition of many component
effects in another respect.   Gaussian signals have the property that when
a signal is passed through a linear filter, the waveform of the response
signal is also a member of a gaussian process.   This is not true for most
nongaussian processes; there is no simple (and often no known) relation
between the nongaussian input to a linear filter and the output from the
filter in response to this input.   The gaussian process is thus in many
ways "natural" to the important topic of electrical noise, since many
transmission or control systems behave approximately as linear
filters.   Analogous situations hold in many other types of physical
systems.

In addition to the widespread occurrence in nature of physical processes which are gaussian or very nearly gaussian, the mathematics of the
gaussian process is much more completely developed than for most other
processes of similar complexity and generality.   In part, this is because
the gaussian process is completely described in terms of second-order
effects: the mean-value function and the autocorrelation function determine completely the character of the gaussian process.   In this section
we shall give a very brief outline of some of the more important mathematical characteristics of gaussian processes.   For more detailed treatments, see any of the references listed at the end of the chapter.

Before turning to general processes, we consider finite families of
random variables.   An extension of the uniqueness property (M1) on
moment-generating functions (Sec. 5-7) asserts that a joint distribution
function for a finite number of random variables is uniquely determined

by the *moment-generating function*

$$M(s_1, s_2, \ldots, s_n) = E[\exp \{s_1X_1 + s_2X_2 + \cdots + s_nX_n\}]$$

or, equivalently, by the *characteristic function*

$$\varphi(t_1, t_2, \ldots, t_n) = E[\exp \{i(t_1X_1 + t_2X_2 + \cdots + t_nX_n)\}]$$

For each $X_k(\cdot)$ we denote the *mean value* by $\mu_k = E[X_k]$, and for each pair $X_j(\cdot)$ and $X_k(\cdot)$, we indicate the *covariance* by

$$K_{jk} = E[(X_j - \mu_j)(\bar{X}_k - \bar{\mu}_k)]$$

Using an argument exactly parallel to that for property ($\varphi 3$) for the autocorrelation function, we find that the quadratic form

$$Q(t_1, t_2, \ldots, t_n) = \sum_{j,k} K_{jk} t_j \bar{t}_k \geq 0$$

so that it is positive semidefinite. If the stronger condition holds that the value is zero only when all t_k are zero (in which case the form is said to be *positive definite*), then the inverse form exists as follows:

$$Q^{-1}(t_1, t_2, \ldots, t_n) = \sum_{j,k} K_{jk}^* \bar{t}_j t_k \geq 0$$

The numbers K_{jk}^* are obtained by inverting the matrix $[K_{jk}]$ of the covariance factors. K_{jk}^* is the element of the jth row and kth column of the inverse matrix.

Definition 7-9a

The class of random variables $\{X_k(\cdot): 1 \leq k \leq n\}$ is said to have a *joint gaussian distribution* (or to be *normally distributed*) iff the characteristic function

$$\varphi(t_1, t_2, \ldots, t_n) = E\left[\exp\left\{i \sum_{k=1}^{n} t_k X_k\right\}\right]$$

$$= \exp\left[i \sum_{k=1}^{n} t_k \mu_k - \tfrac{1}{2}Q(t_1, t_2, \ldots, t_n)\right]$$

If the quadratic form is positive definite, so that the inverse exists, it can be shown (cf. Cramér [1946]) that the joint density function is given by

$$f_{X_1X_2} \cdots X_n(t_1, t_2, \ldots, t_n)$$

$$= \frac{1}{(2\pi)^{n/2} |K^*|^{1/2}} \exp\left[-\tfrac{1}{2}Q^{-1}(t_1 - \mu_1, t_2 - \mu_2, \ldots, t_n - \mu_n)\right]$$

where $|K^*|$ is the determinant of the matrix $[K_{jk}^*]$. If the quadratic form is only positive semidefinite, special arguments must be carried out. The distribution in this case is said to be *singular*. The quadratic form Q, and hence the matrix $[K_{jk}]$, have rank $r < n$. Cramér has shown that in this case the probability mass distribution induced by the random variables is limited to a linear subspace R^r of dimension r, rather than being distributed in n-dimensional space R^n.

Using properties of the characteristic function and properties of convergence [mean²], one may prove the following theorems:

Theorem 7-9A

If the class $\{X_k(\cdot): 1 \leq k \leq n\}$ is normally distributed and if a class $\{Y_i(\cdot): 1 \leq i \leq m\}$ is obtained by a linear transformation,

$$Y_i(\cdot) = \sum_{k=1}^n a_{ik}X_k(\cdot) \qquad i = 1, 2, \ldots, m$$

then the new class is also normally distributed.

Theorem 7-9B

Suppose $\{X_n(\cdot): 1 \leq n < \infty\}$ is a sequence of random variables, each of which is normally distributed. If $X_n(\cdot) \to X(\cdot)$ [mean²] as n becomes infinite, then $X(\cdot)$ is also normally distributed.

Proof of the second theorem utilizes the continuity property of the characteristic function and the Schwarz inequality.

We are now in a position to state a definition of a gaussian random process.

Definition 7-9b

A random process X is said to be *gaussian* (or *normal*) iffi every finite subfamily of random variables from the process is normally distributed.

A random process is *strongly normal* iffi it is normal and either (1) all $X(\cdot, t)$ are real-valued or (2) $E[X(\cdot, s)X(\cdot, t)] = 0$ for $s \neq t$.

Theorems 7-9A and 7-9B imply that derivatives and integrals (in the mean-square calculus discussed briefly in Sec. 7-6) of gaussian processes are also gaussian. A specific value $K_X(s, t)$ of the covariance function (Definition 7-6e) is the covariance of the pair of random variables $X(\cdot, s)$ and $X(\cdot, t)$ from the process. The characteristic function, hence the density or distribution function, for any finite subclass $\{X(\cdot, t_1), X(\cdot, t_2), \ldots, X(\cdot, t_n)\}$ from the process is determined by the values $\mu(t_k)$ of the mean-value function and the values $K_X(t_j, t_k)$ of the covariance function, or equivalently, by the $\mu(t_k)$ and the values $\varphi_{XX}(t_j, t_k)$ of the autocorrelation function. It follows immediately that if a normal process is second-order stationary (i.e., stationary in the wide sense), it is also stationary in the strict sense, since distribution functions of all orders are determined by the autocorrelation function.

The following theorem, which we state without proof (cf. Loève [1963, p. 466]), further characterizes the role of the gaussian process.

Theorem 7-9C

A function $g(\cdot, \cdot)$ defined on $T \times T$ is an autocorrelation function (or a covariance function) iffi it is positive semidefinite.

Every autocorrelation function is the autocorrelation function of a strongly normal process. If the autocorrelation function is real-valued, the process may be chosen to be real-valued.

In particular, any second-order stationary process can be represented by a gaussian stationary process having the same mean-value function and autocorrelation function. Such processes play a central role in investigations in which mean-square approximations are used. Even a cursory survey of the literature on random signals and noise in communication theory will provide convincing evidence of the importance of this topic.

Problems

7-1. The transition matrix for a constant Markov chain is

$$\mathcal{P} = \begin{bmatrix} 0.2 & 0.3 & 0.5 \\ 0 & 0.6 & 0.4 \\ 0.8 & 0.2 & 0 \end{bmatrix}$$

Initial probabilities are $\pi_0(1) = \pi_0(2) = 0$, $\pi_0(3) = 1$. Determine the matrix of second-order transition probabilities and determine $\pi_2(1)$, $\pi_2(2)$, $\pi_2(3)$.
        ANSWER: $\pi_2(1) \doteq 0.16$, $\pi_2(2) = 0.36$, $\pi_2(3) = 0.48$

7-2. An electronic computer operates in the binary number system; i.e., it uses the digits 0 and 1. In the process of computation, numbers are transferred from one part of the machine to another. Suppose that at each transfer there is a small probability p that the digit transferred will be changed into its opposite. Let the digits 0 and 1 correspond to states of the system, and let each stage in the chain of transfer operations correspond to one period.

(*a*) Write the transition matrix for the Markov chain describing the probabilities of the states.

(*b*) Draw a transition diagram for the process.

(*c*) Determine the second-order and third-order transition matrices.

(*d*) If the initial probabilities for the two states are $\pi_0(0) = \pi_0(1) = \frac{1}{2}$, show that the probabilities after n transitions are $\pi_n(0) = \pi_n(1) = \frac{1}{2}$. Interpret this result.

7-3. Construct a transition diagram for the chain described in Example 7-2-3. Identify the closed sets, including the absorbing states, if any.
        ANSWER: States 1 and 9 are absorbing

7-4. Consider a finite chain with N states. Show that if $s^j \to s^k$, then state k can be reached in not more than N transitions from the initial state j. (*Suggestion:* Consider the transition diagram.)

7-5. Draw the transition diagram for the chain described in Prob. 7-1. Show that the chain is irreducible and aperiodic.

7-6. Consider a finite, constant Markov chain having N states. Show that if the transition matrix is doubly stochastic, the chain has stationary distribution $\pi_i = 1/N$, $i = 1, 2, \ldots, N$.

7-7. Show that the "success-failure" process (Example 7-2-4) is always ergodic if none of the $\pi(i, j)$ is zero. Determine the values of the stationary probabilities.

ANSWER: $\pi_1 = \pi(2, 1)/[\pi(1, 2) + \pi(2, 1)]$, $\pi_2 = \pi(1, 2)/[\pi(1, 2) + \pi(2, 1)]$

7-8. Assume the validity of Theorem 7-2E. Show that if a finite, constant Markov chain is irreducible and aperiodic, then it must be ergodic (i.e., all states are ergodic). In this case $\lim\limits_{n \to \infty} \pi_n(i, j) = \pi_j = 1/\tau(j) > 0$ for all pairs i, j.

7-9. Obtain the stationary distribution for the chain described in Prob. 7-1, and obtain the mean recurrence times for the various states.

ANSWER: $\pi_1 = {}^{32}\!/_{104}$, $\pi_2 = {}^{40}\!/_{104}$, $\pi_3 = {}^{32}\!/_{104}$

7-10. Modify the chain described in Example 7-2-3 and Fig. 7-2-1 by making the behavior in positions 1 and 9 the same as that in other positions.

(*a*) Show that the resulting chain is irreducible and that every state is periodic with period 2.

(*b*) Show that the chain has a stationary distribution in which the stationary probability for any position is proportional to the number of ways of reaching that position. Determine these probabilities.

7-11. A random process X consists of four sample functions $X(\xi, t)$, $0 \leq t \leq 6$. The sample functions may be described as follows:

$X(\xi_1, t)$ rises linearly from $X(\xi_1, 0) = 1$ to $X(\xi_1, 6) = 5$.

$X(\xi_2, t)$ is a step function, having value 4 for $0 \leq t < 2.5$, value 1 for $2.5 \leq t < 4.5$, and value 2 for $4.5 \leq t \leq 6$.

$X(\xi_3, t)$ is a triangular function, rising linearly from $X(\xi_3, 0) = 0$ to $X(\xi_3, 3) = 4.5$ and then decreasing linearly to $X(\xi_3, 6) = 0$.

$X(\xi_4, t)$ is a linear function, decreasing from $X(\xi_4, 0) = 5$ to $X(\xi_4, 6) = 2$.
Probabilities $p_k = P(\{\xi_k\})$ are $p_1 = 0.2$, $p_2 = 0.4$, $p_3 = 0.3$, and $p_4 = 0.1$.

(*a*) Plot $F(\cdot, 1)$ and $F(\cdot, 3)$.

ANSWER: $F(\cdot, 1)$ has jumps of 0.3, 0.2, 0.4, 0.1 at $k = 1.5$, 1.67, 4.0, 4.5, respectively.

(*b*) Determine $F(x, t_1; y, t_2)$ for

(1) $(x, t_1, y, t_2) = (2, 1, 3, 5)$

(2) $(x, t_1, y, t_2) = (4, 2, 4, 3)$

(3) $(x, t_1, y, t_2) = (4, 1, 3, 4)$ ANSWER: 0.3, 0.7, 0.7

7-12. Radioactive particles obey a Poisson probability law (i.e., the Poisson process is the counting process for the number of arrivals in a given time t). The average rate λ for a radioactive source is known to be 0.25 particle per second. The number of arrivals in each of five nonoverlapping intervals of 8 seconds duration each is noted. What is the probability of one or more observations of exactly two arrival times in 8 seconds?   ANSWER: $1 - (1 - 2e^{-2})^5$

7-13. Let N be the Poisson process.

(a) Evaluate $\mu[N(\cdot, t)/t]$ and $\sigma^2[N(\cdot, t)/t]$ for $t > 0$.   ANSWER: λ, λ/t

(b) Discuss with the aid of the Chebyshev inequality (or other suitable relation) the problem of determining the parameter λ for the process.

7-14. Let N be the Poisson process, and for a fixed $T > 0$, define the increment process X by the relation $X(\xi, t) = N(\xi, t + T) - N(\xi, t)$.

(a) Show that $E[X(\cdot, t)] = \lambda T$.

(b) Show that the autocorrelation function is given by

$$\varphi_{XX}(\tau) = \begin{cases} \lambda^2 T^2 + \lambda(T - |\tau|) & \text{for } |\tau| \leq T \\ \lambda^2 T^2 & \text{for } |\tau| \geq T \end{cases}$$

[*Suggestion:* Use the results of Examples 5-3-3 and 5-4-4 and the fact that the increments of the Poisson process are independent and stationary. Use the device $N(\cdot, t + a) - N(\cdot, t) = N(\cdot, t + a) - N(\cdot, t + b) + N(\cdot, t + b) - N(\cdot, t)$ for suitable choices of b.]

7-15. Suppose $Z = X + Y$, where X and Y are random processes. Determine the autocorrelation function for Z in terms of autocorrelation functions and cross-correlation functions for X and Y.

7-16. Suppose X and Y are second-order stationary random processes. Consider $E\{|X(\cdot, t) \pm Y(\cdot, t + \tau)|^2\} \geq 0$. Show $|\text{Re } \varphi_{XY}(\tau)| \leq \frac{1}{2}[\varphi_{XX}(0) + \varphi_{YY}(0)]$.

7-17. Let X be a real-valued, stationary random process. It is desired to estimate $X(\xi, t_3)$ when $X(\xi, t_1)$ and $X(\xi, t_2)$ are known $(t_1 < t_2 < t_3)$. One type of estimate is provided by the so-called *linear predictor*. As an estimate, one takes $Y(\xi, t_1, t_2) = aX(\xi, t_1) + bX(\xi, t_2)$, where a and b are suitable constants. Let $\mathcal{E}(\xi, t_1, t_2, t_3) = Y(\xi, t_1, t_2) - X(\xi, t_3)$. As a measure of goodness of the predictor, we take the mean-square error $E[\mathcal{E}^2(\cdot, t_1, t_2, t_3)]$.

(a) Let $t_2 = t_1 + \alpha$ and $t_3 = t_2 + \tau$. Obtain an expression for the mean-square error in terms of the autocorrelation function for the process X.

     ANSWER: $(a^2 + b^2 + 1)\varphi_{XX}(0) + 2ab\varphi_{XX}(\alpha) - 2a\varphi_{XX}(\alpha + \tau) - 2b\varphi_{XX}(\tau)$

(b) Determine values of the constants a and b to minimize the mean-square error.

Instructions for probs. 7-18 through 7-27

Assume that the random variables and the random processes are real-valued, unless stated otherwise. In working any problem in the sequence, assume the results of previous problems in the sequence; also assume the results of Probs. 5-10 through 5-12.

7-18. Consider the random sinusoid (cf. Example 7-1-1) given by $X(\xi, t) = A(\xi) \cos [\omega t + \theta(\xi)]$, where ω is a constant, $A(\cdot)$ and $\theta(\cdot)$ are independent random variables, and $\theta(\cdot)$ is distributed uniformly on the interval $[0, 2\pi]$. Use the results of Prob. 5-11.

(a) Show that the mean-value function is identically zero.

(b) Determine the autocorrelation function for the process, and show that $\varphi_{XX}(t, t + \tau)$ does not depend upon t.

[*Suggestion:* Use the formula $\cos x \cos y = \frac{1}{2} \cos (x + y) + \frac{1}{2} \cos (x - y)$.]

7-19. Repeat Prob. 7-18 for the case in which $\omega(\cdot)$ is a random variable such that $\{A(\cdot), \omega(\cdot), \theta(\cdot)\}$ is an independent class and $\theta(\cdot)$ is uniformly distributed on the interval $[0, 2\pi]$. (*Suggestion:* Use the results of Prob. 5-12.)

7-20. Suppose X is a random process (real or complex) such that, with probability 1, each sample function $X(\xi, \cdot)$ is periodic with period T [i.e., for each ξ, except possibly for a set of probability 0, $X(\xi, t) = X(\xi, t + T)$ for all t]. Show that

$$\varphi_{XX}(t, t + \tau) = \varphi_{XX}(t, t + \tau + T) = \varphi_{XX}(t + T, t + T + \tau)$$

[*Suggestion:* Use the general definition of mathematical expectation and property (I5) for the abstract integral.]

7-21. Show that if X is a second-order stationary random process and $g(\cdot, \cdot)$ is a Borel function, then $E\{g[X(\cdot, t), X(\cdot, t + \tau)]\}$ is a function of τ but does not depend upon t. (*Suggestion:* Use the Stieltjes integral form of the expectation.)

7-22. Consider the real-valued random process Y defined by

$$Y(\xi, t) = A(\xi) \cos [\omega t + \theta(\xi) + X(\xi, t)]$$

where (1) X is a second-order stationary process and (2) $A(\cdot)$, $\theta(\cdot)$, and X are independent in the sense that, for any Borel function $g(\cdot, \cdot)$, the class $\{A(\cdot), \theta(\cdot), g[X(\cdot, t), X(\cdot, t + \tau)]\}$ is an independent class for each t and τ. The parameter ω is a constant. $\theta(\cdot)$ is uniformly distributed $[0, 2\pi]$.

(*a*) Show that the mean-value function for Y is identically zero.

(*b*) Show that $\varphi_{YY}(t, t + \tau) = \frac{1}{2}E[A^2]E\{\cos [\omega\tau + X(\cdot, t + \tau) - X(\cdot, t)]\} = \varphi_{YY}(\tau)$. (*Suggestion:* Use the results of Probs. 5-12 and 7-21.)

7-23. Let $\{X_i : 1 \leq i \leq n\}$ be a class of second-order stationary random processes. Suppose that, for any pair of these processes and any choice of the pair of parameters s, t, the random variables $X_i(\cdot, s)$ and $X_j(\cdot, t)$ are independent. Let the mean-value function for X_i be μ_i, a constant, and the autocorrelation function for X_i be $\varphi_{ii}(\cdot)$. Derive the mean-value function and the autocorrelation function for

the process $Y = \sum_{i=1}^{n} X_i$, and show that $\mu_Y(t) = \mu_Y$ and $\varphi_{YY}(t, t + \tau) = \varphi_{YY}(\tau)$.

7-24. Suppose X is the random process defined by

$$X(\xi, t) = \sum_{k=1}^{n} A_k(\xi) \cos [k\omega t + \theta_k(\xi)]$$

where ω is a constant and the class $\{A_k(\cdot), \theta_j(\cdot) : 1 \leq k \leq n, 1 \leq j \leq n\}$ is an independent class of random variables, with the further property that each $\theta_j(\cdot)$ is uniformly distributed on the interval $[0, 2\pi]$. Determine the mean-value function and the autocorrelation function, and show that $\mu_X(t) = \mu_X$ and $\varphi_{XX}(t, t + \tau) = \varphi_{XX}(\tau)$. Show that the autocorrelation function is periodic, with period $2\pi/\omega$. (*Suggestion:* Use the results of Prob. 7-23.)

7-25. Repeat Prob. 7-24 in the case that ω is also a random variable, with the condition that the class $\{A_k(\cdot), \omega(\cdot), \theta_j(\cdot) : 1 \leq k \leq n, 1 \leq j \leq n\}$ is an independent class. In this case the autocorrelation function may not be periodic. Note, also, that the results of Prob. 7-23 do not apply.

7-26. Suppose W is a real-valued, second-order stationary random process; suppose $\theta(\cdot)$ and $\beta(\cdot)$ are uniformly distributed on $[0, 2\pi]$; and suppose that, for each Borel function $g(\cdot, \cdot)$ and each pair of real numbers s, t, the class $\{g[W(\cdot, s), W(\cdot, t)],$

$\theta(\cdot), \beta(\cdot)\}$ is an independent class. Let ω and a be constants. Consider the processes

$$X(\xi, t) = W(\xi, t) \cos [\omega t + \theta(\xi)]$$
$$Y(\xi, t) = W(\xi, t) \cos [(\omega + a)t + \theta(\xi)]$$
$$Z(\xi, t) = W(\xi, t) \cos [(\omega + a)t + \theta(\xi) + \beta(\xi)]$$

(a) Show that X, Y, and Z are second-order stationary random processes with respect to mean value and autocorrelation functions.

(b) Show that $\varphi_{XY}(t, t + \tau)$ depends upon t, but that $\varphi_{XZ}(t, t + \tau)$ does not.

(c) Use the results of part b to show that the process $X + Z$ is second-order stationary with respect to mean value and autocorrelation functions but that the process $X + Y$ is not.

7-27. Suppose the real-valued process X is defined by $X(\xi, t) = g[t - \theta(\xi)]$, with $g(\cdot)$ a periodic Borel function, with period T, and $\theta(\cdot)$ a random variable distributed uniformly on the interval $[0, T]$. We suppose further that

$$\int_0^T g^2(t) \, dt < \infty$$

(a) Show that the process X is second-order stationary with respect to mean value and autocorrelation functions.

(b) Show that, for every ξ, the time correlation function must be the same as the probability correlation function, i.e., that

$$A_t[X(\xi, t)X(\xi, t + \tau)] = \varphi_{XX}(\tau) \qquad \text{for any } \xi$$

$$\left[\textit{Suggestion:} \text{ Note that if } f(t) = f(t + T) \text{ for all } t, \text{ we must have } \int_0^T f(t) \, dt = \int_a^{a+T} f(t) \, dt \text{ for every } a \text{ and } A_t[f] = \frac{1}{T} \int_a^{a+T} f(t) \, dt. \right]$$

7-28. A random process X has sample functions which are step functions described as follows. For any given ξ the sample function has value $A_n(\xi)$ for $\alpha(\xi) + nT \leq t < \alpha(\xi) + (n + 1)T$, where (1) $\{A_n(\cdot), \alpha(\cdot): -\infty < n < \infty\}$ is an independent class, (2) all $A_n(\cdot)$ have the same distribution with $\mu[A_n] = \mu$ and $\sigma^2[A_n] = \sigma^2$, and (3) $\alpha(\cdot)$ is uniformly distributed on the interval $(0, T)$. Obtain the autocorrelation function. (*Suggestion:* Show that the appropriate independence conditions are met so that the results of Sec. 7-5 and of Example 7-6-2 may be used.)
ANSWER: $\sigma^2 \cdot (1 - |\tau|/T) + \mu^2$ for $|\tau| \leq T$, and μ^2 for $|\tau| > T$

Selected references

General references

DOOB [1953]: "Stochastic Processes." A treatise written for the mature mathematician. This, however, is one of the standard works on random processes and is an indispensable source for those prepared to read it.

GNEDENKO [1962]: cited at the end of our Chap. 3.

LOÈVE [1963]: cited at the end of our Chap. 4.

PARZEN [1962]: "Stochastic Processes." A readable treatment of the subject, which demands less mathematical preparation than the works of Loève or Doob, yet presents a generally sound development. Considerable attention to applications.

ROSENBLATT [1962]: "Random Processes." An exposition of the fundamental ideas of random processes, which is mathematically sound but not so complete or advanced as the work of Loève or of Doob. A good companion to the work of Parzen [1962].

Markov chains

BHARUCHA-REID [1960]: "Elements of the Theory of Markov Processes and Their Applications." A treatise which considers much more general Markov processes than the simple case of constant chains. Develops fundamental theory and gives considerable attention to applications in a variety of fields.

FELLER [1957], chaps. 15, 16. Cited at the end of our Chap. 1. Has an excellent treatment of the fundamental theory of constant chains, with some attention to applications.

HOWARD [1960]: "Dynamic Programming and Markov Processes." Uses the z transform (a type of generating function) to deal with finite chains. Provides some interesting applications to economic problems.

KEMENY, MIRKIL, SNELL, AND THOMPSON [1959]: "Finite Mathematical Structures." A lucid exposition of a wide range of topics coming under the general heading of "finite mathematics." Chapter 6 has a readable treatment of finite, constant chains.

KEMENY AND SNELL [1960]: "Finite Markov Chains." A detailed treatment of the finite chains, with considerable attention to applications.

Poisson processes

GNEDENKO [1962], sec. 51. Provides a derivation similar to that given in our Sec. 7-4. Cited at the end of our Chap. 3.

PARZEN [1962], chap. 4. Provides an extensive discussion of the Poisson and related processes, with axiomatic derivations. Cited above.

Correlation functions, covariance functions

DOOB [1953]: cited above.

LANING AND BATTIN [1956]: "Random Processes in Automatic Control." One of several books dealing with the application of the theory of random processes to problems in filtering and control. Although much of the mathematics is developed heuristically, the work presents major ideas in an important field of application.

LEE [1960]: "Statistical Theory of Communication," chap. 2. An extensive text by one of the pioneer teachers in the field. Chapter 2 gives a very detailed discussion of time correlation functions and their role in signal analysis.

LOÈVE [1963], chap. 10, on second-order properties. Contains a detailed discussion of the covariance function and develops the mean-square calculus. The work is written for graduate-level mathematical courses. Cited at the end of our Chap. 4.

MIDDLETON [1960]: "An Introduction to Statistical Communication Theory." An extensive treatise on the application of random processes to communication theory.

YAGLOM [1962]: "An Introduction to the Theory of Stationary Random Functions." A revision and translation of a well-known work in Russian. Provides a readable treatment of topics in random processes important in statistical communication theory, with more attention to mathematical foundation than is customary in engineering literature.

Gaussian processes

CRAMÉR [1946]: chap. 24.   Gives a careful development of the joint gaussian distribution.   This famous work is written for the mathematician but is widely consulted by others.   Cited at the end of our Chap. 5.

DAVENPORT AND ROOT [1958]: "Introduction to Random Signals and Noise," chaps. 7, 8.   Discusses physically and mathematically some common types of noise processes and gives a good account of the central features of the gaussian process important for applications.

DOOB [1953], chap. 2, sec. 3.   Provides a discussion of wide-sense and strict-sense stationary properties in gaussian processes.   Cited above.

PARZEN [1962].   "Stochastic Processes," chap. 3.   Gives a readable discussion of the topic.   Cited above.

APPENDIX A. SOME ELEMENTS OF COMBINATORIAL ANALYSIS

The fundamental task in the classical theory of probability is to count the number of ways that an event can occur. The discipline of counting (which can become quite sophisticated) is commonly known as *combinatorial analysis*. We shall discuss only the most elementary principles and results in this theory.

A fundamental principle of counting

A basic strategy of counting is the following. Break a selection process S into a finite sequence of selections $S_1, S_2, \ldots, S_N$ such that the number of ways of making selection S_k does not depend upon the results of the previous selections in the sequence. Then the number of ways of realizing S is the product of the numbers of ways of realizing the various selections in the sequence.

A simple set representation of this situation can be made in terms of the cartesian product of sets. Suppose we let S be a set of points—one point for each way of carrying out the selection process. Similarly, we let S_k be a set of points—one point for each way of carrying out the kth step. The selection of a point in S is equivalent to a sequence of selections of points in the spaces S_k, $k = 1, 2, \ldots, N$. Thus we may think of S as the cartesian product space $S_1 \times S_2 \times \cdots \times S_N$. Determination of a sequence of components is tantamount to selection of a point in S. If we let $n(S)$ represent the number of points in space S and $n(S_k)$ be the number of points in space S_k, we have the fundamental relation for product spaces

$$n(S) = n(S_1)n(S_2) \cdots n(S_N) = \prod_{k=1}^{n} n(S_k)$$

Permutations and combinations

In selecting (or displaying) a finite set of distinct objects chosen from a given set, we may take two points of view:

1. We may be interested only in the membership of the set selected, without regard to the order in which the objects are selected or displayed. Each distinguishable set in this case is called a *combination*. Two sets having the same number of objects represent different combinations iffi (if and only if) there is at least one element in one of the sets which is different from every element in the other set.

2. We may be interested in the *order* in which the elements are arranged, as well as the combination which is ordered. Each ordered arrangement of a given combination is called a *permutation*. Two permutations of the same combination differ iffi they differ in the element found in at least one place in the ordering scheme. Two sets represent the same combination iffi for any one permutation of the first set there is a permutation of the second set which does not differ from the given permutation of the first set.

Some basic formulas

1. The number $P_r{}^n$ of *permutations* of r objects (i.e., the number of ordered r-tuples) selected from a set of n objects is given by

$$P_r{}^n = n(n - 1)(n - 2) \cdots (n - r + 1) = \frac{n!}{(n - r)!}$$

2. The number $C_r{}^n$ of *combinations* of r objects selected from a set of n objects is given by

$$C_r{}^n = \frac{n!}{r!(n-r)!} = P_r{}^n/r!$$

3. Let $\begin{pmatrix} n \\ n_1, n_2, \ldots, n_k \end{pmatrix}$ be the number of ways that n objects may be placed in k cells, with n_1 objects in the first, n_2 objects in the second, etc. $(n_1 + n_2 + \cdots + n_k = n)$. The order of placing the objects in any cell is immaterial. Then we have

$$\begin{pmatrix} n \\ n_1, n_2, \ldots, n_k \end{pmatrix} = \frac{n!}{n_1! n_2! \ldots n_k!}$$

Derivation of the basic formulas

The formulas above may be derived by using the basic counting strategy.

(1) (a) Select from the n objects one object for the first position (there are n possibilities).
   (b) Select from the $n - 1$ remaining objects one object for the second position (there are $n - 1$ possibilities, regardless of the result of the previous step).
   (c) Continue the process until finally one selects from the $n - r + 1$ remaining objects one object for the rth place (there are $n - r + 1$ possibilities for this step, regardless of the results of the previous steps).

Then, by the fundamental counting rule,

$$P_r{}^n = n(n - 1)(n - 2) \cdots (n - r + 2)(n - r + 1)$$

(2) Obtain a permutation in two steps:
   (a) Select a combination of r objects ($C_r{}^n$ possibilities).
   (b) Select a permutation of the r objects (there are $P_r{}^r = r!$ possibilities for each combination chosen).

By the fundamental counting rule

$$P_r{}^n = C_r{}^n r! \qquad \text{or} \qquad C_r{}^n = \frac{P_r{}^n}{r!}$$

(3) Obtain a permutation of the entire n objects in the following steps:
   (a) Select a combination of the desired type; there are $\begin{pmatrix} n \\ n_1, n_2, \ldots, n_k \end{pmatrix}$ possibilities.
   (b) Select a permutation of the n_1 objects in the first cell (there are $n_1!$ possibilities).
   (c) In a similar manner, for each of the k cells, obtain a permutation of the objects in that cell (there are $n_i!$ possibilities in the ith cell).

By the fundamental counting rule, the number of permutations of the entire set of n objects must be given by

$$P_n{}^n = n! = \begin{pmatrix} n \\ n_1, n_2, \ldots, n_k \end{pmatrix} n_1! n_2! \cdots n_k!$$

so that the desired result follows immediately.

The binomial coefficient

By an inductive proof it may be shown that

$$(p + q)^n = \sum_{k=0}^{n} C_k^n p^k q^{n-k}$$

so that the number C_k^n appears as a coefficient in the *binomial expansion*, and is hence called a *binomial coefficient*.   Since this coefficient appears naturally in a wide variety of applications, we note some properties.

If we put $p = q = \frac{1}{2}$, we obtain

$$\sum_{k=0}^{n} C_k^n = 2^n$$

In terms of combinations, this says that the number of combinations of n or fewer elements (including zero) from a set of n elements is 2^n.   This is the well-known fact that a space with n elements has 2^n subsets (including the whole space and the empty set).

It is sometimes convenient to extend the range of r in the expression C_r^n by putting

$$C_r^n = 0 \quad \text{for } r > n \text{ or } r < 0$$

By examining the basic formulas, the following relations may be derived easily:

$$C_r^n = C_{n-r}^n \qquad C_r^n = C_{r-1}^{n-1} + C_r^{n-1}$$

$$C_{r+1}^n = \frac{n - r}{r + 1} \cdot C_r^n \qquad \text{for } r \neq -1$$

$$C_{r+1}^n > C_r^n \qquad \text{for } 0 \leq r < \frac{n + 1}{2}$$

$$C_{r+1}^n < C_r^n \qquad \text{for } n \geq r > \frac{n + 1}{2}$$

A wide variety of other formulas may be developed.

Treatments of combinatorial analysis are found in many books on probability. For a simple but good introductory treatment, one may consult Goldberg [1960]; for a much more extensive treatment, see Feller [1957, chaps. 2 through 4].

APPENDIX B. SOME TOPICS IN SET THEORY

Sets and their representation

In any investigation involving set theory, one usually postulates some *basic set*, or *universal set*, which consists of all elements to be considered for that problem. This may be the set R of all real numbers, or it may be an abstract set of elements (finite or infinite in number). As a general model, we consider

A *basic set* S of *elements* ξ:
Set membership is indicated by $\xi \in S$ (this is read, "ξ belongs to S," "ξ is a member of S," or "ξ is in S").
Various *subsets* of the basic space are represented by:
Capital letters, usually near the beginning of the alphabet (A, B, C, etc.).
Bracket notation with a list of elements in the set; e.g., $\{\xi_1, \xi_2, \xi_3, \xi_4\}$ indicates the set whose members are the elements listed.
Bracket notation with a defining proposition: if $\pi(\cdot)$ is a proposition which for any ξ is either true or false, the set $\{\xi: \pi(\xi)\}$ is the set of ξ such that $\pi(\xi)$ is a true statement.
The *empty set* $\emptyset$ is the set which has no elements.

It is necessary to make a clear distinction between an element and a set of elements. In particular, one must distinguish between the single element ξ and the set $\{\xi\}$, which has as member only the single element ξ. In developing the basic probability model in Chap. 2, we let the element ξ represent one of the possible outcomes of a trial or experiment; an event is then a set of such elements. The reader should be warned, however, of an unfortunate anomaly in terminology found in much of the literature. In his fundamental work, Kolmogorov [1933, 1956] used the term elementary event for the elementary outcome ξ; he uses no specific term for the event $\{\xi\}$. Although he does not confuse logically the elementary outcome ξ with the event $\{\xi\}$, his terminology is inconsistent at this point. We shall attempt to be consistent in our usage. A little care will prevent confusion in reading the literature which follows Kolmogorov's usage.

Basic set operations

From given sets, new sets may be obtained by several basic operations.

The *union* of two or more sets. $A \cup B$ indicates the set formed by taking all elements that are in A *or* in B (or both). The idea of the union of sets may be extended to several sets in a straightforward manner.

The *intersection* of two or more sets. $A \cap B$, or AB, indicates the set formed by taking all elements that are in both A *and* B. The idea of intersection may be extended to several sets.

The *complement* of a set. A^c indicates the set of those elements of S which are *not* in A.

In addition to the three operations taken as basic, we add two others which are commonly employed.

The *difference* of two sets $A - B$ is the set formed by taking all those elements in A which are not in B. Thus $A - B = AB^c$.

The *disjunctive union* or *symmetric difference* $A \oplus B$ of two sets is the set of all elements which are in A but not in B or in B but not in A.

We have $A \oplus B = AB^c \cup A^c B = (A - B) \cup (B - A)$. Properties of this combination are most readily studied with the aid of the indicator function for sets, introduced in Sec. 2-7.

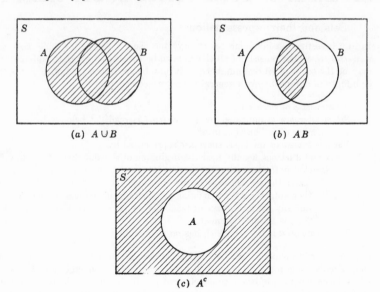

(a) $A \cup B$

(b) AB

(c) A^c

Fig. B-1   Venn diagrams illustrating basic set relations of union, intersection, and complement.

It is often helpful to make a geometrical representation of sets and their combinations by use of the *Venn diagram* (also called the Euler diagram, or the Euler-Venn diagram).   The basic set S is represented as a region on the plane (or as a discrete set of points located on a given region of the plane).   Subsets of the basic space are represented by subregions (or the discrete sets of points located in the subregions).   Figure B-1 shows several Venn diagrams.   The shaded area in Fig. B-1a represents $A \cup B$; that in Fig. B-1b represents AB; the shaded area in Fig. B-1c indicates A^c.   Venn diagrams are used in figures throughout the text.   A special form of the Venn diagram referred to as a *minterm map* is developed in Sec. 2-7 (see Figs. 2-7-2 and 2-7-3 for examples).

Basic set relationships

Three types of relationship between sets are useful in developing a theory of sets and set operations.

1. *Inclusion.*   A set A is included in (or contained in) set B (written $A \subset B$ or $B \supset A$) iffi each element of set A is also an element of set B.   It should be noted that the relation $\subset$ is always a relation between sets.   It is never correct (or even meaningful) to write $\xi_1 \subset \xi_2$ (where ξ_1 and ξ_2 are elements) or $\xi \subset A$.   In the last case we have $\{\xi\} \subset A$ iffi $\xi \in A$.

2. *Equality.*   Sets A and B are equal iffi both $A \subset B$ and $B \subset A$.   In this case we write $A = B$.

3. *Disjointedness.*   Two sets A and B are disjoint iffi they have no elements in common; this is equivalent to the condition that the intersection is empty ($AB = \emptyset$). In the case of the union of disjoint sets, it is convenient to have a special symbol. In this book we use the symbol $A \uplus B$ to indicate the union of two sets, with the further stipulation that the sets are disjoint.

Again, we may use the Venn diagram to give a simple geometrical representation of the basic relationships. In Fig. B-2, we have $A \subset B$, $AB = A$, $AC = \emptyset$, and $BC = \emptyset$.

The characterization of sets in terms of propositions implies the following important set relations:

$$\{\xi: \pi_1(\xi) \ or \ \pi_2(\xi)\} = \{\xi: \pi_1(\xi)\} \cup \{\xi: \pi_2(\xi)\}$$
$$\{\xi: \pi_1(\xi) \ and \ \pi_2(\xi)\} = \{\xi: \pi_1(\xi)\} \cap \{\xi: \pi_2(\xi)\}$$
$$\{\xi: \pi_1(\xi) \ not \ true\} = \{\xi: \pi_1(\xi)\}^c$$

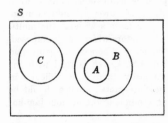

Fig. B-2   Venn diagram illustrating relations of inclusion and disjointedness.

$A \subset B \quad AB = A \quad AC = \emptyset \quad BC = \emptyset$

To say that proposition $\pi_1(\cdot)$ implies proposition $\pi_2(\cdot)$ means that if $\pi_1(\xi)$ is true for any given ξ, the proposition $\pi_2(\xi)$ is also true. From this it is apparent that

$$\pi_1(\cdot) \text{ implies } \pi_2(\cdot) \text{ iffi } \{\xi: \pi_1(\xi)\} \subset \{\xi: \pi_2(\xi)\}$$

A clear grasp of these connections between set relations and logical relations among propositions is an important aid to understanding many relations among events (as sets of elementary outcomes).

Algebra of set operations

A series of simple theorems may be proved which make it possible to construct an algebra of set operations. A basic list is given below. Unless special sets S or $\emptyset$ are designated or unless there are restrictive conditions, the relations are set identities, in the sense that they hold for any subsets of the basic set S.

(1) $A \cup B = B \cup A$ $AB = BA$

(2) $A \cup (B \cup C) = (A \cup B) \cup C$ $A(BC) = (AB)C$

(3) $A \cup BC = (A \cup B)(A \cup C)$ $A(B \cup C) = AB \cup AC$

(4) $A \cup \emptyset = A$ $AS = A$

(5) $A \cup A^c = S$ $AA^c = \emptyset$

(6) $(A^c)^c = A$

(7) $A = B$ iffi $A^c = B^c$

(8) $(A \cup B)^c = A^c B^c$ $(AB)^c = A^c \cup B^c$

(9) $A \subset B$ iffi $AB = A$

(10) $A \subset B$ iffi $A^c \cup B = S$ $A \supset B$ iffi $A^c B = \emptyset$

(11) $A \subset B$ iffi $B^c \subset A^c$

(12) $\displaystyle\bigcup_{k=1}^{\infty} A_k = \biguplus_{k=1}^{\infty} B_k$ where $B_k = A_k \displaystyle\bigcap_{j=1}^{k-1} A_j{}^c$

(Note that use of the symbol $\uplus$ indicates the assertion "the B_k are disjoint.")

Most of these rules can be given simple geometrical interpretations which are an aid to visualizing and remembering the rules. On the other hand, utilization of these rules makes it possible to modify set expressions by "algebraic" manipulation of the symbols according to these and other rules derived therefrom.

One peculiarity of notational usage in utilizing the symbolic rules of algebra should be pointed out. We use $A \cap B$ or AB alternatively to indicate set intersection, just as we use $a \cdot b$ or ab alternatively to indicate multiplication of numbers. In writing sums of products, some problems of grouping may arise. The expression $a \cdot b \cdot c + d \cdot e$ is ambiguous. It could mean $(a \cdot b \cdot c) + (d \cdot e)$. It could just as well mean $a \cdot (b \cdot c + d \cdot e)$ or one of the other possible expressions derived by a different grouping of characters. When we write $abc + de$, however, common usage reads this as $(abc) + (de)$. Adjacent placing of several factors without the multiplication sign is intended to indicate a grouping of the factors which are juxtaposed. In a similar manner we shall deal with set intersection. Whereas the expression $A \cap B \cap C \cup D \cap E$ is ambiguous, the expression $ABC \cup DE$ is read with the following grouping: $(A \cap B \cap C) \cup (D \cap E)$. Ordinarily, there is no difficulty in utilizing this scheme. Where there is danger of misinterpretation, use should be made of parentheses, brackets, or braces as grouping symbols in the manner familiar from ordinary algebra.

Cartesian product of sets

Consider two basic sets S and U with $\xi \in S$ and $\eta \in U$. A new type of set may be constructed by considering all ordered pairs (ξ, η), where $\xi \in S$ and $\eta \in U$. Each such pair is considered an element in the *(cartesian) product set* $S \times U$. The fundamental model for this construction is the plane, considered as the cartesian product of two real lines. Real numbers are represented as points on the real lines (coordinate axes). Pairs of real numbers may then be considered as points on the plane. If R_1 represents one coordinate axis (i.e., one coordinate of a point on the plane is a number represented by a point on this line) and R_2 represents the other coordinate axis, the plane then becomes the cartesian product space $R_1 \times R_2$. In keeping with this geometrical model, the sets S and U in the abstract case are called *coordinate sets* (or spaces). The concept of a cartesian product may be extended in a natural way to the product of several coordinate sets. If S, U, and V are such sets, one may make several constructions. One may take the cartesian product $S \times U$ to form a new space. Then this new space may be combined with V to form $(S \times U) \times V$. Or U and V could be combined to form $U \times V$, and then a second combination $S \times (U \times V)$ could be made. Or one could simply take $S \times U \times V$, in which case each element is a triple consisting of one element each from S, U, and V in that order. For most purposes, it is convenient and useful to identify the three spaces in a natural way. If $\xi \in S$, $\eta \in U$, and $\zeta \in V$, then points from the three product spaces are $[(\xi, \eta), \zeta]$, $[\xi, (\eta, \zeta)]$, and (ξ, η, ζ), respectively. It is apparent now that these can be considered identical.

A wide variety of subsets can be described by conditions on the coordinates of the points in a cartesian product space. One of the simplest is the *rectangle set*. Suppose the two basic spaces are S_1 and S_2 and that subsets in these sets are indicated by corresponding subscripts. The set $A_1 \times A_2$ is the set of all pairs (ξ, η) such that $\xi \in A_1$ and $\eta \in A_2$. The sets A_1 and A_2 may be quite complicated subsets of the coordinate spaces. These are not the most general subsets of the product space, however. For example, the union of two or more rectangle sets is not necessarily a rectangle set. Examples of rectangle sets on the plane are found in Figs. 3-5-2, 3-7-1, and 3-7-2.

A number of rules of combination for rectangle sets are easily proved and are often useful in dealing with these sets. Some of the more useful are

$$(A_1 \cup B_1) \times A_2 = (A_1 \times A_2) \cup (B_1 \times A_2)$$
$$(A_1 \cap B_1) \times A_2 = (A_1 \times A_2) \cap (B_1 \times A_2)$$
$$(A_1 \cup B_1) \times (A_2 \cup B_2) = (A_1 \times A_2) \cup (A_1 \times B_2) \cup (B_1 \times A_2) \cup (B_1 \times B_2)$$
$$(A_1 \cap B_1) \times (A_2 \cap B_2) = (A_1 \times A_2) \cap (B_1 \times B_2)$$
$$A_1 \times A_2 = (A_1 \times S_2) \cap (S_1 \times A_2)$$
$$(A_1 \times A_2)^c = (A_1{}^c \times S_2) \cup (S_1 \times A_2{}^c)$$

These may be interpreted in terms of sets on the plane. The coordinate sets A_1, B_1 are sets on one of the coordinate axes, and the coordinate sets A_2, B_2 are sets on the other coordinate axis. The validity may be argued from definitions, without regard to geometrical interpretations, however.

Classes of sets

The basic concept of a set is that of a collection of elements. It is desirable to deal also with collections of sets, that is, sets of sets. In order to prevent verbal confusion, it is customary to use the term *class* of sets (or perhaps *family* of sets) to indicate such a collection. If A, B, C are members of a given class α, we say $A \in \alpha$, $B \in \alpha$, and $C \in \alpha$. The sets A and B may stand in the relationship $A \subset B$. We do not say $A \in B$ or $A \subset C$. It is, however, correct to say $\alpha \subset \mathcal{B}$, when $\mathcal{B}$ is a class such that every set in α is also in $\mathcal{B}$.

The term *countable class* may refer to either a finite class or a countably infinite class. A class is countably infinite iffi its members can be put into a one-one correspondence with the positive integers (natural numbers). A class is countable iffi its index set J is countable; it is finite iffi its index set J is finite.

A convenient notation for classes of sets uses subscripts to identify various sets in the class and a bracket notation as follows:

$$\alpha = \{A_i : i \in J\} \qquad \text{where } J \text{ is a suitable } \textit{index set}$$

This is read, "α is the class of all those sets A_i such that i is a member of the index set J." Subclasses of the set α may be indicated by the choice of suitable subsets of the index set J. Thus

$$\alpha_0 = \{A_i : i \in J_0\} \subset \alpha \qquad \text{iffi } J_0 \subset J$$

The first statement means that every set in class α_0 is also in class α; the second statement means that every index i in J_0 is also in J.

The concepts of union and intersection may be extended to general classes, as follows:

$\bigcup\limits_{i \in J} A_i$ is the set of all elements in at least one of the sets A_i in the class $\{A_i : i \in J\}$.

$\bigcap\limits_{i \in J} A_i$ is the set of all elements in every set of the class $\{A_i : i \in J\}$.

It is apparent that if the class contains only two sets, these definitions coincide with those for the union and intersection of two sets.

The rules on complements under (8) in the list of set operations above may be extended to general classes as follows:

$$[\bigcup_{i \in J} A_i]^c = \bigcap_{i \in J} A_i^{\,c} \qquad [\bigcap_{i \in J} A_i]^c = \bigcup_{i \in J} A_i^{\,c}$$

An alternative notation is frequently useful. If $\alpha = \{A_i : i \in J\}$ is any class of sets, we may denonte $\bigcup_{i \in J} A_i$ by $\bigcup_{\alpha} A$ and $\bigcap_{i \in J} A_i$ by $\bigcap_{\alpha} A$. Other variations may be used. These are generally described when used or are clear from the context.

These basic rules are known as *DeMorgan's rules.*

A class is said to be a *disjoint class* if no two sets in the class have any point in common. This implies that the intersection of every subclass of sets must be empty. In the case of disjoint classes, we may use a special symbol for the union, as follows:

$$\underset{i \in J}{\biguplus} A_i \text{ implies the stipulation that the class is disjoint}$$

It is not incorrect to use the ordinary symbol for the union of a disjoint class. The use of the special symbol, however, indicates the union plus the assertion that the class is disjoint. An example of this usage is given (somewhat prematurely) in proposition (12) among the theorems on Algebra of Set Operations.

Sequences of sets

Sequences of sets are countable classes which are indexed by the integers or by some subset of the integers such as the positive integers or by the integers $\{1, 2, \ldots, n\}$ for some finite n. The members of a sequence can be put in natural order, according to the ordering of the integers in the index set. We speak of infinite sequences and finite sequences of sets in an obvious way. It is often convenient to designate sequences by one of the following notational schemes, according to the situation:

$A_1, A_2, A_3, \ldots$
$\ldots, A_{-2}, A_{-1}, A_0, A_1, A_2, \ldots$
$\{A_i : -\infty < i < \infty\}$
$\{A_i : 0 \leq i < \infty\}$

. . . .

In the case of a sequence, the concept of a limit can be introduced in a natural and useful manner. Suppose $\alpha = \{A_i : 1 \leq i < \infty\}$. The *superior limit* of the sequence, written lim sup A_i, is the set of all those elements ξ which belong to A_i for an infinite number of the i in the index set. The *inferior limit*, written lim inf A_i, is the set of all those elements ξ which belong to all but a finite number of the A_i. It should be apparent that any ξ in $A_* = $ lim inf A_i must also be in $A^* = $ lim sup A_i: that is, $A_* \subset A^*$.

If ξ belongs to an infinity of the A_i, for every n it must belong to the union of all those A_i for which $i \geq n$. If it belongs to such a union for every n, it must clearly belong to an infinity of the A_i. Hence

$$A^* = \lim \sup A_i = \bigcap_{n=1}^{\infty} \bigcup_{i=n}^{\infty} A_i$$

If ξ belongs to all but a finite number of the A_i, there must be some n such that ξ belongs to A_i for all $i \geq n$; that is, there must be some n such that ξ belongs to the intersection of all the A_i for which $i \geq n$. On the other hand, if ξ belongs to such an intersection, it must belong to all but a finite number of the A_i. Thus

$$A_* = \lim \inf A_i = \bigcup_{n=1}^{\infty} \bigcap_{i=n}^{\infty} A_i$$

If the superior limit and the inferior limit are the same set, the common set is called the *limit* of the sequence, written $\lim A_i$. If the sequence is *monotone*, that is, either *monotone-increasing* ($A_i \subset A_{i+1}$, all i) or *monotone-decreasing* ($A_i \supset A_{i+1}$, all i), the limit always exists and is given by the following formulas:

Monotone-increasing: $\quad \lim A_i = \bigcup_{i=1}^{\infty} A_i$

Monotone-decreasing: $\quad \lim A_i = \bigcap_{i=1}^{\infty} A_i$

These formulas can be derived from the formulas for $\lim \sup A_i$ and $\lim \inf A_i$ by the use of the special conditions, with their implications for the unions and intersections.

Fields and sigma fields

Various operations on sets (notably the formation of unions, of intersections, and of complements) have been introduced. These operations produce resultant sets which are usually different sets, although in some cases they may be the same set. In the discussion of events as sets in Chap. 2, it is shown that the union or intersection of countable classes of events or the complement of an event must be an event for the model to be satisfactory. This means that the class $\mathcal{E}$ of events, as a class of sets, must have the property that the formation of unions or intersections of countable subclasses and the formation of complements of individual sets must produce resultant sets within the class $\mathcal{E}$ of events.

Definition.
A class $\mathcal{A}$ of subsets of the basic space S is said to be *closed under an operation* on sets iff the application of this operation to appropriate sets or subclasses of sets always yields a set in $\mathcal{A}$.

Thus, the class $\mathcal{E}$ of events must be closed under the operations of forming countable unions, countable intersections, and complements (three distinct operations). Various types of classes may be defined by specifying closure under certain collections of operations.

We consider two such types of class. First we make the

Definition.
A nonempty class $\mathcal{A}$ of sets is called a *field* of sets iff it is closed under i) finite unions and ii) complements.

This means that if $A_1, A_2, \ldots, A_n$ are all members of $\mathcal{A}$ so is the set $A = \bigcup_{i=1}^{n} A_i$. Also, $A_i{}^c$ is a member of $\mathcal{A}$ for any i. Since $\bigcap_{i=1}^{n} A_i = [\bigcup_{i=1}^{n} A_i{}^c]^c$, it follows that a field is closed under finite intersections. It is clear that the empty set $\emptyset$ and the basic space S always belong to a field. Other names for a field are an *algebra* of sets or an *additive class* of sets.

If we have a whole family of fields of sets, the intersection of these classes will be nonempty and will form a field. If we begin with any nonempty class $\mathcal{C}$ of sets and consider all the fields which include $\mathcal{C}$, their intersection will be the minimal field which contains $\mathcal{C}$.

Definition.
The minimal field $\mathfrak{F}$ which includes a nonempty class $\mathcal{C}$ is called the *field generated by* $\mathcal{C}$.

If $\mathcal{C}$ is itself a field, then it is identical with its generated field.

Fields play an important role in developing basic probability theory. One often begins by assigning probabilities to members of a field, then extending the definition to a more general class $\mathcal{E}$ of events. We consider, next, such classes.

Definition.
A nonempty class $\mathfrak{F}$ of sets is called a *sigma field* iff it is closed under i) countable unions and ii) complements.

Just as in the case of a field, it follows that a sigma field is closed under the formation of countable intersections. These are the properties needed for a class $\mathcal{E}$ of events. In a formal development of probability theory, the class $\mathcal{E}$ of events is taken to be a sigma field of subsets of the basic space S. Again, we have for any nonempty class $\mathcal{C}$ a minimal sigma field of sets which contains $\mathcal{C}$, called the *sigma field generated* by $\mathcal{C}$. Alternate terms found in the literature for sigma fields are *Borel fields* and *completely additive classes*.

Borel sets

The discussion above considers sigma fields on any abstract space. In dealing with real-valued or vector-valued random variables, we are concerned with sets of numbers viewed as points on the real line, or with sets of points on euclidean spaces of various dimensions. The Borel sets on the line appear as the sigma field generated by the intervals on the real line. Actually, any one of several classes of intervals may be taken as the generating class. One may begin with a class composed of all finite intervals of any specified one of the types (a, b), $(a, b]$, $[a, b)$, or $[a, b]$, where a and b are any two real numbers with $a \leq b$. One may also begin with the class of semi-infinite intervals $(-\infty, a]$, or with the class consisting of intervals $(-\infty, a)$.

Suppose one of the classes of intervals is taken as a starting point. The generated class then contains all intervals in any of the other classes, since an interval of any one type can be obtained by a countable number of operations of union, intersection, and complementation upon sets of any one of the other kinds (see Example 3-1-5 for an illustration of this conversion). Simple arguments show that the generated classes for each of the interval types must be identical.

Extensions to sets in euclidean spaces of higher dimension are straightforward. A variety of types of intervals may serve as the basis for the generating class, but they lead to the same generated completely additive class. It is apparent, in the case of Borel sets in the plane, that this class contains all rectangle sets whose coordinate sets are Borel sets on the real line. Also, the class of Borel sets in the plane has as members all sets obtained by performing a countable number of operations of taking unions, intersections, and complements of rectangle sets of the kind just described.

APPENDIX C. MEASURABILITY OF FUNCTIONS

The idea of a function is discussed in Sec. 3-1. There it is shown that the inverse function produces a correspondence between sets in the range T and sets in the domain space S. This correspondence preserves unions, intersections, and complements in the sense discussed in Theorem 3-1A. Then the random variable $X(\cdot)$ is introduced as a function whose domain is the basic space S and whose range is the real line R, or a suitable subset thereof. The random variable cannot be defined as a completely arbitrary function, however. It is necessary to be able to make appropriate probability statements. These may be made if, for each semi-infinite interval $I_t = (-\infty, t]$ on the real line, the inverse image $X^{-1}(I_t)$ is an event (i.e., a subset of the basic space for which probability is defined). This condition implies the more general condition stated in the theorem below. The proof of this theorem is instructive as an example of the role of **sigma fields** of sets.

Theorem 3-1B

The requirement that a random variable $X(\cdot)$ satisfy the condition that $X^{-1}(I_t)$ be an event for each real t implies the condition that $X^{-1}(M)$ be an event for each Borel set M.

PROOF To prove the theorem, we define several classes of sets and show that appropriate relationships hold between them.

Let ε be the class of events (a sigma field of subsets of S)

$\mathcal{J}$ be the class of semi-infinite intervals I_t on the real line R

$\mathcal{B} = \mathcal{a}(\mathcal{J})$ be the class of Borel sets on the real line R, which is the sigma field generated by class $\mathcal{J}$

$\varepsilon_B = X^{-1}(\mathcal{B})$ be the class of inverse images of the Borel sets

$\varepsilon_I = X^{-1}(\mathcal{J})$ be the class of inverse images of the semi-infinite intervals I_t

$\varepsilon_0 = \mathcal{a}(\varepsilon_I)$ be the sigma field generated by the inverse images of the intervals

We prove the theorem by showing $\varepsilon_B = \varepsilon_0 \subset \varepsilon$.

1. Because of the preservation of unions and complements under the inverse transformation (Theorem 3-1A), it follows that ε_B is a sigma field.
2. Since $\varepsilon_I \subset \varepsilon$ by the definition of a random variable, it follows that $\varepsilon_0 = \mathcal{a}(\varepsilon_I)$ must be a subclass of the events ε.
3. Since $\mathcal{J} \subset \mathcal{B}$, it follows that $\varepsilon_I \subset \varepsilon_B$.
4. By the definition for the sigma field generated by a given class, it follows that $\varepsilon_0 \subset \varepsilon_B$.
5. Since $\varepsilon_0 \subset \varepsilon_B$, every set in ε_0 must be of the form $E_M = X^{-1}(M)$, where M is a Borel set (that is, $M \in \mathcal{B}$). Let $\mathcal{B}_0$ be the class of sets on the real line such that $\varepsilon_0 = X^{-1}(\mathcal{B}_0)$. Then $\mathcal{B}_0 \subset \mathcal{B}$.
6. Since ε_0 is a sigma field, it follows easily that $\mathcal{B}_0$ is a sigma field.
7. Since ε_I contains the inverse images of all intervals I_t, it follows that $\mathcal{B}_0$ contains all intervals I_t. Hence $\mathcal{B} \subset \mathcal{B}_0$, so that $\mathcal{B}_0 = \mathcal{B}$. Thus $\varepsilon_0 = \varepsilon_B$, and both are subclasses of ε, as was to be shown. ■

The condition that $X^{-1}(M)$ is an event for each Borel set M is called a *measurability condition*. It is motivated by the desire to be able to assign a measure to each set $X^{-1}(M)$. The domain of a measure function is a sigma field. A measurability condition holds with respect to such a class.

Definition C-1

Let $f(\cdot)$ be a function whose domain is a space S and whose range is the real line (or a subset thereof). Let ε be a sigma field of subsets of the space S. Then the function $f(\cdot)$ is said to be *measurable* (with respect to the class ε) iff $f^{-1}(M) \in \varepsilon$ for each Borel set M.

A random variable $X(\cdot)$ is thus a measurable function with respect to the class of events. Probability measure is defined on this class of events. This makes it possible to assign probabilities to each event of the type $\{\xi : X(\xi) \in M\}$, where M is a Borel set.

A second type of measurable function which plays an important role in the development of probability theory in this book is the so-called *Borel function* (or sometimes *Baire function*). A real-valued Borel function of a single real variable is one whose domain is the real line and whose range is in the real line (i.e., it is a real-valued function of a real variable); the Borel function is measurable with respect to the class of Borel sets. Thus the inverse image of any Borel set in the range space is a Borel set in the domain space. Such functions are considered in Sec. 3-8 (Definition 3-8b). In the case of functions of several real variables, measurability is with respect to the Borel sets on the multidimensional euclidean space, which is the domain space. The class of Borel functions is quite broad and includes most of the functions ordinarily encountered in practice. It certainly includes the continuous functions, since the inverse image of an interval is a union of intervals. This class then includes sectionally continuous functions if reasonable care is used to define the functions at the discontinuities. Many other functions in the class of Borel functions may be described. For most practical purposes, the class is so broad that there is little need to verify that a given function under consideration is in fact a Borel function. It is for this reason that considerations of measurability can be ignored in the usual works directed toward applications.

APPENDIX D.   PROOFS OF SOME THEOREMS

D-1   Classes of events equal with probability 1

Theorem 3-10B

Let α and $\mathfrak{B}$ be finite or countably infinite classes such that $\alpha = \mathfrak{B}$ $[P]$.   Then the following three conditions must hold:

1. $\underset{i\in J_1}{\cup} A_i = \underset{i\in J_1}{\cup} B_i$ $[P]$ for every $J_1 \subset J$
2. $\underset{i\in J_2}{\cap} A_i = \underset{i\in J_2}{\cap} B_i$ $[P]$ for every $J_2 \subset J$
3. $A_i{}^c = B_i{}^c$ $[P]$ for each $i \in J$

PROOF   Condition 3 follows from condition 3 of Theorem 3-10A.
Condition 1 is established by the following calculations:

$$P(\underset{i}{\cup} A_i) = P[\underset{i}{\cup} A_i(B_i \cup B_i{}^c)] = P(\underset{i}{\cup} A_iB_i)$$

since $P(A_iB_i{}^c) = 0$ for every i (Theorem 3-10A).
Interchanging the role of A_i and B_i gives the same expression for $P(\underset{i}{\cup} B_i)$.

Now

$$
\begin{aligned}
P[(\underset{i}{\cup} A_i) \cap (\underset{j}{\cup} B_j)] &= P[(\underset{i}{\cup} A_iB_i) \cup (\underset{j\neq i}{\cup} A_iB_j)] \\
&= P\{\underset{i}{\cup} [A_iB_i \cup (\underset{j\neq i}{\cup} A_iB_j)]\} \\
&= P\{\underset{i}{\cup} [A_iB_i \cup (\underset{j\neq i}{\cup} A_iB_j(A_iB_i)^c)]\} \\
&= P(\underset{i}{\cup} A_iB_i)
\end{aligned}
$$

since $P[A_iB_j(A_iB_i)^c] = P[A_iB_j(A_i{}^c \cup B_i{}^c)] = 0$ for every i, j.
We have thus established the equalities required by the definition.
Condition 2 follows by considering complements, since

$$\underset{i}{\cap} A_i = \underset{i}{\cap} B_i \ [P] \qquad \text{iffi} \qquad \underset{i}{\cup} A_i{}^c = \underset{i}{\cup} B_i{}^c \ [P]$$

The theorem follows from conditions 1 and 3, which are established. ∎

Theorem 3-10C

If a finite or countably infinite class α has the properties

(1) $\underset{i\in J}{\sum} P(A_i) = 1$ and   (2) $P(A_iA_j) = 0$ for $i \neq j$

then there exists a partition $\mathfrak{B}$ such that $\alpha = \mathfrak{B}$ $[P]$.

PROOF   Suppose the events of the class α are numbered to form a sequence $A_1, A_2, \ldots, A_n, \ldots$ First, consider the case in which the union of the A_i is the whole space.   Put

$$B_1 = A_1, B_2 = A_2A_1{}^c, \ldots, B_n = A_nA_1{}^c \cdots A_{n-1}{}^c, \ldots$$

Then it is easy to see that the class $\mathfrak{B} = \{B_i : i \in J\}$ is a partition, for the sets are disjoint and their union is the whole space.   We note, further, that

$$P(B_n) = P(A_nA_1{}^c \cdots A_{n-1}{}^c) = P(A_n) \qquad \text{since } P(A_nA_i) = 0, i \neq n$$

and

$$P(A_n B_n) = P(B_n) \qquad \text{since } B_n \subset A_n$$

Thus the class $\mathfrak{B}$ is a partition which is equal with probability 1 to the class $\mathfrak{a}$. If the union of the class $\mathfrak{a}$ is not the whole space, we simply consider the set A_0, which is the complement of this union and whose probability is zero, and take as B_1 the union of A_1 and A_0. It is evident that in this case $B_1 = A_1$ $[P]$. ∎

D-2   Random variables equal with probability 1

Theorem 3-10D

Consider two simple random variables $X(\cdot)$ and $Y(\cdot)$ with ranges T_1 and T_2, respectively. Let $T = T_1 \cup T_2$, $t_i \in T$. Put $A_i = \{\xi : X(\xi) = t_i\}$ and $B_i = \{\xi : Y(\xi) = t_i\}$. Then $X(\cdot) = Y(\cdot)$ $[P]$ iffi $A_i = B_i$ $[P]$ for each i.

PROOF   We note that if t_i is not in T_1, $A_i = \emptyset$, and if t_i is not in T_2, $B_i = \emptyset$.
1. Given $X(\cdot) = Y(\cdot)$ $[P]$. To show $A_i = B_i$ $[P]$. Put $D = \{\xi : X(\xi) \neq Y(\xi)\}$. $P(D) = 0$.
Let $D_i = A_i B_i{}^c \uplus A_i{}^c B_i$. Now on D_i we must have $X \neq Y$, so that $D_i \subset D$. This implies $P(D_i) = 0$ and hence that $A_i = B_i$ $[P]$. The argument holds for any i.
2. If $A_i = B_i$ $[P]$, then $P(D_i) = 0$. Since $D = \cup_i D_i$, we have $P(D) \leq \sum_i P(D_i)$
$= 0$, so that $X(\cdot) = Y(\cdot)$ $[P]$. ∎

Theorem 3-10E

Consider two random variables $X(\cdot)$ and $Y(\cdot)$. Put $E_M = \{\xi : X(\xi) \in M\} = X^{-1}(M)$ and $F_M = \{\xi : Y(\xi) \in M\} = Y^{-1}(M)$ for any Borel set M. Then, (A) $X(\cdot) = Y(\cdot)$ $[P]$ iffi (B) $E_M = F_M$ $[P]$ for each Borel set M.

PROOF

1. To show (A) implies (B) we consider

$$S_1 = \{\xi : X(\xi) = Y(\xi)\} \qquad \text{and note that} \qquad S_1 = S \; [P]$$

from which it follows that $P(A) = P(AS_1) + P(AS_1{}^c) = P(AS_1)$ for every event A. Moreover, a little reflection shows that $E_M S_1 = F_M S_1 = E_M F_M S_1$ for any Borel set M. We thus obtain the fact that $P(E_M) = P(F_M) = P(E_M F_M)$, which is the defining condition for $E_M = F_M$ $[P]$.
2. To show (B) implies (A), we consider first the case that $X(\cdot)$ and $Y(\cdot)$ are simple functions. Define the sets T_1, T_2, T, A_i, and B_i as in the previous theorem. Then $A_i = X^{-1}(\{t_i\})$ and $B_i = Y^{-1}(\{t_i\})$. By hypothesis, $A_i = B_i$ $[P]$, which, by Theorem 3-10D, above, implies $X(\cdot) = Y(\cdot)$ $[P]$.
In the general case, $X(\cdot)$ is the limit of a sequence $X_n(\cdot)$, and $Y(\cdot)$ is the limit of a sequence $Y_n(\cdot)$, $n = 1, 2, \ldots$. Under hypothesis (B) and the construction of the approximating functions $X_n(\cdot)$ and $Y_n(\cdot)$, $X_n{}^{-1}(\{t_{in}\}) = X^{-1}[M(i, n)]$ and $Y_n{}^{-1}(\{t_{in}\}) = Y^{-1}[M(i, n)]$ are equal with probability 1. Thus $X_n(\cdot) = Y_n(\cdot)$ $[P]$.
If we put $C_n = \{\xi : X_n(\xi) \neq Y_n(\xi)\}$, we must have $P(C_n) = 0$. Let $C = \bigcup_{n=1}^{\infty} C_n$. Then $P(C) = 0$. On the complement C^c we have $X_n(\xi) = Y_n(\xi)$ for all n, so that the two sequences approach the same limit for each ξ in C^c. This means that $X(\xi) = Y(\xi)$ for all ξ in C^c. Since $P(C^c) = 1$, the theorem is proved. ∎

Theorem 3-10F

Suppose $X(\cdot)$ and $Y(\cdot)$ are two nonnegative random variables. Then (A) $X(\cdot) = Y(\cdot)$ [P] iffi (B) there exist nondecreasing sequences of simple random variables $\{X_n(\cdot): 1 \leq n < \infty\}$ and $\{Y_n(\cdot): 1 \leq n < \infty\}$ satisfying the three conditions:

1. $X_n(\cdot) = Y_n(\cdot)$ [P], $1 \leq n < \infty$
2. $\lim\limits_{n \to \infty} X_n(\cdot) = X(\cdot)$
3. $\lim\limits_{n \to \infty} Y_n(\cdot) = Y(\cdot)$

PROOF

1. To show (A) implies (B), we form the simple functions according to the scheme described in Sec. 3-3. Let the points in the subdivision for the nth approximating functions be t_{in}. Put $E(i, n) = \{\xi: X_n(\xi) = t_{in}\}$ and $E'(i, n) = \{\xi: Y_n(\xi) = t_{in}\}$. By Theorem 3-10$E$ (proved above), we must have $E(i, n) = E'(i, n)$ [P] for each i, n, so that $X_n(\cdot) = Y_n(\cdot)$ [P] for every n. By Theorem 3-3A, properties 2 and 3 must hold.

2. To show (B) implies (A), let $D_n = \{\xi: X_n(\xi) \neq Y_n(\xi)\}$. Then $P(D_n) = 0$, so that, if $D = \bigcup\limits_{n=1}^{\infty} D_n$, we must have $P(D) = 0$. On D^c, $X_n(\xi) = Y_n(\xi)$ for each ξ, n, so that $\lim\limits_{n \to \infty} X_n(\xi) = \lim\limits_{n \to \infty} Y_n(\xi)$ for each $\xi \in D^c$. Thus $X(\cdot) = Y(\cdot)$ on D^c, so that the set on which these fail to be equal must be a subset of D and thus have probability zero. ∎

D-3   Proof of Theorem 6-1F on the law of large numbers

Theorem 6-1F

Suppose the class of random variables $\{X_i(\cdot): 1 \leq i < \infty\}$ is pairwise independent and each random variable in the class has the same distribution with $\mu[X_i] = \mu$. The variance may or may not exist. Then $\lim\limits_{n \to \infty} P(|A_n - \mu| < \epsilon) = 1$ for any $\epsilon > 0$.

PROOF   The proof makes use of a classical device known as the *method of truncation*. Let δ be any positive number. For each pair of positive integers i, n, define the random variables $Z_{in}(\cdot)$ and $W_{in}(\cdot)$ as follows:

$Z_{in}(\xi) = X_i(\xi)$ and $W_{in}(\xi) = 0$ for $|X_i(\xi)| \leq n\delta$
$Z_{in}(\xi) = 0$ and $W_{in}(\xi) = X_i(\xi)$ for $|X_i(\xi)| > n\delta$

First we show the $Z_{in}(\cdot)$ are pairwise independent for any given n. Let

$B_n = [-\delta n, \delta n] = \{t: -\delta n \leq t \leq \delta n\}$

For any Borel set M, we must have

$$Z_{in}^{-1}(M) = \begin{cases} X_i^{-1}(M) & \text{for } 0 \notin M \\ X_i^{-1}(MB_n) \uplus X_i^{-1}(B_n^c) & \text{for } 0 \in M \end{cases}$$

For any distinct pair of integers i, j, the pairwise independence of the variables $X_i(\cdot)$ and $X_j(\cdot)$ ensures the independence conditions necessary to apply Theorem 2-6G.

We may thus assert the independence of $Z_{in}^{-1}(M)$ and $Z_{jn}^{-1}(N)$ for any Borel sets M and N, which is the condition for independence of the random variables $Z_{in}(\cdot)$ and $Z_{jn}(\cdot)$. We may assert further that

$$X_i(\cdot) = Z_{in}(\cdot) + W_{in}(\cdot) \qquad |X_i(\cdot)| = |Z_{in}(\cdot)| + |W_{in}(\cdot)|$$
$$\mu = \mu[Z_{in}] + \mu[W_{in}] = \mu_n + \mu_n' \qquad \text{with } \mu_n \to \mu \text{ as } n \to \infty$$
$$E[|X_i|] = b = E[|Z_{in}|] + E[|W_{in}|] = b_n + b_n' \qquad \text{with } b_n' \to 0 \text{ as } n \to \infty$$
$$Z_{in}^2(\cdot) \leq n\delta|Z_{in}(\cdot)| \quad \text{so that} \quad \sigma^2[Z_{in}] \leq E[Z_{in}^2] \leq n\delta E[|Z_{in}|]$$
$$= n\delta b_n \leq n\delta b$$

$$P(W_{in} \neq 0) = P(|X_i| \geq n\delta) \leq \frac{1}{n\delta}E[|W_{in}|] = \frac{b_n'}{n\delta}$$

Hence, given any $\epsilon > 0$ and $\delta > 0$, there is an n_0 such that

$$|\mu - \mu_n| < \frac{\epsilon}{2} \qquad \text{and} \qquad P(W_{in} \neq 0) < \frac{\delta}{n} \qquad \text{for all } n \geq n_0$$

For simplicity of writing, put $A_n^*(\cdot) = (1/n) \sum_{i=1}^{n} Z_{in}(\cdot)$. By the Chebyshev inequality, $P(|A_n^* - \mu_n| \geq \epsilon/2) \leq 4b\delta/\epsilon^2$.

Now $|\mu - \mu_n| < \epsilon/2$ and $|A_n^* - \mu| \geq \epsilon$ implies

$$|A_n^* - \mu_n| \geq \frac{\epsilon}{2}$$

so that in this case

$$P(|A_n^* - \mu| \geq \epsilon) \leq \frac{4b\delta}{\epsilon^2}$$

Also, for n sufficiently large,

$$P(\bigcup_{i=1}^{n} \{W_{in} \neq 0\}) \leq \sum_{i=1}^{n} P(W_{in} \neq 0) < \delta$$

Thus, for any ϵ, δ, arbitrarily chosen, there is an n_1 such that, for all $n > n_1$,

$$P(|A_n - \mu| \geq \epsilon) \leq P(|A_n^* - \mu| \geq \epsilon) + P(\bigcup_{i=1}^{n} \{W_{in} \neq 0\})$$
$$\leq \delta\left(1 + \frac{4b}{2}\right)$$

The arbitrariness of δ implies the limit asserted. ∎

APPENDIX E. INTEGRALS OF COMPLEX-VALUED RANDOM VARIABLES ˙

Suppose $Z(\cdot) = X(\cdot) + iY(\cdot)$, where $X(\cdot)$ and $Y(\cdot)$ are real-valued random variables. Then $X(\cdot) = \mathrm{Re}\, Z(\cdot)$ and $Y(\cdot) = \mathrm{Im}\, Z(\cdot)$.

Definition E-1

If E is any event,

$$\int_E Z \, dP = \int_E X \, dP + i \int_E Y \, dP$$

provided that both integrals on the right-hand side of the equation exist.

It is apparent that $\mathrm{Re} \int_E Z \, dP = \int_E \mathrm{Re}\, Z \, dP$ and $\mathrm{Im} \int_E Z \, dP = \int_E \mathrm{Im}\, Z \, dP$.

We may use properties of integrals in the real case to obtain corresponding properties in the complex case.

(I1) *Linearity with respect to the integrand*

$$\int_E (aZ + bW) \, dP = a \int_E Z \, dP + b \int_E W \, dP$$

PROOF Straightforward use of algebra to expand $(a_1 + ia_2)(X + iY) + (b_1 + ib_2)(U + iV)$, separation of the real and imaginary parts, use of linearity in the real case, and recombination serves to show the validity of the expression. ■

(I2) *Linearity with respect to measure.* Because measure is nonnegative, this property extends immediately.

(I3) *Additivity.* If $\{E_i : i \in J\}$ is a finite or countably infinite partition of E, then

$$\int_E Z \, dP = \sum_{i \in J} \int_{E_i} Z \, dP$$

PROOF The proof consists of separating the real and imaginary parts, using additivity in the real case, and recombining. ■

(I4) $\int_E Z \, dP$ exists iffi $\int_E |Z| \, dP$ does, and $\left| \int_E Z \, dP \right| \leq \int_E |Z| \, dP$.

PROOF To simplify writing, we express the integrals for the whole space; exactly similar arguments hold for any event E.

$\int Z \, dP$ exists iffi $\int X \, dP$ and $\int Y \, dP$ both exist. The latter two integrals exist iffi $\int |X| \, dP$ and $\int |Y| \, dP$ exist. Since $|Z| \leq |X| + |Y|$, $|X| \leq |Z|$, and $|Y| \leq |Z|$, we must have by properties of real integrals $\int |Z| \, dP \leq \int |X| \, dP + \int |Y| \, dP$, $\int |X| \, dP \leq \int |Z| \, dP$, and $\int |Y| \, dP \leq \int |Z| \, dP$. Suppose $\int Z \, dP = re^{i\theta}$ with $r \geq 0$. The inequality is trivially true if $r = 0$. For $r > 0$, consider $\int e^{-i\theta} Z \, dP = r > 0$. Then $|\int Z \, dP| = r = \mathrm{Re} \int e^{i\theta} Z \, dP = \int \mathrm{Re}\, e^{i\theta} Z \, dP \leq \int |\mathrm{Re}\, e^{i\theta} Z| \, dP \leq \int |e^{-i\theta} Z| \, dP = \int |Z| \, dP$.

From the inequalities thus established, the theorem follows immediately. ■

Properties (I5) and (I6) extend in an immediate and obvious fashion. Property (I7) applies to real-valued functions and hence has no complex extension.

(I8)   *Product rule for independent random variables.*   If each of the classes $\{X(\cdot),$ $U(\cdot)\}$, $\{X(\cdot),\ V(\cdot)\}$, $\{Y(\cdot),\ U(\cdot)\}$, and $\{Y(\cdot),\ V(\cdot)\}$ is an independent class and if $Z(\cdot) = X(\cdot) + iY(\cdot)$ and $W(\cdot) = U(\cdot) + iV(\cdot)$, then

$$\int ZW\ dP = \int Z\ dP \int W\ dP$$

PROOF   Expand the integrands, separate real and imaginary parts, and apply the theorem for the real case.   Upon collection of the resulting integral products, the theorem follows easily.   ∎

APPENDIX F. SUMMARY OF PROPERTIES AND KEY THEOREMS

Basic probability

Definition 2-3a

A *probability system* (or probability space) consists of the triple

1. A *basic space* S of *elementary outcomes* (elements) ξ
2. A *class* ε *of events* (a sigma field of subsets of S)
3. A *probability measure* $P(\cdot)$ defined for each event A in the class ε and having the following properties:

(P1) $P(S) = 1$ (probability of the sure event is unity)

(P2) $P(A) \geq 0$ (probability of an event is nonnegative)

(P3) If $\alpha = \{A_i: i \in J\}$ is a countable partition of A (i.e., a mutually exclusive class whose union is A), then

$$P(A) = \sum_{i \in J} P(A_i) \qquad \text{(additivity property)}$$

Further properties

(P4) $P(A^c) = 1 - P(A)$

(P5) $P(\emptyset) = 0$

(P6) If $A \subset B$, then $P(A) \leq P(B)$.

(P7) $P(A \cup B) = P(A) + P(A^cB) = P(B) + P(AB^c)$
$= P(A^cB) + P(AB^c) + P(AB)$
$= P(A) + P(B) - P(AB)$

(P8) Let $\mathcal{B} = \{B_i: i \in J\}$ be any countable class of mutually exclusive events. If the occurrence of the event A implies the occurrence of one of the B_i (i.e., if $A \subset \uplus B_i$), then $P(A) = \sum_{i \in J} P(AB_i)$.

(P9) If $\{A_n: 1 \leq n < \infty\}$ is a decreasing or an increasing sequence of events whose limit is the event A, then $\lim_{n \to \infty} P(A_n) = P(A)$.

(P10) Let $\alpha = \{A_i: i \in J\}$ be any countable class of events, and let A be the union of the class. Then

$$P(A) \leq \sum_{i \in J} P(A_i) \qquad \text{(subadditivity property)}$$

Conditional probability

Definition 2-5a

If E is an event with positive probability, the *conditional probability of the event* A, *given* E, written $P(A|E)$, is defined by the relation

$$P(A|E) = \frac{P(AE)}{P(E)}$$

(CP1) *Product rule for conditional probability*

$$P(A_1A_2 \cdots A_n) = P(A_1)P(A_2|A_1)P(A_3|A_1A_2) \cdots P(A_n|A_1A_2 \cdots A_{n-1})$$

(CP2) Let $\mathcal{B} = \{B_i: i \in J\}$ be any countable class of mutually exclusive events, each with positive probability. If the occurrence of the event A implies the occurrence of one of the B_i (i.e., if $A \subset \biguplus_{i \in J} B_i$), then

$$P(A) = \sum_{i \in J} P(A|B_i)P(B_i)$$

(CP3) *Bayes' rule.* Let $\mathfrak{B} = \{B_i : i \in J\}$ be any countable class of mutually exclusive events, each with positive probability. Let A be any event with positive probability such that $A \subset \bigcup_{i \in J} B_i$. Then

$$P(B_i|A) = \frac{P(A|B_i)P(B_i)}{P(A)} = \frac{P(A|B_i)P(B_i)}{\sum_{j \in J} P(A|B_j)P(B_j)}$$

(CP4) $P(A_1 A_2 \cdots A_n | E) = P(A_1|E)P(A_2|A_1 E) \cdots P(A_n | A_1 \cdots A_{n-1} E)$

(CP5)   Let $\mathfrak{B} = \{B_i : i \in J\}$ be any finite or countably infinite class of mutually exclusive events, each with positive probability. If the occurrence of the event A implies the occurrence of one of the B_i (i.e., if $A \subset \biguplus_{i \in J} B_i$), then

$$P(A|E) = \sum_{i \in J} P(A|B_i E)P(B_i|E)$$

Stochastic independence

Definition 2-6a

Two events A and B are said to be (*stochastically*) *independent* iffi the following *product rule* holds:

$$P(AB) = P(A)P(B)$$

Definition 2-6b

A class of events $\mathfrak{a} = \{A_i : i \in J\}$, where J is a finite or an infinite index set, is said to be an *independent class* iffi the product rule holds for every finite subclass of $\mathfrak{a}$.

Theorem 2-6E

Suppose $\mathfrak{a} = \{A_i : i \in I\}$ is any class of events. Let $\{\mathfrak{a}_j : j \in J\}$ be a family of finite subclasses of $\mathfrak{a}$ such that no two have any member event A_i in common. Let B_j be the intersection of all the sets in $\mathfrak{a}_j$. Put $\mathfrak{B} = \{B_j : j \in J\}$. Then $\mathfrak{a}$ is an independent class iffi every class $\mathfrak{B}$ so formed is an independent class.

Theorem 2-6F

If $\mathfrak{a} = \{A_i : i \in J\}$ is an independent class, so also is the class $\mathfrak{a}'$ obtained by replacing the A_i in any subclass of $\mathfrak{a}$ by either $\emptyset$, S, or A_i^c. The particular substitution for any given A_i may be made arbitrarily, without reference to the substitution for any other member of the subclass.

Theorem 2-6G

Suppose $\mathfrak{a} = \{A_i : i \in I\}$ and $\mathfrak{B} = \{B_j : j \in J\}$ are countable disjoint classes whose members have the property that any A_i is independent of any B_j; that is, $P(A_i B_j) =$

$P(A_i)P(B_j)$ for any $i \in I$ and $j \in J$. Then the events

$$A = \biguplus_{i \in I} A_i \text{ and } B = \biguplus_{j \in J} B_j \text{ are independent}$$

Theorem 2-8A

Suppose $\{A_1, A_2, \ldots, A_n, B_1, B_2, \ldots, B_m\}$ is an independent class, and let

$$F = f(A_1, A_2, \ldots, A_n) \qquad \text{and} \qquad G = g(B_1, B_2, \ldots, B_m)$$

be boolean functions of the indicated events. Then F and G are independent events.

Boolean functions of events

Definition 2-7d

A *boolean function* $F = f(A, B, \ldots)$ *of a finite class of sets* is a set obtained by a finite number of applications of the operations of union, intersection, and complementation to the members of the class.

Theorem 2-7A   Minterm Expansion Theorem

Any boolean function $F = f(A_{N-1}, A_{N-2}, \ldots, A_1, A_0)$ may be expressed as the disjoint union of an appropriate subclass of the minterms m_i generated by the class $\{A_{N-1}, A_{N-2}, \ldots, A_1, A_0\}$. In symbols, $F = \biguplus_{i \in J_F} m_i$, where J_F is a suitable index set.

Random variables and events

Theorem 3-1A

Suppose $X(\cdot)$ is a mapping carrying elements ξ of the domain S into elements t in the range T. Then

    1. The inverse image of the union (intersection) of a class of the t sets is the union (intersection) of the class of inverse images of the separate t sets.

    2. The inverse image of the complement of a t set is the complement of the inverse image of the t set.

    3. If a class of t sets is a disjoint class, the class of the inverse images is a' disjoint class.

    4. The relation of inclusion is preserved by the inverse mapping.

Theorem 3-1B

If the function $X(\cdot)$ is such that $\{\xi : X(\xi) \leq t\}$ is an event for each real t, then $X^{-1}(M)$ is an event for each Borel set M.

Definition 3-1a

A real-valued function $X(\cdot)$ from the basic space S to the real line R is called a (real-valued) *random* variable iffi for each real t it is true that $\{\xi : X(\xi) \leq t\}$ is an event.

Definition 3-2a

The probability measure $P_X(\cdot)$ defined on the class of Borel sets $\mathcal{B}_R$ by

$$P_X(M) = P[X^{-1}(M)] \qquad \text{for each } M \in \mathcal{B}_R$$

is called the *probability measure induced by the random variable* $X(\cdot)$.

Discrete random variables

Definition 3-3a

A random variable $X(\cdot)$ whose range T consists of a finite set of values is called a *simple random variable*. If the range T consists of a countably infinite set of distinct values, the function is referred to as an *elementary random variable*. The term *discrete random variable* is used to indicate the fact that the random variable is either simple or elementary.

Definition 3-3b

The discrete random variable $X(\cdot)$ is said to be in *canonical form* iffi it is written

$$X(\cdot) = \sum_{i \in J} t_i I_{A_i}(\cdot)$$

where $T = \{t_i : i \in J\}$ is a set of distinct constants and $\{A_i : i \in J\}$ is a partition.

Definition 3-3c

The discrete random variable $X(\cdot)$ is said to be in *reduced canonical form* iffi it is written

$$X(\cdot) = \sum_{i \in J} t_i I_{B_i}(\cdot)$$

where $T' = \{t_i : i \in J\}$ is a set of distinct, nonzero constants and $\{B_i : i \in J\}$ is a disjoint class.

Theorem 3-3A

A bounded, nonnegative random variable $X(\cdot)$ can be represented as the limit of a nondecreasing sequence of simple functions. The convergence of this sequence is uniform in ξ over the whole basic space S.

Probability distribution functions

Definition 3-4a

For any real-valued random variable $X(\cdot)$, we define the *distribution function* $F_X(\cdot)$ by the expression

$$F_X(t) = P(\{\xi : X(\xi) \le t\}) = P(X \le t)$$

for each real t.

Definition

The function

$$u(t) = \begin{cases} 0 & \text{for } t < 0 \\ 1 & \text{for } t > 0 \end{cases}$$

is called the *unit step function*. The function has been left undefined at $t = 0$. If we define the function to be continuous from the right, we use the symbol $u_+(\cdot)$.

(F1) $F_X(t)$ is monotonically increasing with increasing t.

(F2) $F_X(-\infty) = \lim_{t \to -\infty} F_X(t) = 0$ and $F_X(\infty) = \lim_{t \to \infty} F_X(t) = 1$

(F3) $P(a < X \le b) = P(X \in (a, b]) = F_X(b) - F_X(a)$

(F4) $F_X(\cdot)$ has a jump discontinuity of magnitude $\delta > 0$ at $t = a$ iffi $P(X = a) = \delta$. $F_X(\cdot)$ is continuous at $t = a$ iffi $P(X = a) = 0$.

(F5) $F_X(\cdot)$ is continuous from the right.

Definition 3-4b

If the probability measure $P_X(\cdot)$ induced by the real-valued random variable $X(\cdot)$ is such that it assigns zero probability to any point set of Lebesgue measure (generalized length) zero, the probability measure and the probability distribution are said to be *absolutely continuous* (with respect to Lebesgue measure). In this case, the random variable is also said to be *absolutely continuous*.

Definition 3-4c

If the probability measure $P_X(\cdot)$ induced by the real-valued random variable $X(\cdot)$ is absolutely continuous, the function $f_X(\cdot)$ defined on the real line such that

$$\int_M f_X(u) \, du = P_X(M) \qquad \text{for each Borel set } M$$

is called the *probability density function* for $X(\cdot)$.

(f1) $f_X(\cdot) \ge 0$ almost everywhere

(f2) $\displaystyle\int_{-\infty}^{\infty} f_X(u) \, du = 1$

(f3) $\displaystyle\int_{-\infty}^{t} f_X(u) \, du = F_X(t)$

(f4) $f_X(t) = \dfrac{d}{dt} F_X(t)$ at points of continuity of $f_X(\cdot)$

Definition 3-5a

The probability measure $P_{XY}(\cdot)$ defined on the Borel sets in the plane is called the *joint probability measure* induced by the joint mapping $(t, u) = Z(\xi) = [X, Y](\xi)$. The probability mass distribution is called the *joint distribution*. The probability measures $P_{XY}(\cdot \times R_2) = P_X(\cdot)$ and $P_{XY}(R_1 \times \cdot) = P_Y(\cdot)$ are called the *marginal probability measures* induced by $X(\cdot)$ and $Y(\cdot)$, respectively. The corresponding probability mass distributions are called the *marginal distributions*.

Definition 3-6a

The function $F_{XY}(\cdot, \cdot)$ defined by

$$F_{XY}(t, u) = P(X \le t, Y \le u)$$

is called the *joint distribution function* for $X(\cdot)$ and $Y(\cdot)$. The special cases $F_{XY}(\cdot, \infty)$ and $F_{XY}(\infty, \cdot)$ are called the *marginal distribution functions* for $X(\cdot)$ and for $Y(\cdot)$, respectively.

Definition 3-6b

If the joint probability measure $P_{XY}(\cdot)$ induced by $X(\cdot)$ and $Y(\cdot)$ is absolutely continuous, a function $f_{XY}(\cdot, \cdot)$ exists such that $\iint\limits_{Q} f_{XY}(t, u) \, dt \, du = P_{XY}(Q)$ for each Borel set Q on the plane. The function $f_{XY}(\cdot, \cdot)$ is called a *joint probability density function* for $X(\cdot)$ and $Y(\cdot)$.

(f1) $\quad f_{XY}(\cdot, \cdot) \geq 0 \qquad$ almost everywhere

(f2) $\quad \displaystyle\iint_{-\infty}^{\infty} f_{XY}(x, y) \, dx \, dy = 1$

(f3) $\quad \displaystyle\int_{-\infty}^{t} \int_{-\infty}^{u} f_{XY}(x, y) \, dy \, dx = F_{XY}(t, u)$

(f4) $\quad f_{XY}(t, u) = \dfrac{\partial^2 F}{\partial u \, \partial t} = \dfrac{\partial^2 F}{\partial t \, \partial u} \qquad$ (under suitable regularity conditions)

Independent random variables

Definition 3-7a

The random variables $X(\cdot)$ and $Y(\cdot)$ are said to be (*stochastically*) *independent* iffi, for each choice of Borel sets M and N, the events $X^{-1}(M)$ and $Y^{-1}(N)$ are independent events.

Definition 3-7b

A class $\{X_i(\cdot): i \in J\}$ of random variables is said to be an *independent class* iffi, for each class $\{M_i: i \in J\}$ of Borel sets, arbitrarily chosen, the class of events $\{X_i^{-1}(M_i): i \in J\}$ is an independent class.

Theorem 3-7A

Random variables $X(\cdot)$ and $Y(\cdot)$ are independent iffi

$$P_{XY}(M \times N) = P_X(M)P_Y(N)$$

for each Borel set M in R_1 and N in R_2.

Theorem 3-7B

Any two real-valued random variables $X(\cdot)$ and $Y(\cdot)$ are independent iffi, for all semi-infinite, half-open intervals M_t and N_u, defined by $M_t = \{\xi: X(\xi) \leq t\}$ and $N_u = \{\xi: Y(\xi) \leq u\}$,

$$P_{XY}(M_t \times N_u) = P_X(M_t)P_Y(N_\cdot)$$

By definition, the latter condition is equivalent to the condition

$$P(X \in M_t, \, Y \in N_u) = P(X \in M_t)P(Y \in N_u)$$

for all such half-open intervals.

Theorem 3-7C

Two random variables $X(\cdot)$ and $Y(\cdot)$ are independent iffi their distribution functions satisfy the *product rule*

$$F_{XY}(t, u) = F_X(t)F_Y(u).$$

If the density functions exist, then independence of the random variables is equivalent to the product rule $f_{XY}(t, u) = f_X(t)f_Y(u)$ for the density functions.

Theorem 3-7D

Suppose $X(\cdot)$ and $Y(\cdot)$ are independent random variables, each of which is nonnegative. Then there exist nondecreasing sequences of nonnegative simple random variables $\{X_n(\cdot): 1 \leq n < \infty\}$ and $\{Y_m(\cdot): 1 \leq m < \infty\}$ such that

$$\lim_{n \to \infty} X_n(\xi) = X(\xi) \qquad \text{and} \qquad \lim_{m \to \infty} Y_m(\xi) = Y(\xi) \qquad \text{for all } \xi$$

and $\{X_n(\cdot), Y_m(\cdot)\}$ is an independent pair for any choice of m, n.

Functions of random variables

Definition 3-8a

If $g(\cdot)$ is a real-valued function of a single real variable t, the function $Z(\cdot) = g[X(\cdot)]$ is defined to be the function on the basic space S which has the value $v = g(t)$ when $X(\xi) = t$.

Definition 3-8b

Let $g(\cdot)$ be a real-valued function, mapping points in the real line R_1 into points in the real line R_2. The function $g(\cdot)$ is called a *Borel function* iffi, for every Borel set M in R_2, the inverse image $N = g^{-1}(M)$ is a Borel set in R_1. An exactly similar definition holds for a *Borel function* $h(\cdot, \cdot)$, mapping points in the plane $R_1 \times R_2$ into points on the real line R_3.

Theorem 3-8A

If $W(\cdot)$ is a random vector and $g(\cdot)$ is a Borel function of the appropriate number of variables, then $Z(\cdot) = g[W(\cdot)]$ is a random variable measurable $\varepsilon(W)$.

Theorem 3-8B

If $\{X_i(\cdot): i \in J\}$ is an independent class of random variables and if, for each $i \in J$, $W_i(\cdot)$ is $\varepsilon(X_i)$ measurable, then $\{W_i(\cdot): i \in J\}$ is an independent class of random variables.

Almost-sure relationships

Definition 3-10a

Two *events* A and B are said to be *equal with probability 1*, designated in symbols by

$$A = B \ [P] \qquad \text{iffi} \qquad P(A) = P(B) = P(AB)$$

We also say A and B are *almost surely equal*.

Definition 3-10b

Two classes of events $\mathfrak{a}$ and $\mathfrak{B}$ are said to be *equal with probability 1* or *almost surely equal,* designated $\mathfrak{a} = \mathfrak{B}$ $[P]$, iffi their members may be put into a one-to-one correspondence such that $A_i = B_i$ $[P]$ for each corresponding pair.

Definition 3-10d

A *property* of a random variable or a *relationship* between two or more random variables is said to *hold with probability 1* (indicated by the symbol $[P]$ after the appropriate expression) iffi the elements ξ for which the property or relationship fails to hold belong to a set D having 0 probability.   In this case we may also say that the property or the relationship *holds almost surely.*

A property or relationship is said to *hold with probability 1 on* (*event*) E (indicated by "$[P]$ on E") iffi the points of E for which the property or relationship fails to hold belong to a set having 0 probability.   We also use the expression "*almost surely on E.*"

Mathematical expectation

Definition 5-1a

If $X(\cdot)$ is a real-valued random variable and $g(\cdot)$ a Borel function, the *mathematical expectation* $E[g(X)] = E[Z]$ *of the random variable* $Z(\cdot) = g[X(\cdot)]$ is given by

$$E[Z] = E[g(X)] = \int Z \, dP = \int g(X) \, dP$$

(E1) $E[aI_A] = aP(A)$, where a is a real or complex constant.

(E2)   *Linearity.*   If a and b are real or complex constants, $E[aX + bY] = aE[X] + bE[Y]$.

(E3)   *Positivity.*   If $X(\cdot) \geq 0$ $[P]$,   then $E[X] \geq 0$.   If $X(\cdot) \geq 0$ $[P]$,   then $E[X] = 0$ iffi $X(\cdot) = 0$ $[P]$.   If $X(\cdot) \geq Y(\cdot)$ $[P]$, then $E[X] \geq E[Y]$.

(E4) $E[X]$ exists iffi $E[|X|]$ does, and $|E[X]| \leq E[|X|]$.

(E5)   *Schwarz inequality.* $|E[XY]|^2 \leq E[|X|^2]E[|Y|^2]$

In the real case, equality holds iffi there is a real constant λ such that

$$\lambda X(\cdot) + Y(\cdot) = 0 \ [P].$$

(E6)   *Product rule for independent random variables.*   If $X(\cdot)$ and $Y(\cdot)$ are independent, integrable random variables, then $E[XY] = E[X]E[Y]$.

(E7)   If $g(\cdot)$ is a nonnegative Borel function and if $A = \{\xi\colon g(X) \geq a\}$, then $E[g(X)] \geq aP(A)$.

(E8)   If $g(\cdot)$ is a nonnegative, strictly increasing, Borel function of a single real variable and c is a nonnegative constant, then

$$P(|X| \geq c) \leq E[g(|X|)]/g(c)$$

(E9)   *Jensen's inequality.*   If $g(\cdot)$ is a convex Borel function and $X(\cdot)$ is a real random variable whose expectation exists, then

$$g(E[X]) \leq E[g(X)]$$

Mean value

Definition 5-3a

If $X(\cdot)$ is a real-valued random variable, its mean value, denoted by one of the symbols μ, μ_X, or $\mu[X]$, is defined by $\mu[X] = E[X]$.

Variance

Definition 5-4a

Consider a real-valued random variable $X(\cdot)$ whose square is integrable. The *variance of $X(\cdot)$*, denoted $\sigma^2[X]$, is given by

$$\sigma^2[X] = E[(X - \mu)^2]$$

where $\mu = \mu[X]$ is the mean value of $X(\cdot)$.

(**V1**) $\sigma^2[X] = E[X^2] - E^2[X] = E[X^2] - (\mu_X)^2$
(**V2**) $\sigma^2[aX] = a^2\sigma^2[X]$
(**V3**) $\sigma^2[X + a] = \sigma^2[X]$
(**V4**) $\sigma^2[X \pm Y] = \sigma^2[X] + \sigma^2[Y] \pm 2\{E[XY] - E[X]E[Y]\}$
(**V5**) If $\{X_i(\cdot): 1 \leq i \leq n\}$ is a class of pairwise independent random variables

and $X(\cdot) = \sum_{i=1}^{n} \delta_i X_i(\cdot)$, where each δ_i has one of the values $+1$ or -1, then

$$\sigma^2[X] = \sum_{i=1}^{n} \sigma^2[X_i] + \sum_{i \neq j} \delta_i\delta_j\{E[X_iX_j] - E[X_i]E[X_j]\}$$

(**V6**) Consider the random variable $X(\cdot)$, with mean $\mu[X] = \mu$ and standard deviation $\sigma[X] = \sigma$. Then the random variable

$$Y(\cdot) = \frac{X(\cdot) - \mu}{\sigma}$$

has mean $\mu[Y] = 0$ and standard deviation $\sigma[Y] = 1$.

Theorem 5-4B   Chebyshev Inequality

Let $X(\cdot)$ be any random variable whose mean μ and standard deviation σ exist. Then

$$P(|X - \mu| \geq k\sigma) \leq \frac{1}{k^2}$$

or equivalently,

$$P(|X - \mu| \geq k) \leq \sigma^2/k^2$$

Moment-generating function

Definition 5-7a

If $X(\cdot)$ is a random variable and s is a parameter, real or complex, the function of s defined by

$$M_X(s) = E[e^{sX}]$$

is called the *moment-generating function for* $X(\cdot)$.   If $s = iu$, where i is the imaginary unit having the formal properties of the square root of -1, the function $\varphi_X(\cdot)$ defined by

$$\varphi_X(u) = M_X(iu) = E[e^{iuX}]$$

is called the *characteristic function for* $X(\cdot)$.

(M1)   Consider two random variables $X(\cdot)$ and $Y(\cdot)$ with distribution functions $F_X(\cdot)$ and $F_Y(\cdot)$, respectively.   Let $M_X(\cdot)$ and $M_Y(\cdot)$ be the corresponding moment-generating functions for the two variables.   Then $M_X(iu) = M_Y(iu)$ for all real u iffi $F_X(t) = F_Y(t)$ for all real t.

(M2)   If $E[|X|^n]$ exists, the nth-order derivative of the characteristic function exists and

$$\varphi_X{}^{(n)}(0) = \frac{d^n}{du^n} M_X(iu) \Big|_{u=0} = i^n E[X^n]$$

(M3)   If the region of convergence for $M_X(\cdot)$ is a proper strip in the s plane (which will include the imaginary axis), derivatives of all orders exist and

$$M_X{}^{(n)}(s) = \frac{d^n}{ds^n} M_X(s) = \int t^n e^{st} \, dF_X(t)$$
$$M_X{}^{(n)}(0) = E[X^n]$$

(M4)   If $Z(\cdot) = aX(\cdot) + b$, then $M_Z(s) = e^{bs} M_X(as)$.

(M5)   If $X(\cdot)$ and $Y(\cdot)$ are *independent* random variables and if $Z(\cdot) = X(\cdot) + Y(\cdot)$, then $M_Z(s) = M_X(s) M_Y(s)$ for all s.

Types of convergence

Definition 6-3a

The sequence is said to *converge with probability 1 to* $X(\cdot)$, indicated

$$X_n(\cdot) \to X(\cdot) \quad [P]$$

iffi there is a set E with $P(E^c) = 0$ such that $X_n(\xi) \to X(\xi)$ for each $\xi \in E$.

Definition 6-3b

The sequence is said to *converge almost uniformly to* $X(\cdot)$, indicated

$$X_n(\cdot) \to X(\cdot) \quad [\text{a. unif}]$$

iffi, to each $\epsilon > 0$, there corresponds a set E with $P(E^c) < \epsilon$ such that $X_n(\xi)$ converges uniformly to $X(\xi)$ for all $\xi \in E$.

Definition 6-3c

The sequence is said to *converge in probability to* $X(\cdot)$, indicated

$$X_n(\cdot) \to X(\cdot) \quad [\text{in prob}]$$

iffi

$$\lim_{n \to \infty} P(\{\xi : |X(\xi) - X_n(\xi)| \geq \epsilon\}) = 0 \quad \text{for each } \epsilon > 0$$

Definition 6-3d

The sequence is said to *converge in the mean of order p* ($p \geq 1$), indicated

$$X_n(\cdot) \to X(\cdot) \qquad [\text{mean}^p]$$

iffi

$$\lim_{n \to \infty} \int |X - X_n|^p \, dP = 0$$

Theorem 6-3D

[a. unif] $\Rightarrow$ [P] $\Rightarrow$ [in prob] $\Leftarrow$ [mean^p]

Theorem 6-3E

If $Y(\cdot)$ is a nonnegative random variable such that $Y^p(\cdot)$ is integrable ($p \geq 1$) and $|X_n(\cdot)| \leq Y(\cdot)$ [P], then, for $p = 1$, [P] $\Rightarrow$ [a. unif], and for $p \geq 1$, [in prob] $\Rightarrow$ [mean^p], so that in this case

[a. unif] $\Leftrightarrow$ [P] $\Rightarrow$ [in prob] $\Leftrightarrow$ [mean^p]

Theorem 6-3F

A sequence of random variables $\{X_n(\cdot): 1 \leq n < \infty\}$ satisfies the condition

$$(A) \quad X_n(\cdot) \to X(\cdot) \qquad [\text{in prob}]$$

iffi

(B) Each subsequence has a further subsequence which converges to $X(\cdot)$ with probability 1.

Expectations for random processes

Definition 7-6a

A process X is said to be of *order p* if, for each $t \in T$, $E[|X(t)|^p] < \infty$ (p is a positive integer).

Definition 7-6b

The *mean-value function* for a process is the first moment

$$\mu_X(t) = E[X(t)]$$

Definition 7-6e

The *covariance function* $K_X(\cdot, \cdot)$ of a process X, if it exists, is the function defined by

$$K_X(s, t) = E[(X(s) - \mu_X(s))\overline{(X(t) - \mu_X(t))}]$$

The bar denotes the complex conjugate.   The *autocorrelation function* $\varphi_{XX}(\cdot, \cdot)$ of a process, if it exists, is the function defined by

$$\varphi_{XX}(s, t) = E[X(s)\overline{X(t)}]$$

Definition 7-6f

The *cross-correlation functions* for two random processes X and Y are defined by

$$\varphi_{XY}(s, t) = E[X(s)\overline{Y(t)}]$$
$$\varphi_{YX}(s, t) = E[Y(s)\overline{X(t)}]$$

Properties of correlation functions

Definition 7-6g

A function $g(\cdot, \cdot)$ defined on $T \times T$ is *positive semidefinite* (or nonnegative definite) iffi, for every finite subset T_n contained in T and every function $h(\cdot)$ defined on T_n. it follows that

$$\sum_{s,t \in T_n} g(s, t)h(s)\overline{h(t)} \geq 0$$

(φ1) $\varphi_{XY}(s, t) = \overline{\varphi_{YX}(t, s)}$
(φ2) $|\varphi_{XY}(s, t)|^2 \leq \varphi_{XX}(s, s)\varphi_{YY}(t, t)$
(φ3)   The autocorrelation function $\varphi_{XX}(\cdot, \cdot)$ is positive semidefinite.
(φ4)   The random process X is continuous [mean²] at t iffi the autocorrelation function $\varphi_{XX}(\cdot, \cdot)$ is continuous at the point t, t.
(φ5)   If $\varphi_{XX}(s, t)$ is continuous at all points t, t, it is continuous for all s, t.

(φ6)   If $\dfrac{\partial^2}{\partial s\, \partial t} \varphi_{XX}(s, t)$ exists for all points t, t, then $X'(\cdot, t)$ exists for all t.

(φ7)   If $X'(\cdot, s)$ exists for all s and $Y'(\cdot, t)$ exists for all t, then the following correlation functions and partial derivatives exist and the equalities indicated hold:

$$\varphi_{X'Y}(s, t) = \frac{\partial}{\partial s} \varphi_{XY}(s, t)$$

$$\varphi_{XY'}(s, t) = \frac{\partial}{\partial t} \varphi_{XY}(s, t)$$

$$\varphi_{X'Y'}(s, t) = \frac{\partial^2}{\partial s\, \partial t} \varphi_{XY}(s, t)$$

Stationary random processes

Definition 7-7a

A random process X is said to be *stationary* iffi. for every choice of any finite number n of elements $t_1, t_2, \ldots, t_n$ from the parameter set T and of any h such that $t_1 + h, t_2 + h, \ldots, t_n + h$ all belong to T, we have the *shift property*
$$F(\cdot, t_1; \cdot, t_2; \cdots ; \cdot, t_n) = F(\cdot, t_1 + h; \cdot, t_2 + h; \cdots ; \cdot, t_n + h)$$

Definition 7-7b

A random process X is said to be *second-order stationary* if it is of second order and if its first and second distribution functions have the shift property

$$F(x, t) = F(x) \qquad \text{and} \qquad F(x, s; y, t) = F(x, 0; y, t - s)$$

If the process is of second order and $\varphi_{XX}(t, t + \tau) = \varphi_{XX}(0, \tau)$ for all t, τ, the process is said to be *stationary in the wide sense*.

Bibliography

This bibliography provides a consolidated listing of those books and articles which are included in one or more of the lists of selected references at the ends of the chapters. Boldfaced numbers in square brackets following a bibliographical entry indicate the chapter or chapters in which the work is included as a selected reference. Citations in the text are by author and date (e.g., Brunk [1965]); hence dates are listed immediately after the author's name in this bibliography.

ABRAMSON, NORMAN [1963]: "Information Theory and Coding," McGraw-Hill Book Company, New York [5].

ARLEY, NIELS, AND K. RANDER BUCH [1950]: "Introduction to the Theory of Probability and Statistics" (transl. from the Danish), John Wiley & Sons, Inc., New York [5].

BHARUCHA-REID, A. T. [1960]: "Elements of the Theory of Markov Processes and Their Applications," McGraw-Hill Book Company, New York [7].

BOGDANOFF, JOHN L., AND FRANK KOZIN (eds.) [1963]: "Proceedings of the First Symposium on Engineering Applications of Random Function Theory and Probability," John Wiley & Sons, Inc., New York.

BRUNK, H. D. [1965]: "An Introduction to Mathematical Statistics," 2d ed., Blaisdell Publishing Company, New York [2, 3, 5, 6].

CHERNOFF, H. [1952]: A Measure of Asymptotic Efficiency for Tests of a Hypothesis Based on Sums of Observations, *Ann. Math. Statistics*, vol. 23, pp. 493–507 [6].

CRAMÉR, HARALD [1946]: "Mathematical Methods of Statistics," Princeton University Press, Princeton, N.J. [5, 6, 7].

DAVENPORT, WILBUR B., JR., AND WILLIAM L. ROOT [1958]: "Introduction to Random Signals and Noise," McGraw-Hill Book Company, New York [7].

DAVID, F. N. [1962]: "Games, Gods, and Gambling," Hafner Publishing Company, Inc., New York [1].

DOOB, J. L. [1953]: "Stochastic Processes," John Wiley & Sons, Inc., New York [7].

FANO, ROBERT M. [1961]: "Transmission of Information," The M.I.T. Press, Cambridge, Mass., and John Wiley & Sons, Inc., New York [5].

FELLER, WILLIAM [1957]: "An Introduction to Probability Theory and Its Applications," vol. 1, 2d ed., John Wiley & Sons, Inc., New York [1, 7].

FISZ, MAREK [1963]: "Probability Theory and Mathematical Statistics" (transl. from the Polish), 3d ed., John Wiley & Sons, Inc., New York [2, 3, 6].

GNEDENKO, B. V. [1962]: "The Theory of Probability" (transl. from the Russian by B. D. Sekler), Chelsea Publishing Company, New York [3, 5, 6, 7].

GOLDBERG, SAMUEL [1960]: "Probability: An Introduction," Prentice-Hall, Inc., Englewood Cliffs, N.J. [1, 2, 3].

HALMOS, PAUL R. [1950]: "Measure Theory," D. Van Nostrand Company, Inc., Princeton, N.J. [6].

HOWARD, RONALD A. [1960]: "Dynamic Programming and Markov Processes," The M.I.T. Press, Cambridge, Mass., and John Wiley & Sons, Inc., New York [7].

KEMENY, JOHN G., HAZLETON MIRKIL, J. LAURIE SNELL, AND GERALD L. THOMPSON, [1959]: "Finite Mathematical Structures," Prentice-Hall, Inc., Englewood Cliffs, N.J. [7].

———— AND J. LAURIE SNELL [1960]: "Finite Markov Chains," D. Van Nostrand Company, Inc., Princeton, N.J. [7].

KOLMOGOROV, A. N. [1956]: "Foundations of the Theory of Probability," Chelsea Publishing Company, New York (a translation of the monograph "Grundbegriffe der Wahrscheinlichkeitsrechnung," 1933) [2].

LANING, J. HALCOMBE, JR., AND RICHARD H. BATTIN [1956]: "Random Processes in Automatic Control," McGraw-Hill Book Company, New York [7].

LEE, Y. W. [1960]: "Statistical Theory of Communication," John Wiley & Sons, Inc., New York [7].

LLOYD, DAVID K., AND MYRON LIPOW [1962]: "Reliability: Management, Methods, and Mathematics," Prentice-Hall, Inc., Englewood Cliffs, N.J. [3].

LOÈVE, MICHEL [1963]: "Probability Theory," 3d ed., D. Van Nostrand Company, Inc., Princeton, N.J. [4, 5, 6, 7].

MCCORD, JAMES R., III, AND RICHARD M. MORONEY, JR. [1964]: "Introduction to Probability Theory," The Macmillan Company, New York [2, 3, 6].

MIDDLETON, DAVID [1960]: "An Introduction to Statistical Communication Theory," McGraw-Hill Book Company, New York [7].

MUNROE, M. E. [1953]: "Introduction to Measure and Integration," Addison-Wesley Publishing Company, Inc., Cambridge, Mass. [2, 4, 6].

NATIONAL BUREAU OF STANDARDS [1964]: Milton Abramowitz and Irene A. Stegun (eds), "Handbook of Mathematical Functions with Formulas, Graphs, and Mathematical Tables," Applied Mathematics Series, no. 55 [3].

PARZEN, EMANUEL [1960]: "Modern Probability Theory and Its Applications," John Wiley & Sons, Inc., New York [2, 3, 5].

———— [1962]: "Stochastic Processes," Holden-Day, Inc., San Francisco [7].

PFEIFFER, PAUL E. [1964]: "Sets, Events, and Switching," McGraw-Hill Book Company, New York [2].

PIERCE, J. R. [1961]: "Symbols, Signals and Noise: The Nature and Process of Communication," Harper & Row, Publishers, Incorporated, New York.

ROSENBLATT, MURRAY [1962]: "Random Processes," Oxford University Press, New York [7].

SHANNON, C. E. [1948]: A Mathematical Theory of Communication, *Bell System Tech. J.*, vol. 27, pp. 379–423, 623–656 [5].

——— [1957]: Certain Results in Coding Theory for Noisy Channels, *Information and Control*, vol. 1, pp. 6–25 [6].

USPENSKY, J. V. [1937]: "Introduction to Mathematical Probability," McGraw-Hill Book Company, New York [1].

WADSWORTH, GEORGE A., AND JOSEPH G. BRYAN [1960]: "Introduction to Probability and Random Variables," McGraw-Hill Book Company, New York [3, 5].

WAX, NELSON (ed.) [1954]: "Selected Papers on Noise and Stochastic Processes," Dover Publications, Inc., New York.

WIDDER, DAVID VERNON [1941]: "The Laplace Transform," Princeton University Press, Princeton, N.J. [5].

WOZENCRAFT, JOHN M., AND BARNEY REIFFEN [1961]: "Sequential Decoding," The M.I.T. Press, Cambridge, Mass., and John Wiley & Sons, Inc., New York [6].

YAGLOM, A. M. [1962]: "An Introduction to the Theory of Stationary Random Functions," Prentice-Hall, Inc., Englewood Cliffs, N.J. [7].

Index